LIGHT
SCATTERING
by small particles

by
H. C. van de Hulst
Leiden Observatory

Dover Publications, Inc.
New York

Published in Canada by General Publishing Company,
Ltd., 30 Lesmill Road, Don Mills, Toronto, Ontario.
Published in the United Kingdom by Constable and Com-
pany, Ltd., 10 Orange Street, London WC2H 7EG.

This Dover edition, first published in 1981, is an un-
abridged and corrected republication of the work originally
published in 1957 by John Wiley & Sons, Inc., N.Y.

International Standard Book Number: 0-486-64228-3
Library of Congress Catalog Card Number: 81-68483

Manufactured in the United States of America
Dover Publications, Inc.
180 Varick Street
New York, N.Y. 10014

CONTENTS

Part I BASIC SCATTERING THEORY

CONTENTS

Part II SPECIAL TYPES OF PARTICLES

CONTENTS

CONTENTS

CONTENTS

CONTENTS

CONTENTS

PREFACE

The scattering of electromagnetic waves by a homogeneous sphere is a problem with a known solution. I first met this problem when I needed some numbers and curves in an astrophysical investigation. I soon learned that it is a long way from the formulae containing the solution to reliable numbers and curves. Subsequent conversations and correspondence with other research workers, notably in chemistry, showed that the same difficulty was felt in other fields.

The studies on which the present book is based were started in 1945 in an attempt to compile the data available in the literature and to fill in the gaps, where needed. Several related problems, such as the scattering by cylinders, were added to the original topic.

For clearer presentation, the problems of mathematical physics dealing with the scattering properties of single particles (part II) have been separated from the problems arising in specific fields of application (part III). The properties of the particles that should be known in order to describe the optical properties of a medium consisting of such particles have been defined in general terms (part I).

New formulae or numerical results are contained in almost all chapters. They are noted in the references at the end of each chapter. The reference lists have steadily grown in the course of the years; they probably are fairly complete, but no systematic bibliographical study has been made.

Although the book has a mathematical character, requirements of mathematical rigor do not dominate the presentation. Arguments based on physical intuition are given wherever they illuminate the subject more clearly than a mathematical derivation. Simple results that arise under special sets of assumptions are often derived both ways. In view of the wishes expressed by several colleagues, I have not shrunk from a certain inconsistency in the level of presentation. For instance, chapter 17, which comes closest to an actual research report, contains less explanation of elementary detail than some earlier chapters, which may be consulted by research workers without special mathematical training.

Personal acknowledgment of the support received from numerous friends and colleagues in writing this book is impossible. I wish to express my thanks both for the information they have contributed and for their inspiring questions.

H. C. van de Hulst

Leiden, The Netherlands
March, 1957

PART I

Basic Scattering Theory

1. INTRODUCTION

1.1. Scattering, Absorption, Extinction

This book is a treatise on the *scattering* of light. Hardly ever is light observed directly from its source. Most of the light we see reaches our eyes in an indirect way. Looking at a tree, or a house, we see diffusely reflected sunlight. Looking at a cloud, or at the sky, we see scattered sunlight. Even an electric lamp does not send us light directly from the luminous filament but usually shows only the light that has been scattered by a bulb of ground glass. Everyone engaged in the study of light or its industrial applications meets the problem of scattering.

Scattering is often accompanied by *absorption*. A leaf of a tree looks green because it scatters green light more effectively than red light. The red light incident on the leaf is absorbed; this means that its energy is converted into some other form (what form of energy is irrelevant for our purpose) and is no longer present as red light. Absorption is preponderant in materials such as coal and black smoke; it is nearly absent (at visual wavelengths) in clouds.

Both scattering and absorption remove energy from a beam of light traversing the medium: the beam is attenuated. This attenuation, which is called *extinction*, is seen when we look directly at the light source. The sun, for instance, is fainter and redder at sunset than at noon. This indicates an extinction in the long air path, which is strong in all colors but even stronger in blue light than in red light. Whether scattering or absorption is mainly responsible for this extinction cannot be judged from this observation alone. Looking sideways at the air, through which the sun shines, we see that actually blue light is scattered more strongly. Measurements show that all light taken away from the original beam reappears as scattered light. Therefore, scattering, and not absorption, causes the extinction in this example.

Other terminology is sometimes used but is not recommended. Here the word absorption is used in the sense of extinction as defined above.[1] Actual absorption is then designated as "pure absorption" or "true absorption." Throughout this book terms will be used as defined above, so that

$$\text{Extinction} = \text{scattering} + \text{absorption}.$$

[1] E.g., in the term "interstellar absorption."

3

1.2. Subject Limitations

Only a few of the multitude of scattering phenomena are treated in this book.

A first restriction is that we shall always assume that the scattered light has the same frequency (i.e., the same wavelength) as the incident light[2]. Effects like the Raman effect, or generally any quantum transitions, are excluded.

1.21   Independent Scattering

A second, most important limitation is that independent particles are considered. The distinction is roughly this: the scattering by well-defined separate particles, such as occur in a fog, is within the province of this book, whereas the scattering by a diffuse medium, as for instance a solution of a high polymer, is not discussed.

A more precise distinction may be made. If light traverses a perfectly homogeneous medium, it is not scattered. Only inhomogeneities cause scattering. Now, in fact, any material medium has inhomogeneities as it consists of molecules, each of which acts as a scattering center, but it depends on the arrangement of these molecules whether the scattering will be very effective.   In a perfect crystal at zero absolute temperature the molecules are arranged in a very regular way, and the waves scattered by each molecule interfere in such a way as to cause no scattering at all but just a change in the overall velocity of propagation. In a gas, or fluid, on the other hand, statistical fluctuations in the arrangement of the molecules cause a real scattering, which sometimes may be appreciable. In these examples, whether or not the molecules are arranged in a regular way, the final result is a *cooperative effect* of all molecules. The scattering theory then has to investigate in detail the phase relations between the waves scattered by neighboring molecules. Any such problem, in which the major difficulty is in the precise description of the cooperation between the particles, is called a problem of dependent scattering and is not treated in this book.[3]

Frequently, however, the inhomogeneities are alien bodies immersed in the medium. Obvious examples are water drops and dust grains in atmospheric air and bubbles in water or in opal glass. If such particles are sufficiently far from each other, it is possible to study the scattering

---

[2] This may technically be called coherent scattering.   However, this term is often used with a different connotation: an assembly of particles is said to scatter incoherently if the positions of the individual particles vary sufficiently (sec. 1.21).

[3] See the references at the end of this chapter.   References appear throughout at the end of each chapter.

by one particle without reference to the other ones. This will be called *independent scattering*; it is the exclusive subject of this book.

It may be noted that waves scattered by different particles from the same incident beam in the same direction still have a certain phase relation and may still interfere. The fact that the wavelength remains the same means that the scattered waves must be either in phase and enhance each other or out of phase and destroy each other, or any intermediate possibility. The assumption of independent scattering implies that there is no systematic relation between these phases. A slight displacement of one particle or a small change in the scattering angle may change the phase differences entirely. The net effect is that *for all practical purposes the intensities scattered by the various particles must be added* without regard to phase. It thus seems that the scattering by different particles is incoherent, although in the strict sense this is not true. An exception must be made for virtually zero scattering angles. In these directions no scattering in the ordinary sense can be observed. (See chap. 4.)

What distance between particles is sufficiently large to ensure independent scattering? Early estimates have shown that a mutual distance of 3 times the radius is a sufficient condition for independence. This may not be a general rule, but a more precise discussion is beyond the scope of this book. In most practical problems the particles are separated by much larger distances. Even a very dense fog consisting of droplets 1 mm in diameter and through which light can penetrate only 10 meters has about 1 droplet in 1 cm^3, which means that the mutual distances are some 20 times the radii of the drops. The same is true for many colloidal solutions.

1.22. Single Scattering

A third limitation is that the effects of multiple scattering will be neglected. Practical experiments most often employ a multitude of similar particles in a cloud or a solution. The obvious relations for a thin and tenuous cloud containing M scattering particles are that the intensity scattered by the cloud is M times that scattered by a single particle, and the energy removed from the original beam (extinction) is also M times that removed by a single particle. This simple proportionality to the number of particles holds only if the radiation to which each particle is exposed is essentially the light of the original beam.

Each particle is also exposed to light scattered by the other particles, whereas the light of the original beam may have suffered extinction by the other particles. If these effects are strong, we speak of *multiple scattering* and a simple proportionality does not exist. This situation may be illustrated by a white cloud in the sky. Such a cloud is like a dense fog; its droplets may be considered as independent scatterers. Yet the total intensity scattered by the cloud is not proportional to the number of droplets contained in it, for not each droplet is illuminated by full sunlight. Drops within the cloud may receive no direct sunlight at all but only diffuse light which has been scattered by other drops. Most of the light that emerges from a cloud has been scatttered by two or more droplets successively. It is estimated (for a very thick cloud) that about 10 per cent emerges after a single scattering.

Multiple scattering does not involve new physical problems, for the assumption of independence, which states that each droplet may be thought to be in free space, exposed to light from a distant source, holds true whether this source is the sun or another droplet. Yet the problem of finding the intensities inside and outside the cloud is an extremely difficult mathematical problem. This problem has been studied extensively in many ramifications. It is usually called the problem of radiative transfer. Common applications are the transfer of radiation in a stellar atmosphere and the scattering of neutrons in an atomic pile. The cases treated so far refer to rather simple forms both of the single scattering pattern (isotropic scattering, Rayleigh scattering) and of the entire cloud (infinite or finite slab with plane boundaries, sphere). The reader is referred to the literature for further details.

A simple and conclusive test for the absence of multiple scattering is to double the concentrations of particles in the investigated sample. If the scattered intensity is doubled, only single scattering is important. Another criterion may be the extinction. The intensity of a beam passing through the sample is reduced by extinction to $e^{-\tau}$ of its original value. Here τ is the optical depth of the sample along this line. If $\tau < 0.1$ single scattering prevails; for $0.1 < \tau < 0.3$ a correction for double scattering may be necessary. For still larger values of the optical depth the full complexities of multiple scattering become a factor. They may not prevent a determination of the scattering properties of a single particle, but they certainly make the interpretation much less clear. Caution is invariably required when the optical depth is not small in all directions through the sample.

Concluding this section it may be noted that this book treats only the very simplest case occurring in the theory of many particles. This leaves room for a thorough treatment of the scattering theory for one particle.

1.3. Historical Review

A proper understanding of the subject will be helped greatly by a review of its history, even though this has to be brief and can only show some of the highlights.

The nature of light was the subject of speculation and research by nearly all the great scientists of the seventeenth century. Snell's law, Newton's rings, Huygens' principle, and Fermat's principle date from this era. The general feeling was that light was something in the ether as sound is in air, but the phenomenon of polarization seemed to present insurmountable difficulties in such a concept, so that the century closed with the problem of the nature of light unsolved. Nor did the eighteenth century add much to the solution of this problem.

The decisive steps were taken by Young and Fresnel at the beginning of the nineteenth century. Young studied *diffraction* phenomena and showed that the pattern of maxima and minima in the shadow space behind a hair was caused by interference of waves coming from both sides of the hair. The nature of these waves remained obscure to him. Fresnel showed that these waves originate from the undisturbed wave front at either side of the obstacle. In this explanation Fresnel drew upon the old principle of Huygens that each point of a wave front may be considered as a center of secondary waves. By combining this principle with Young's principle of interference, Huygens' rule that the envelope of the secondary waves forms a new wave front had a natural explanation. If part of the original wave front is blocked by an obstacle, the system of secondary waves is incomplete so that diffraction phenomena occur. The exact agreement obtained between theory and experiments in a number of difficult problems left no doubt that Fresnel's explanation was correct. It will also form the basis of many problems discussed in this book.

The final explanation of *polarization* was given by Young's suggestion that the ether must exhibit transverse vibrations, like a rigid solid. It was a happy circumstance that in the same period Malus had discovered that polarization occurs at reflection and Brewster had measured the intensities of the polarized components at any angle of incidence. Then Fresnel, taking up Young's idea of transverse vibration, was able to derive these intensity rules theoretically on the basis of the simple boundary condition that the tangential component of the amplitude of vibration must be continuous.

One of the outstanding achievements of the later nineteenth century was Maxwell's electromagnetic theory of light, by which electric and optical phenomena were linked together. The modern way of expressing the boundary condition is, therefore, to say that the tangential component of the electric field must be continuous. Yet this improvement is not always essential to our problem. Many scattering problems involving polarization might as well be formulated in Fresnel's terminology as in modern terms by means of electric and magnetic fields.

The nineteenth century, particularly its second half, was the era of the great mathematical physicists: Poisson, Cauchy, Green, Kirchhoff, and the paragons Stokes and Rayleigh, if a very incomplete enumeration may suffice. With the exception of Stokes's discussion of the nature of natural and partly polarized light as a superposition of many polarized waves (sec. 5.13. of this book), no fundamental problems in optics were solved. The quest was for new skill in the mathematical formulation of complex phenomena rather than for added physical insight into simple phenomena. Coordinate systems in which the wave equation is separable were found. Fresnel's version of Huygens' principle was given a mathematical basis by Kirchhoff; Bessel functions and related functions were made into a powerful tool. A problem typical for this era was the scattering of light by a homogeneous sphere, one of the main topics of this book. It was among the more difficult problems and, though many special cases had been solved before, its full solution was formulated by Mie only in 1908.

This period came to an end with the rise of quantum mechanics. Debye was possibly the last one to study scattering problems of this kind with the devotion, insight, and mathematical technique displayed by the masters of the nineteenth century. Soon afterwards most of the mathematical physicists of top rank began devoting their time to studies of quantum mechanics or other fields of current interest. The scattering problems discussed in this book became a subject for applied scientists, interested in numerical results, or for students writing doctor's theses, of whom the writer was one. Formulae and numerical results were gradually collected, but few important ideas were added during this period.

The final stage of this brief and oversimplified history of the subject under review is its rather curious comeback in more recent years. The new interest devoted to it springs from very diverse sources, ranging from unemployment, which helped to start the New York Mathematical Tables Project, to the invention of radar and the development of quantum mechanics. Also new research in astronomy and chemistry prompted more extensive calculations than had been made before. It is important to see what quantum mechanics has to do with it. The analogy between traveling electrons and waves of light, or sound, had been an important help in the early development of quantum mechanics. So it was evident that scattering of an electron by an atom should have points of analogy to scattering of light, or sound, by a solid particle. In the late thirties quantum mechanics had developed so far that the need for accurate computations of scattering cross sections was felt. To this end new methods were devised, which partly were variants of the methods established in the domain of optics thirty or more years before and partly were of a new character. This has stimulated new studies of optical scattering problems. The method of phase shifts and the variational methods were new and have since found their application also in optical problems.

1.4. Sketch of the Book

The book has one theme: single scattering by independent particles. This means that only those experimental conditions are considered in which the particles are so far from each other that each of them is exposed to a parallel beam of light (i.e., light from a distant source) and has sufficient room to form its own scattering pattern, undisturbed by the presence of other particles. The book consists of three parts, which treat three distinct phases of the subject.

Part I. This part gives general theorems for particles that may have arbitrary size, shape, and composition. It is shown that the scattering by any finite particle is fully characterized by its four amplitude functions, S_1, S_2, S_3, and S_4, which are complex functions of the directions of incidence and of scattering. Knowledge of these functions suffices for computing the intensity and polarization of scattered light, the total cross sections of the particle for scattering, absorption, and extinction, and the radiation pressure exerted on the particle. For homogeneous spheres only two such functions $S_1(\theta)$ and $S_2(\theta)$ are needed, where θ is the scattering angle.

Chapter 2 presents all that can be expressed in terms of intensities without the introduction of phases and complex numbers. Chapter 3 introduces phases and complex amplitudes. Chapter 4 gives the main theorems derived for a single particle as well as a medium of independent particles; sec. 4.42 is the one most often referred to, as it gives the formulae for homogeneous spheres. The simplifications resulting if arbitrary particles are distributed in orientations with certain symmetry relations are summarized in chap. 5.

Part II. This is the main part of the book. It specifies the amplitude functions for a great variety of special particles. These chapters contain many cross references, as one special case and another special case often have a common limiting case. The final aim is, in each case, to derive the amplitude functions and the cross sections. Roughly, we can distinguish three groups of chapters.

Chapters 6 to 8 discuss particles that do not have a special form. They are, respectively, very small, very "soft," and very large.

Chapters 9 to 14 treat homogeneous spheres of arbitrary size. The two parameters in all formulae are $x = 2\pi a/\lambda$ (a = radius, λ = wavelength) and m, the refractive index. The rigorous solution (Mie) is derived, its limiting forms discussed, and many numerical results reproduced in tables and graphs. A survey of these chapters can best be gained from the Contents and from sec. 10.1.

Chapters 15 to 17 are devoted to particles of other regular forms, namely long circular cylinders, some miscellaneous geometrical forms, and large bodies with a smoothly curved surface.

Part III. This part gives selected applications in many domains of chemistry, physics, meteorology, and astronomy. Whereas the earlier parts are reasonably complete, this part is meant to give only typical examples of practical problems in which the preceding theories have proved important. Some common features of such applications are discussed in chap. 18, and chaps. 19 to 21 give examples from different fields. A scientist who is not directly concerned with the mathematical complexities of the subject may find it useful to turn to the chapter on his own field first in order to find his way into the formulae, graphs, and tables of the preceding parts.

References

The "dependent" scattering by polymers, etc., not treated in this book, is surveyed by

G. Oster, *Chem. Revs.*, **43,** 319 (1946).

B. H. Zimm, P. Doty, and R. Stein, *Theory and Application of Light Scattering,* New York, John Wiley & Sons, in preparation.

The "multiple" scattering, also excluded, is discussed by

S. Chandrasekhar, *Radiative Transfer,* Oxford, Oxford Univ. Press, 1950.

V. Kourganoff, *Basic Methods in Transfer Problems,* Oxford, Oxford Univ. Press, 1952.

H. C. van de Hulst, *The Atmospheres of the Earth and Planets* (2nd ed.), chap. 3, G. P. Kuiper, ed., Chicago, Univ. of Chicago Press, 1952.

An excellent book on the history is

E. T. Whittaker, *A History of the Theories of Aether and Electricity* (2nd ed.), part I, London, Longmans, Green & Co., 1952.

2. CONSERVATION OF ENERGY AND MOMENTUM

2.1. Scattering Diagram and Phase Function

In accordance with the limitations imposed upon our subject in the first chapter (sec. 1.2) we shall consider a single particle of arbitrary size and form, illuminated by a very distant light source. We shall inquire into the properties of the scattered light at a large distance from the particle. This implies the assumption that other particles leave sufficient room about this particle for the distant scattered field to be established (assumption of independence, sec. 1.21).

The most important property of the scattered wave is its *intensity*. By intensity I we shall understand the energy flux per unit area; its c.g.s. units are erg per cm^2 per sec. In optics this is called the irradiance. The incident wave and the scattered wave at any point in the distant field are unidirectional, i.e., each confined to one direction or to a very small solid angle around this direction. The term intensity as used in this book refers to the total energy flux in this solid angle. The waves are also assumed to be monochromatic, i.e., confined to one frequency or to a small frequency interval. The intensity refers to the total energy flux in this interval. Changing to other units (m.k.s. units: watt per m^2) makes no difference in the formulae, except where I is expressed directly in terms of electric and magnetic fields. With the same exception and with the exception of the formulae giving the radiation pressure we can read everywhere for I the illuminance, i.e., luminous flux per unit area (units: lumen per m^2 = lux).

Neither the incident nor the scattered light is completely characterized by its intensity; the additional properties are *polarization and phase*. The phases cannot be measured directly, but they are of importance in the correct formulation of the scattering of polarized light. So, throughout the second and main part of this book we shall work with scattering functions $S_1(\theta, \varphi)$ and $S_2(\theta, \varphi)$, which are complex numbers and describe amplitude *and* phase of the scattered waves. In the present chapter the relations are derived that can be formulated in terms of intensities without reference to phase. The phases are introduced in chap. 3 and applied to scattering theory in chaps. 4 and 5.

The scattered wave at any point in the distant field has the character

11

of a spherical wave, in which energy flows outward from the particle. The direction of scattering, i.e., the direction from the particle to this point, is characterized by the angle θ which it makes with the direction of propagation of the incident light and an azimuth angle φ (Fig. 1).

Let I_0 be the intensity of the incident light, I the intensity of the scattered light in a point at a large distance r from the particle, and k the wave number defined by $k = 2\pi/\lambda$, where λ is the wavelength in the surrounding medium. Since I must be proportional to I_0 and r^{-2} we may write

$$I = \frac{I_0 F(\theta, \varphi)}{k^2 r^2}.$$

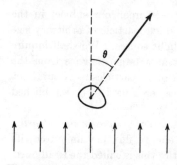

Here $F(\theta, \varphi)$ is a dimensionless function (F/k^2 is an area) of the direction but not of r. It also depends on the orientation of the particle with respect to the incident wave and on the state of polarization of the incident wave.

Fig. 1. Definition of scattering angle. Incident light is from below in this and in other figures.

The relative values of I, or of F, may be plotted in a polar diagram, as a function of θ in a fixed plane through the direction of incidence. This diagram is called a *scattering diagram* of the particle. When $F(\theta, \varphi)$ is divided by $k^2 C_{sca}$, where C_{sca} is the area defined below, another function of direction, the *phase function*,[1] is obtained. The phase function has no physical dimension, and its integral over all directions is 1.

2.2. Conservation of Energy

Let the total energy scattered in all directions be equal to the energy of the incident wave falling on the area C_{sca}. By this definition and by the preceding equation we have

$$C_{sca} = \frac{1}{k^2} \int F(\theta, \varphi) \, d\omega,$$

where $d\omega = \sin \theta \, d\theta \, d\varphi$ is the element of solid angle and the integral is taken over all directions. Likewise, the energy absorbed inside the particle may by definition be put equal to the energy incident on the area C_{abs}, and the energy removed from the original beam may by definition be

---

[1] The word phase has come into this expression via astronomy (lunar phases) and has nothing to do with the phase of a wave.

put equal to the energy incident on the area C_{ext}. The law of conservation of energy then requires that

$$C_{ext} = C_{sca} + C_{abs}$$

The quantities C_{ext}, C_{sca}, C_{abs}, are called the *cross sections* of the particle for extinction, scattering, and absorption, respectively. They have the dimension of area. Generally, they are functions of the orientation of the particle and the state of polarization of the incident light.

Non-absorbing particles have $C_{ext} = C_{sca}$. This cross section will sometimes be denoted by C without suffix.

2.3. Conservation of Momentum; Radiation Pressure

According to Maxwell's theory, light carries momentum as well as energy. The direction is that of propagation and the amount is determined by

$$\text{Momentum} = \frac{\text{energy}}{c},$$

where c is the velocity of light. We shall consider the component of momentum in the direction of propagation of the incident wave (which will further be called the forward direction). The momentum removed from the original beam is proportional to C_{ext}. Of this, the part C_{abs} is not replaced, but the part C_{sca} is partially replaced by the forward component of the momentum of the scattered light. This component in any direction is proportional to $I \cos \theta$. The total forward momentum carried by the scattered radiation is, therefore, proportional to

$$\overline{\cos \theta}\, C_{sca} = \frac{1}{k^2} \int F(\theta, \varphi) \cos \theta \, d\omega.$$

This equation defines the weighted mean of $\cos \theta$ with the scattering function as weighting function. Examples are worked out in secs. 10.62 and 12.5. It follows that the part of the forward momentum that is removed from the incident beam and *not* replaced by the forward momentum of the scattered light is proportional to

$$C_{pr} = C_{ext} - \overline{\cos \theta}\, C_{sca}.$$

For non-absorbing particles

$$C_{pr} = (1 - \overline{\cos \theta})\, C.$$

This momentum is given to the scattering particle. A certain force is, therefore, exerted on the scattering particle in the direction of propagation of the incident wave. This is the well-known phenomenon of *radiation pressure.* The force is equal to the force that would be exerted by the incident light on the area C_{pr} of a black wall. Its magnitude is

$$\text{Force} = I_0 C_{pr}/c.$$

Generally, the particle will also suffer a component of force perpendicular to the direction of propagation of the incident light. Its magnitude may be derived in a similar way. The influence of this component cancels out in a cloud of particles which are oriented at random. The particle is generally also subject to a torque. Its calculation requires not only the distant field but also the field approximated to a higher power of $1/r$.

2.4. Efficiency Factors

Most particles have an obvious geometrical cross section G. A sphere of radius a has, for instance, $G = \pi a^2$. The dimensionless constants

$$Q_{ext} = C_{ext}/G,$$
$$Q_{sca} = C_{sca}/G,$$
$$Q_{abs} = C_{abs}/G,$$
$$Q_{pr} = C_{pr}/G$$

will be called the *efficiency factors* for extinction, scattering, absorption, and radiation pressure, respectively. For quite general particles these factors depend on the orientation of the particle and on the state of polarization of the incident light. For spheres they are independent of both. In all cases we have

$$Q_{ext} = Q_{sca} + Q_{abs}.$$

2.5. Scattering Diagram for Polarized Light

The way in which $F(\theta, \varphi)$ depends on the form and size of the scattering particle is the main subject of this book and will be treated in chaps. 6 to 17. The way in which it depends on the state of polarization of the incident light will be treated here inasmuch as it can be formulated without referring to phase effects. The derivation is found in secs. 4.41, 5.13, and 5.14.

The full relations indicating how the intensity *and* state of polarization of the scattered light depend on the intensity *and* state of polarization of the incident light are contained in the matrix equation

$$\{I, Q, U, V\} = \frac{1}{k^2 r^2} \mathbf{F} \cdot \{I_0, Q_0, U_0, V_0\}.$$

Here I, Q, U, and V are the Stokes parameters of the scattered light, the meaning of which is explained in sec. 5.13, and I_0, Q_0, U_0, and V_0 are the corresponding parameters of the incident light. The matrix

F consists of 16 components, each of them a real function of the directions of incidence and scattering. The first of the four equations contained in this matrix equation reads

$$I = \frac{1}{k^2 r^2} \{F_{11}I_0 + F_{12}Q_0 + F_{13}U_0 + F_{14}V_0\}.$$

On comparing this equation with the one in sec. 2.1, we find that the value of F is

$$F = F_{11} + F_{12}\frac{Q_0}{I_0} + F_{13}\frac{U_0}{I_0} + F_{14}\frac{V_0}{I_0},$$

which specifies the manner in which F depends on the state of polarization of the incident light, defined by the quantities Q_0/I_0, U_0/I_0, and V_0/I_0. For incident natural light the latter quantities are zero, so that

$$F = F_{11}.$$

In the most general case the matrix **F** is asymmetric. For a single particle the number of independent constants is reduced to 7 because 9 relations exist between the 16 elements (sec. 5.14). For a cloud consisting of many particles the number of constants actually is 16. However, symmetry relations reduce the number of independent constants in most practical circumstances (secs. 5.2 and 5.3). For instance, the scattering matrix of a homogeneous spherical particle is characterized by 3 independent constants, i_1, i_2, and δ, which are functions of the angle θ (secs. 4.42 and 9.31). In this case 10 of the 16 constants are zero, and the other 6 are quadratic functions of the complex amplitude functions $S_1(\theta)$ and $S_2(\theta)$. Full formulae are found in sec. 4.42.

2.6. Scattering and Extinction by a Cloud Containing Many Particles

Let a cloud contain many scattering particles and be optically thin so that the incident intensity I_0 for each particle is the same (cf. sec. 1.22). It is then possible to write the equation

$$I_i = \frac{1}{k^2 r^2} F_i(\theta, \varphi) I_0$$

for each particle, each particle being denoted by an index i. The particles need not be similar. By summation we find that a formula of the same form as the first equation of sec. 2.1 holds for the entire cloud and that

$$F(\theta, \varphi) = \sum_i F_i(\theta, \varphi).$$

This addition is based on the assumption that phase effects may be neglected (sec. 1.22).

This formula may be applied to a volume element V of an extended medium that contains N identical particles per unit volume, each characterized by the same function $F(\theta, \varphi)$. The number of particles in the element then is NV, and the scattered intensity at a distance r is given by

$$I = \frac{NV}{k^2 r^2} F(\theta, \varphi) I_0.$$

If the projected area of the volume in this direction is A, the radiation is contained in the solid angle A/r^2 so that the average brightness (radiance) of the scattering element is

$$B = \frac{NVF(\theta, \varphi) I_0}{k^2 A}$$

In terms of luminous units I_0 is the illuminance, B is the luminance and is measured in lumen per steradian or candle per m².

Many evident variations on these results may be made. Among them may be mentioned the one that, if polarizations are involved, the summation formula holds for each of the 16 elements F_{ik} separately.

A similar addition formula holds for the cross sections C_{ext}, C_{sca}, and C_{abs}. The somewhat hidden physical ground of this rather obvious assumption is explained in secs. 4.22 and 4.3.

One very common application is the extinction by a cloud of spherical particles of the same composition but of different sizes. Here the efficiency factor $Q_{ext}(a)$ and the extinction cross section $C_{ext}(a) = \pi a^2 Q_{ext}(a)$ are functions of the radius a. Let there be $N(a)\, da$ particles per cm³ with radii in the interval da so that $\int_0^\infty N(a)\, da = N$ is the total number per cm³. The extinction coefficient of the medium, which equals the total cross section per cm³ (sec. 4.3), is then

$$\gamma = \int_0^\infty \pi a^2 Q(a) N(a)\, da.$$

Frequently it is desirable to change from the integration variable a to the variable $x = 2\pi a/\lambda$.

3. WAVE PROPAGATION IN VACUUM

The preceding chapters dealt with intensities only. Throughout this book we shall describe a wave not only by its intensity but also by its phase. This phase plays an important role in chap. 4 (the general extinction formula) and in chap. 5 (the general formulation of polarized light).

As preparation for these and further chapters we shall review the very simplest problem: the phase relations in a plane wave traveling in vacuum. This problem was first successfully discussed by Fresnel.

3.1. Fresnel's Formulation of Huygens' Principle

3.11. Complex Numbers

Complex amplitudes are an indispensable tool in the mathematical physics of wave phenomena. We briefly recall their definition and simple properties. The amplitude a and the phase α of a periodic wave are combined into a complex amplitude:

$$A = ae^{i\alpha} = a \cos \alpha + ia \sin \alpha.$$

Here i is the imaginary unit $\sqrt{-1}$. A complex number may be graphically displayed as a point in a plane (the complex domain) by plotting the real part $(a \cos \alpha)$ horizontally along the "real axis" and the imaginary part $(a \sin \alpha)$ vertically along the "imaginary axis." The factors -1 and i may always be written in an exponential form by means of

$$-1 = e^{i\pi}, \quad i = e^{i\pi/2}.$$

The physical quantity represented by a complex expression containing A as a factor is always assumed to be equal to the real part of that expression. The intensity of the wave is proportional to the square of the amplitude. This square may be written in various ways as

$$|A|^2 = A \cdot A^* = (a \cos \alpha + ia \sin \alpha)(a \cos \alpha - ia \sin \alpha) = a^2.$$

The vertical bars denote the modulus or absolute value, and the asterisk denotes the conjugate complex value (i replaced by $-i$ wherever it occurs explicitly or implicitly).

17

3.12. Derivation of Fresnel's Formula

Let a light source be at infinite distance so that we have a plane wave of constant intensity. Let the wave propagate in the positive z-direction. Further, t = time, λ = wavelength, k = wave number = $2\pi/\lambda$, c = velocity of light, ω = circular frequency = kc. The field of the light wave may then be represented by the complex expression[1]

$$u = e^{-ikz + i\omega t}$$

Here u may represent any component of the electric or magnetic fields;

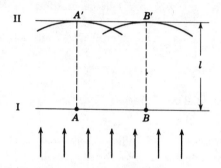

Fig. 2. Explanation of rectilinear propagation by Fresnel.

in the early nineteenth century, when light was still considered an elastic vibration of the ether, it was called the "disturbance."

Polarized light is characterized by two such amplitudes. All arguments in this chapter are valid for these two amplitudes separately to the degree of approximation to which they are valid at all. This means that the state of polarization is preserved. For simplicity the theory is presented with one amplitude only.

The planes z = constant are the planes of constant phase and therefore are called wave fronts. Figure 2 shows two such wave fronts, I and II, at a mutual distance l, which we suppose to be much larger than the wavelength:

$$kl \gg 1.$$

The disturbance at front I may be considered as the cause of the disturbance at front II at a time that is l/c seconds later. Roughly, the disturbance at A' is caused by that at A, the disturbance at B' is caused by

---

[1] The choice of i or $-i$ is arbitrary. The present choice of positive i in the time factor is the classical one and corresponds with the classical form of the complex refractive index (sec. 14.1). This choice is retained throughout this book. In modern books the other choice has become more common.

that at B. This corresponds to rectilinear propagation. However, this rule holds only approximately. If for instance all light at the left side of A is screened off, there is not a sharp shadow edge at A'. Apparently the disturbance at A' is caused to a certain extent by the disturbances at all points close to A. Huygens visualized this idea by assuming that all points of I were centers of secondary spherical waves and that the envelope of these waves determines the new wave front II (Huygens' principle). Huygens could explain the laws of reflection and refraction in this way, but the question *how large* a surrounding of A cooperates in determining the disturbance at A' remained open. Therefore, a quantitative theory could not be derived.

A very plausible solution of this problem was found by Fresnel, and its essential correctness was proved by the successful solution of many diffraction problems. Later it proved to be an approximation (valid whenever $kl \gg 1$) of the rigorous formula. Fresnel assumed that the secondary waves from all points of I should cooperate at A' (or indeed at any point beyond I) according to the principle of *interference*, just discovered by Young. This means that the disturbances due to these waves have to be added, each with its proper phase.

Let dS be a surface element of plane I at a distance r from A'. Then a spherical wave emitted by dS causes the disturbance

$$q \, dS \frac{e^{-ikr}}{r} u_{\mathrm{I}}$$

at A'. Here u_{I} is the disturbance at any point of plane I, and q is a constant to be fixed later. With rectangular coordinates in plane I, centered at A, we have

$$dS = dx \, dy$$

and if we anticipate that only points of I for which

$$x \ll l \qquad \text{and} \qquad y \ll l$$

effectively influence the disturbance at A', we also have

$$r = (l^2 + x^2 + y^2)^{1/2} \approx l + \frac{1}{2l} (x^2 + y^2).$$

The r in the denominator may be replaced by l. The analytic representation of Huygens' principle thus becomes

$$u_{\mathrm{II}} = \frac{q}{l} e^{-ikl} \int_{-\infty}^{\infty} \int_{-\infty}^{\infty} u_{\mathrm{I}} e^{\frac{-ik(x^2+y^2)}{2l}} \, dx \, dy.$$

Here u_{I} has been retained under the integrals because in diffraction problems in general u_{I} is a function of x and y, e.g., zero where the light is screened and 1 where it is not screened.

In the propagation treated in this section u_{I} is independent of x and y. The remaining integrals are of the type

$$\int_{-\infty}^{\infty} e^{\frac{-ikx^2}{2l}}\, dx = \left(\frac{2\pi l}{k}\right)^{1/2} e^{-i\pi/4} = \sqrt{(l\lambda)}\, e^{-i\pi 4}.$$

(If the limits are not ∞ this is called Fresnel's integral.)   The relation reduces then to

$$u_{\mathrm{II}} = \frac{q}{l}\, e^{-ikl}\, u_{\mathrm{I}} \cdot - il\lambda.$$

The factors l cancel out, as they should, for we know from the first formula in this section that the result should be

$$u_{\mathrm{II}} = e^{-ikl}\, u_{\mathrm{I}}.$$

This finally fixes the constant q:

$$q = \frac{i}{\lambda}$$

The full result is: *The disturbance caused by an area dS of a wave front with disturbance u_{I} at a point at the distance r, which is in a direction not too far from the direction of propagation, is*

$$\frac{i}{r\lambda}\, u_{\mathrm{I}}\, e^{-ikr}\, dS.$$

This formula suffices for a quite accurate solution of most diffraction problems. The derivation presented here closely resembles the one given by Fresnel in 1818.

3.13. Quantitative Examples

The applications of this formula are so well known that we may refer to textbooks on physical optics for further details. Straightforward applications in this book are found, e.g., in sec. 8.2 (diffraction by opaque bodies) and in sec. 11.3 (diffraction by transparent spheres).

Most textbooks emphasize the *relative* contributions of the various surface elements dS to the final result. For instance, if, as in Fig. 2, the disturbance at a distance l from the wave front I is sought, this wave front may be subdivided into successive zones, for which the phase of exp $(-ikr)$ is such that surface elements in those zones give contributions of alternating sign to the final amplitude. These are the Fresnel zones. The central zones contribute most effectively. The outer zones are less effective because their phases change so rapidly that their contributions tend to cancel out. The contribution of all points outside A lags behind in phase, because $BA' > AA'$. The average phase lag is $\pi/2$; it is compensated by the factor i in Fresnel's formula at the end of sec. 3.12.

It may be stressed that the same formula admits of quite simple quantitative applications. The quantitative definition of the effective area, useful for precise computations of intensity, is the following: *The amplitude at a distance l beyond a plane wave front is such as if an area $l\lambda$ of the wave front contributes with equal phase and the remaining part of the wave front not at all.*

Some examples may illustrate the application of this rule.

(a) *A lens without aberration* is placed in a parallel beam of light. By what factor does the intensity at its focus exceed the intensity in the undisturbed beam? *Answer:* The light at the focus comes with the same phase from the entire area S of the lens at a distance f, where f is the focal distance. If the lens were absent the light would come effectively from the area λf at the same distance. So the amplitude is increased by the factor $S/\lambda f$, the intensity by the factor $S^2/\lambda^2 f^2$. This formula is correct for any form of the lens and can be found in a much more elaborate manner from the theory of Fraunhofer diffraction near the focus.

(b) *Rectilinear propagation.* A pencil of light of length l can exist only if its width at its base is large compared to $\sqrt{(\lambda l)}$. Since l has to be $\gg \lambda$ for the preceding theory to have any sense at all, the required width is certainly $> \lambda$ but in cases of long pencils considerably greater. More precisely; a pencil of width of the order of $p\lambda$ can lead an independent existence over a length of the order of $p^2\lambda$. This implies quite clearly that *it is impossible to trace rays by geometrical optics through a particle with a size of the order of λ or smaller.* Only for very much larger particles is this method permitted. For applications see secs. 8.1 and 13.24.

3.2. Converging and Diverging Beams

3.21. Amplitude Outside a Focal Line

The preceding derivation was made for the propagation in a plane wave. The writer does not know whether Fresnel himself has considered converging and diverging beams. The extension of his reasoning to this case is so simple and yet so important that it will be given here. It will be applied in sec. 12.22.

A small section of a curved wave front is generally characterized by its two principal radii of curvature. The normals to the wave front converge at a distance f_1 from the wave front in one principal plane and at a distance f_2 in the perpendicular principal plane. The pencil of light coming from this wave front is astigmatic, with focal lines at distances f_1 and f_2 from the front. Figure 3a presents an example of a converging beam with positive f_1 and f_2. If one or both of the focal lines are behind the front, the corresponding values of f_1 and/or f_2 are negative.

Figure 3b shows the situation in one of the principal planes. F represents the focus so that $FA = FB = f$. By Fresnel's use of Huygens' principle we determine the disturbance at P, where $AP = l$. The path PB is longer than PA if P is before F; it is shorter than PA if P is beyond F. In either circumstance we have

$$PB - PA = \tfrac{1}{2}x^2 \left(\frac{1}{l} - \frac{1}{f} \right).$$

The formula holds also for a diverging beam (negative f). Always the integration by means of Huygens' principle leads to an integral of the same type as in sec. 3.12. The only difference with plane waves is the replacement of the factor $1/l$ in the exponent by $1/l - 1/f$. Since this is true for the two principal planes, the result is

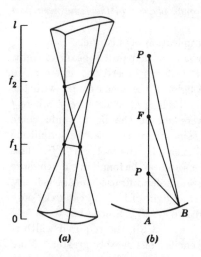

Fig. 3. Geometry of astigmatic beam for explanation of phase relation.

$$u_P = u_A \left(1 - \frac{l}{f_1} \right)^{-1/2} \left(1 - \frac{l}{f_2} \right)^{-1/2}$$

This simple formula tells all we need, both for the intensity and the phase at P. The following comments may be made:

(a) The intensity at P is

$$I_P = I_A \left(1 - \frac{l}{f_1} \right)^{-1} \left(1 - \frac{l}{f_2} \right)^{-1}.$$

Looking at the geometry of Fig. 3a we find that this is in agreement with the simple rule: *The intensity is inversely proportional to the cross section of the beam as computed by geometrical optics* (rectilinear propagation). The same rule follows directly from the conservation of energy in the beam.

(b) The phase factors are 1 or $e^{-i\pi/2}$, depending on whether the sign of $f - l$ is positive or negative. This sign changes from positive to negative if we pass a focal line. The resulting rule is: *The phase in an astigmatic beam advances by $\tfrac{1}{2}\pi$ at the passage of a focal line.*

After the passage of two focal lines the phase has advanced by π, which means a simple change of sign of the disturbance, or amplitude. This holds also for a beam with one focus (where the two focal lines may be thought to coincide). An "advance" in phase of $\pi/2$ means an apparently shorter path by the amount $\lambda/4$. This phase shift is essential, e.g., in the rainbow theory and in other examples of scattering functions computed by geometrical optics (secs. 12.22 and 13.2).

(c) The area of the front at A effective in determining the disturbance at P was found to be λl in the case of a plane wave. This changes in the case of an astigmatic beam to

$$\sqrt{\frac{\lambda}{\left|\frac{1}{l}-\frac{1}{f_1}\right|}} \cdot \sqrt{\frac{\lambda}{\left|\frac{1}{l}-\frac{1}{f_2}\right|}} \cdot$$

The two factors give the dimensions in the principal planes. This may also be expressed as the rule: *The effective area is λl times the square root of the increase of intensity from the front to the point P.*

If the disturbance at a very large distance, l, in the beam is sought, we find that the intensity decreases as $1/l^2$ and that the effective area of the initial front determining this intensity is $\lambda\sqrt{f_1 \cdot f_2}$| independent of l.

3.22. Amplitude in or near a Focal Line

The formulae just derived do not hold in a focus or focal line or close to a focus or focal line. The reason is that the large concentration of light has to come from a large effective area which may be too big for the preceding approximations to hold. Right at the focus or focal line the formulae indicate an infinite area. The dimension of the effective area becomes infinite in one direction if either $l = f_1$ or $l = f_2$. It becomes infinite in all directions if we want to calculate the intensity in a focus, $l = f_1 = f_2$.

Evidently an infinite effective area is unrealistic, so that our formulae must be modified to take account of the real properties of the wave front. The infinity is removed by one of two circumstances.

(a) *If a beam has a finite cross section, determined by a diaphragm, the dimension of the effective area that tends to become infinite* (i.e., one or both of the factors of the expression given in c above) *has to be replaced by the actual width of the beam.*

This is true for the classical (Fraunhofer) diffraction phenomena, e.g., the diffraction in a good telescope. The intensity at the exact position of the focus then is $I_F = I_A \cdot \dfrac{(\text{area})^2}{(\lambda l)^2}$, as follows from the proportionality of the intensity to the square of the effective area. See example a in sec. 3.13.

.(b) If a beam has aberrations, the assumption of a perfectly circular cross section of the wave front breaks down, and a more precise integration by the Fresnel formula gives a finite result. The integration over the wave front then has to include higher-order terms than x^2 and y^2. The exact result depends on the nature of these higher-order terms. This subject is an important one in modern lens design. In the context of this book it proves to be the fundamental clue to the theories of rainbow and glory, both of which may be mathematically characterized as "focal lines at infinity" (secs. 13.2 and 13.3).

3.3. Rigorous Diffraction Theory

The treatment of Huygens' principle in this chapter is in close accord-ance with Fresnel's first successful formulation and directly in line with Huygens' intuitive course of thoughts.

The formulation given some forty years later than Fresnel by Kirchhoff is less intuitive but more rigorous. It has opened a whole field of investigation in mathematical physics that has branched off somewhat from the main topic of this book. Readers acquainted with the latter type of diffraction problems may have some difficulty in establishing the connection with the topic of this book. The brief survey in this section may help them find their way.

The distinction can best be explained by examining the various uses of the word *diffraction*.

(a) *Diffraction = small deviation from rectilinear propagation.*

The original meaning of the word diffraction is a small deviation from rectilinear propagation, or, more generally, from ray propagation according to the theory of geometrical optics. Such deviations occur if an obstacle is placed in the beam of light. They are small only if the dimensions of the obstacle are large compared to the wave length. The appropriate theory is that of Fresnel, discussed in the preceding sections. This theory has rigorous validity only as an asymptotic theory for very large sizes, very small angles. Diffraction in this sense includes both the so-called Fresnel diffraction patterns (not near a focus) and the Fraunhofer diffraction patterns (near a focus, or at ∞ in a parallel beam).

The word diffraction in this sense is encountered in many common physical terms, such as the "diffraction pattern of a telescope," the "diffraction theory of aberrations." It will serve in this sense also in this book, viz., as the scattering law that holds asymptotically for very big particles and for very small angles. The following rules hold for diffraction in this sense.

1. The diffraction laws are the same for scalar waves (e.g., sound) and for light.

2. For the latter the diffraction effects are independent of polarization, and the state of polarization of the incident light is preserved.

3. The diffraction depends not on the material of an (opaque) body but only on the form of its geometrical shadow area. More details are found in sec. 8.2. The generalization to transparent particles, for which rules 1 and 2 still hold, is found in sec. 11.3.

(b) *Diffraction = wave motion in the presence of an obstacle of given size, form, and composition.*

It has become fairly common to use the word diffraction for the most complete generalization of the diffraction phenomenon described above. All problems discussed in the main part of this book are diffraction problems in the general sense. They require a complete solution of Maxwell's equations with given boundary conditions. Only simple geometrical forms are amenable to such a solution (see sec. 16.1). When the word diffraction occurs in this sense it may usually be replaced by the word scattering. The writer has generally preferred to do so, e.g., in the title of this book. This has made it possible to reserve the term diffraction in this book for the more restricted meaning of *a* on p. 24, in which it refers to asymptotic formulae for very large particle sizes and very small scattering angles.

(c) *Diffraction formula = an integral relation holding for a function satisfying the wave equation.*

A formula which can replace that of Fresnel for arbitrary angles was given by Helmholtz and Kirchhoff. This formula (see sec. 17.23 for a special case) gives the field in a point P in terms of the fields and the derivatives of the fields on a closed surface S surrounding P.

This well-known formula has the following properties:

1. It was originally derived for scalar waves only. A new somewhat similar formulation had to be given for electromagnetic waves. In the latter we have to note precisely the polarization of the waves because E and H do not enter the formulae in the same manner. A diffraction theory based on this formula gives different results for different polarization.

2. It is not evidently an expression of Huygens' intuitive principle.

3. Although it is rigorous, it does not give a rigorous solution of any of the extended diffraction problems mentioned in *b* above, for in such problems the fields are unknown in all space and we cannot find a surface

S on which they are known. Kirchhoff's own application was to take S along the opaque screen and its holes and then around P at a very large distance. He assumed that the field at the dark side of the opaque screens would be zero, and in the holes equal to the field of the incident wave. This is not true, except asymptotically for very small wavelengths, and in that circumstance Kirchhoff's solution is simplified to Fresnel's.

4. Kirchhoff's formula solves one important puzzle. Neither Huygens nor Fresnel had been able to show why the construction of the wave front could not have been made in exactly the same manner in the reverse direction, i.e., opposite to the direction of propagation. When the transition from Kirchhoff's formula to that discussed in sec. 3.12 is made and the terms containing the fields and the derivative of the field on S are kept separate, they give equal results with a factor 1 and cos θ, respectively. This means that the combined factor for nearly forward directions is nearly 2, the combined factor for nearly backward directions is nearly 0.

(d)   Diffraction = scattering by a flat particle.

This is meant to include the scattering by flat screens with holes, half planes, and so on. Sometimes the theory devoted to diffraction problems of this kind is called planar diffraction theory. Historically, the interest in plane screens comes from experimental conditions and from the fact that, in diffraction theories in the sense of *a*, a thick body and a thin body with the same form of shadow give the same diffraction pattern. It thus seemed natural to choose a mathematically thin screen as the simplest problem to discuss.

It was soon realized that thinness and opacity are compatible only for totally reflecting screens. A black and thin screen does not exist in Maxwell's theory (cf. sec. 14.1). Thus a problem was defined which is a special case of the scattering problem defined in *b* p. 25 It is the problem to solve Maxwell's equations with special boundary conditions on the faces of the screen When these conditions have been correctly formulated, the problem is solvable for strips and holes of arbitrary size. It is mathematically possible to replace the condition of total reflection by that of a given surface impedance.

The literature on these problems is very extensive: first, as they are technically important in radio engineering; second, because the "edge" conditions proved not so simple as had been anticipated and because wrong solutions were often made; third, because they proved a fertile field for trying out the variational technique. This method is based on the fact that it is possible to formulate integral equations in such a way that a trial solution with a limited accuracy will give another solution with a higher order of accuracy. Some numerical results are quoted in secs. 16.22 and 16.23.

The condition of thinness is met with sufficient accuracy for microwaves screened by metal plates (thickness $\ll \lambda$). It is not at all met by light waves screened by metal plates, e.g., in a spectrograph slit. Here a quite different problem arises, since the slit "edge" must be described as a thick body with a radius of curvature $\gg \lambda$. This problem is discussed in chap. 17.

References

The historical account is again splendidly given in Whittaker's book (see ref. at end of chap. 1). The Fresnel zones belong to the elementary theory of physical optics (see any good textbook). The phase shift π in passing through a focus (sec. 3.21) was discovered and studied by Gouy in 1890. The situation at the focus (sec. 3.22) was first studied by

F. Reiche, *Ann. Physik*, **29**, 65 and 401 (1909).

P. Debye, *Ann. Physik*, **30**, 755 (1909).

and has, as "the diffraction theory of aberrations," gained an important place in the study of lens design, e.g.,

J. Picht, *Optische Abbildung*, Braunschweig, Vieweg, 1931.

For a more recent study of the same problem in quite general terms, see

I. Kay and J. B. Keller, *J. Appl. Phys.*, **25**, 876 (1954).

Kirchhoff's theory is found in most textbooks; its generalization for electromagnetic waves was given by Kottler and others. Excellent reviews of this mathematically difficult subject (as well as the scalar diffraction theory) are:

B. B. Baker and E. T. Copson, *The Mathematical Theory of Huygens' Principle*, Oxford, Clarendon Press, (2nd ed.), 1950.

C. J. Bouwkamp, *Reps. Progr. in Phys.* **17**, 35 (1954).

4. WAVE PROPAGATION IN A MEDIUM CONTAINING SCATTERERS

This chapter contains the general relations in which the phase of the scattered wave occurs.

A complex amplitude function $S(\theta, \varphi)$ describing the amplitude and phase of a scattered *scalar wave* is defined in sec. 4.1. The amplitude function in the forward direction ($\theta = 0$) is $S(0)$. Its value is decisive for the extinction, as is shown in two ways: in sec. 4.2 with a view to a single particle, and in sec. 4.3 with a view to the medium as a whole. In the latter case also the phase velocity in the medium is found.

The formulation in these sections holds for *light waves* in the particular (but quite usual) case that $S(0)$ does not depend on polarization. The complete formulation for polarized light, which involves four amplitude functions, is given in sec. 4.4. A condition is that the mutual distances of the particles are $\gg \lambda$. The connection with the theory of molecular optics, in which this is not so, is clarified in sec. 4.5.

4.1. Amplitude Function of a Single Particle

Let a fixed particle of arbitrary shape and composition be illuminated by a plane scalar wave of infinite extent from the negative z-direction. The origin of coordinates is chosen somewhere in the particle. The "disturbance" of the incident light may be written as

$$u_0 = e^{-ikz+i\omega t}.$$

The historical term "disturbance" is used here in order to emphasize the far-reaching analogy between all kinds of waves. The results in secs. 4.2 and 4.3 hold for any kind of scalar waves (sound waves, electron waves, etc.) and also for electromagnetic waves (light waves) under the condition that the functions $S_1(0)$ and $S_2(0)$, which are defined in sec. 4.41, are the same and the functions $S_3(0)$ and $S_4(0)$ are zero. These conditions are fulfilled by homogeneous spherical particles. Under these conditions the results of secs. 4.2 and 4.3 are correct for light with an arbitrary state of polarization.

The scattered wave in the distant field is a spherical, outgoing wave with amplitude inversely proportional to the distance r. We may therefore write it in the form

$$u = S(\theta, \varphi) \frac{e^{-ikr+i\omega t}}{ikr},$$

thereby defining the *amplitude function* $S(\theta, \varphi)$ of the scattering particle. A factor i is added in the denominator for later convenience, and a factor k is added to make $S(\theta, \varphi)$ a pure number. Please note the choice of sign of i made in sec. 3.12. Combining the two preceding equations we have

$$u = S(\theta, \varphi) \frac{e^{-ikr+ikz}}{ikr} u_0.$$

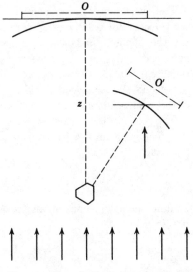

The amplitude function is, in general, complex and may also be written as

$$S(\theta, \varphi) = s \cdot e^{i\sigma},$$

where s is positive and σ is real, and both are functions of θ and φ. The phase σ depends on the choice of the origin and on the sign conventions that have to be made if polarized waves are involved. A negative $\sigma - (\pi/2)$ denotes a phase lag of the scattered wave. The amplitude s is independent of these choices. The intensity is proportional to the square of the amplitude. As a direct consequence we have

Fig. 4. Hypothetical experiment to measure the extinction by one particle.

$$I_{sca} = \frac{s^2(\theta, \varphi)}{k^2 r^2} I_0.$$

Any point in space is traversed by two wave systems: the incident and the scattered wave. In a mathematical point the flow of energy in these two waves cannot be separated, but by analyzing the light falling on an area O' (Fig. 4), we can distinguish between the two waves, each traveling in its own direction and with its own intensity. Physically this means that a telescope with the objective O' has to be used, which has sufficient resolving power to separate the images of the primary source (incident light) and of the secondary source (scattering particle). This distinction can be made even for very small θ by using a large objective O' at a very large distance (O' proportional to $r^{1/2}$). So, the definition of $S(\theta, \varphi)$ is physically meaningful for all directions.

4.2. The Fundamental Extinction Formula

4.21. A Single Particle

In the forward direction, $\theta = 0$, the same definition holds for reasons of continuity: superposed on the incident plane wave is a spherical wave, the amplitude and phase of which are characterized by the amplitude function

$$S(0) = s(0) \, e^{i\sigma(0)}.$$

The phase $\sigma(0)$ is independent of the choice of origin and of sign conventions for polarized light, provided that the same conventions hold for the incident and scattered waves. An experiment by which this spherical wave can be observed is impossible, for a telescope with objective O (Fig. 4) sees the primary and secondary source in the same direction: their images coincide. We shall calculate the total intensity of this combined image as seen in a large telescope at a very large distance.

The plane of the objective is $z =$ constant. Let (x, y, z) be a point in this plane, within the boundaries of O. For this point x and y are $\ll z$, so that

$$r = z + \frac{x^2 + y^2}{2z}.$$

Adding the amplitudes u_0 and u of the incident and scattered wave we obtain

$$u_0 + u = u_0 \left\{ 1 + \frac{S(0)}{ikz} e^{-ik(x^2+y^2)/2z} \right\}.$$

The large distance implies that the second term between brackets is $\ll 1$. The intensity incident on any point of O is found by squaring the modulus of this expression:

$$|u_0 + u|^2 = 1 + \frac{2}{kz} \operatorname{Re} \left\{ \frac{S(0)}{i} e^{-ik(x^2+y^2)/2z} \right\}.$$

Integrating this intensity over the entire objective with area O we find the total intensity of the combined image:

$$O - C,$$

where O is the integral of the first term and C the integral of the second term. The interpretation of this result is that the total light that enters the telescope is reduced by the presence of the particle. The amount of the reduction is such *as if* an area C of the objective had been covered up. This is the phenomenon of *extinction* described in chap. 2. Therefore we shall further write C as C_{ext}, according to the notation introduced earlier.

The double integral over $dx\,dy$ by which C_{ext} is defined contains two Fresnel integrals, each giving a factor $(2\pi z/ik)^{1/2}$, if the limits are extended to ∞ (see sec. 3.12). The result is

$$C_{ext} = \frac{4\pi}{k^2}\,\mathrm{Re}\left\{S(0)\right\}.$$

This is the fundamental extinction formula. The preceding derivation shows that the actual extinction process is not a blocking of the wave but a subtle interference phenomenon. The scattered wave removes some of the energy of the original wave by interference. The "active area" of the integrals is of the order of $z\lambda$. The telescope will register the full extinction only if its diameter is much larger than $(z\lambda)^{1/2}$. With increasing z the linear size of the active area increases as $z^{1/2}$. So the forward scattered wave has an ever weaker influence extending over ever wider circles in such a way that the total energy suppressed remains constant and equal to the energy incident on the area C_{ext}.

4.22.   A Cloud of Many Particles

Let us consider a cloud consisting of many independently scattering particles, not necessarily similar, each of them characterized by their amplitude function $S_i(\theta, \varphi)$. The index i denotes any individual particle. The precise amplitude function for the entire cloud would have to be derived by first referring all amplitude functions to a common origin and then adding them. The transformation to a common origin would involve large phase shifts depending on the precise positions of the particles. These shifts have a random nature and change rapidly even during one experiment. So the interference phenomena that would be expressed by the addition of the amplitudes are not noticed in practice. Consequently, the intensities and not the amplitudes have to be added. This is expressed in the formula

$$I(\theta, \varphi) = \sum_i I_i(\theta, \varphi).$$

By integration over all directions we find as a consequence

$$C_{sca} = \sum_i C_{i,sca}.$$

The situation for $\theta = 0$ is different. A change of origin does not change the phase, and the interference takes place irrespective of the precise position of the particles. Repeating the argument of the preceding section for a cloud of particles we find

$$S(0) = \sum_i S_i(0),$$

and as a consequence

$$C_{ext} = \sum_i C_{i,ext}.$$

It has thus been found that, for quite different reasons, the scattering and extinction cross sections of the single particles must be added to give the corresponding cross sections for the entire cloud. By subtraction, the same rule holds also for the absorption cross sections.

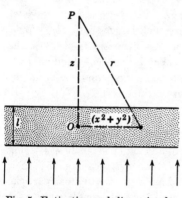

Fig. 5. Extinction and dispersion by a plane-parallel slab containing many particles.

4.3. Extinction and Dispersion in a Medium Containing Scatterers

Let us consider the experimental arrangement shown in Fig. 5. A plane-parallel slab containing many scattering particles, which are all identical and identically oriented so that they all have the amplitude function $S(\theta, \varphi)$, is illuminated from below. Let the slab have a height l and let there be N particles per unit volume. The field at P is influenced by scattering from all particles in the slab, but the *forward* traveling wave in P is coherently influenced only by the particles in the "active" volume of the slab, which coincides with the few central Fresnel zones as seen from P. If $PO = z$ is large enough, only small angles are involved. The total amplitude in P then is

$$u = u_0 \left\{ 1 + S(0) \sum \frac{1}{ikr} e^{-ik(x^2+y^2)/2r} \right\},$$

where the summation is extended over all particles in the active volume. If these particles are numerous, the Σ sign may be replaced by

$$\int N \, dx \, dy \, dz.$$

Direct integration gives

$$u = u_0 \left\{ 1 - \frac{2\pi}{k^2} NlS(0) \right\}.$$

This result may formally be represented as the influence of a complex refractive index of the medium as a whole. If the slab is replaced by a slab of homogeneous material with the complex refractive index $\tilde{m}$, close to 1, the amplitude of the wave is changed by the slab in the proportion

$$e^{-ikl(\tilde{m}-1)} = 1 - ikl(\tilde{m} - 1).$$

Consequently, the formal refractive index of the medium has the value

$$\tilde{m} = 1 - iS(0) \cdot 2\pi Nk^{-3}.$$

If we write the real and imaginary parts of $\tilde{m}$ separately by the formula

$$\tilde{m} = n - in'$$

(sec. 14.1), we find that this important formula comprises two effects: The real part,

$$n = 1 + 2\pi Nk^{-3} \operatorname{Im} \{S(0)\},$$

determines a phase lag (or advance) of the wave traveling through the medium. This is the phenomenon of dispersion: the phase velocity of the wave through the medium is changed to c/n. The imaginary part,

$$n' = 2\pi Nk^{-3} \operatorname{Re} \{S(0)\},$$

determines a decrease of the intensity. In general, the absorption coefficient in a medium with a complex refractive index is $\gamma = 2 kn'$ (sec. 14.1). Here the same formal relation gives the total extinction coefficient:

$$\gamma = 4\pi Nk^{-2} \operatorname{Re} \{S(0)\}.$$

With the result of sec. 4.21, this may also be written in the form

$$\gamma = NC_{ext}.$$

So the extinction cross sections of all particles in a unit volume have to be added to give γ.

We shall not try to give a complete analysis of the assumptions underlying this result. One of these is that $\tilde{m}$ is close to 1 (cf. sec. 4.5). It seems obvious that under certain conditions the point P may also be taken inside the slab. The successive effect of many thin slabs is then identical to the effect of one thick slab with the refractive index $\tilde{m}$. Consequently, for any value of l the influence of the slab on the amplitude is expressed by the factor

$$e^{-ikl(\tilde{m}-1)}.$$

A warning is in order against incorrect use of this refractive index. The formal "absorption" is not absorption in the proper sense but consists of scattering in all directions plus, possibly, absorption inside the scattering particles. The light "reflected" back from the slab cannot be derived from a formula based on this refractive index but should be computed by means of the scattering function for $\theta = \pi$.

4.4. Extinction and Dispersion of Polarized Light

4.41. General Case

In the preceding sections of this chapter polarization effects were left out for simplicity. The two basic formulae, the *fundamental extinction formula* (sec. 4.2) and the *complex dispersion formula* (sec. 4.3) as they stand, are correct: (1) for any type of scalar waves, (2) for light waves under simple conditions; see sec. 4.1. and below. The general formulation which includes all polarization effects is derived as follows.

As explained in the next chapter, the scattering in any direction is described by four amplitude functions, S_1, S_2, S_3, and S_4, all functions of θ and φ, which form a matrix $\mathbf{S}(\theta, \varphi)$ of four elements. The definition of $S(\theta, \varphi)$ in sec. 4.1 thus has to be replaced by

$$\begin{pmatrix} E_l \\ E_r \end{pmatrix} = \begin{pmatrix} S_2 & S_3 \\ S_4 & S_1 \end{pmatrix} \cdot \frac{e^{-ikr+ikz}}{ikr} \begin{pmatrix} E_{l0} \\ E_{r0} \end{pmatrix}.$$

See chap. 5 for details of notation[1]. The corresponding intensity matrix **F** has 16 elements and was mentioned in sec. 2.5.

By making $\theta = 0$ we obtain the four complex numbers $S_1(0)$, $S_2(0)$, $S_3(0)$, $S_4(0)$, forming the matrix $\mathbf{S}(0)$ of four elements. The derivation of sec. 4.3 can now be easily generalized, and the obvious result is that the electric field at a point P beyond the slab is

$$\begin{pmatrix} E_l \\ E_r \end{pmatrix} = \left\{ \begin{matrix} 1 - qS_2(0), & -qS_3(0) \\ -qS_4(0), & 1 - qS_1(0) \end{matrix} \right\} \begin{pmatrix} E_{l0} \\ E_{r0} \end{pmatrix},$$

where $q = 2\pi N l k^{-2}$.

This equation describes a combination of effects which include the following: (1) different phase velocities of plane-polarized light vibrating in different planes (double refraction); (2) different phase velocities of circularly polarized light of different sense of rotation (rotation of the plane of polarization); (3) different extinction of plane-polarized light vibrating in different planes (linear dichroism); (4) different extinction of circularly polarized light of different sense of rotation (circular dichroism).

These effects might be further formalized by introducing a matrix

---

[1] We follow Chandrasekhar's notation, where l and r stand for the last letters of the words parallel and perpendicular.

of four elements $\tilde{\mathbf{m}}$ instead of the scalar $\tilde{m}$ in sec. 4.3, but this is not needed in practice. The one special case that is needed throughout the major part of this book is formulated in sec. 4.42.

If many particles of different kinds or in different orientations form a cloud, or medium, the rules of sec. 4.22 may be applied. The scattering properties in an arbitrary direction are found by making separate sums of each component of the **F** matrix. The extinction and dispersion, likewise, are found by summing each component of the matrix **S**(0). These rules will seldom be used in complete generality; important applications are found in media where particles are oriented at random, or have certain symmetry properties; see secs. 5.2 to 5.4.

4.42. Spherical Particles

The formulae for spherical particles, which are used in 95 per cent of all applications, are very simple, even if arbitrary polarization of the incident light is admitted.

Spherical particles have $S_3 = S_4 = 0$. So two complex amplitude functions occur for any direction; these functions are $S_1(\theta)$ and $S_2(\theta)$; they depend only on the scattering angle θ. The matrix equation for an arbitrary direction $\theta \neq 0$ now gives the two relations

$$E_r = S_1(\theta) \frac{e^{-ikr+ikz}}{ikr} E_{r0},$$

$$E_l = S_2(\theta) \frac{e^{-ikr+ikz}}{ikr} E_{l0},$$

where r and l refer to the electric fields perpendicular to and parallel with the plane of scattering.

By taking the squares of the moduli we obtain for perpendicular polarization:

$$I = \frac{i_1}{k^2 r^2} I_0,$$

for parallel polarization:

$$I = \frac{i_2}{k^2 r^2} I_0,$$

for incident natural light:

$$I = \frac{\frac{1}{2}(i_1 + i_2)}{k^2 r^2} I_0,$$

where

$$i_1 = |S_1(\theta)|^2 \quad \text{and} \quad i_2 = |S_2(\theta)|^2.$$

These three equations are special cases of the relations that connect

the Stokes parameters of the incident and scattered light. In full they
are (sec. 5.14):

$$I_l = \frac{i_2}{k^2 r^2} I_{l0},$$

$$I_r = \frac{i_1}{k^2 r^2} I_{r0},$$

$$U = \frac{\sqrt{(i_1 i_2)}}{k^2 r^2} \{U_0 \cos \delta - V_0 \sin \delta\},$$

$$V = \frac{\sqrt{(i_1 i_2)}}{k^2 r^2} \{U_0 \sin \delta + V_0 \cos \delta\},$$

where δ is defined by

$$S_1(\theta) = \sqrt{i_1} \cdot e^{i\sigma_1}, \ S_2(\theta) = \sqrt{i_2} \cdot e^{i\sigma_2}, \ \delta = \sigma_1 - \sigma_2.$$

For forward scattering ($\theta = 0$) spherical particles give an even further
simplification by the fact that $S_1(0) = S_2(0)$. We denote this value by
$S(0)$ without index. The second matrix equation of sec. 4.41 then reduces
to the scalar equation

$$E = \{1 - qS(0)\}E_0,$$

which holds for any component of the electric (and magnetic) field
separately and is identical with the equation derived for scalar waves
(sec. 4.3).

The conclusion is that all results of the preceding sections hold without
restriction for light of arbitrary polarization, provided that the scattering
particles are homogeneous and spherical. There is no double refraction
or dichroism of any kind. The effect of the medium on the traversing
wave is just that of one complex refractive index (sec. 4.3):

$$\tilde{m} = 1 - i\, 2\pi N k^{-3} S(0).$$

The extinction cross section per particle is (sec. 4.2):

$$C_{ext} = 4\pi k^{-2} \operatorname{Re} \{S(0)\}.$$

Dividing by the geometrical cross section πa^2 and introducing the
notation $x = ka$ we obtain the efficiency factor (sec. 2.4):

$$Q_{ext} = \frac{4}{x^2} \operatorname{Re} \{S(0)\}.$$

4.5. Relation to the Classical Theory of Molecular Optics

Readers acquainted with the classical theory of molecular optics
may have an uncomfortable feeling after reading the preceding sections.

The theory of molecular optics explains the refractive index of a medium (e.g., a fluid or gas) from the scattering properties of the molecules. This seems to be exactly the same problem as discussed and solved in sec. 4.3 for particles of any type (which might include molecules). The annoying point is that the results are not the same. The Lorentz-Lorenz formula derived in molecular optics is *not* identical with the general formula for $\tilde{m}$ derived in sec. 4.3. Yet both formulae express the refractive index of the particulate medium as a whole ($\tilde{m}$) in terms of the properties of the individual particles that scatter light.

The solution of this paradox is that both formulae have a different domain of validity. The finesses cannot be discussed here, but the following sketch may indicate the fundamental difference between the theories behind both formulae. For convenience the particles are thought to be in vacuum, their number N per cm^3.

I. *Lorentz-Lorenz formula.* The condition is that the mutual distances of the particles are small compared with the wavelength λ. This necessarily means that the sizes of the particles themselves are $\ll \lambda$. Their scattering is then characterized by the polarizability (which is for convenience thought to be isotropic, sec. 6.11). The Lorentz-Lorenz formula is

$$4\pi\alpha N = \frac{3(\tilde{m}^2 - 1)}{\tilde{m}^2 + 2}.$$

It is based on the assumption that there are many particles in a sphere with radius λ. The field to which a molecule is exposed is due to the scattered fields of a fairly large number of other molecules at all sides. These fields are nearly in phase with the incident field. The statistical effect of these fields gives rise to the Lorentz-Lorenz formula.

Further implications are that the molecules may be densely packed and that α may be of the order of magnitude of the volume V of the individual molecules. This means that $\tilde{m}$ may deviate very appreciably from 1. In fact this makes the Lorentz-Lorenz formula important for practical purposes. The value of $\tilde{m}$ is real if α is real. This means that extinction by the medium occurs only if the particles have true absorption. (There is, however, a second-order imaginary term in α due to scattering; for further details see sec. 6.13).

II. *The formula of sec. 4.3.* The condition is that the mutual distances are large compared to the wavelength. They must also be large compared to the size of the particles (when this would be $> \lambda$). This means that dense packing is excluded. We have seen that under these conditions the wave propagation is governed by the function for forward scattering $S(0)$ of each particle (we consider only the simplest case, mentioned in sec. 4.42) and that the refractive index of the medium is

$$\tilde{m} = 1 - i2\pi N k^{-3} S(0), \quad (k = 2\pi/\lambda).$$

In this case the important part of the field (which has systematic phase relations with the original wave) does not come from particles on all sides but only from particles that are almost directly behind the point considered. The value of $\tilde{m}$ found from this formula is always close to 1, and its deviations from 1 in the imaginary direction may be just as large as those in the real direction even if the particles themselves do not absorb the light but only scatter it.

A further difference between I and II is that the theory under I may be extended to give the amount of radiation reflected from a (sharply bounded) slab of the particulate medium but that a similar extension of II is impossible. This is because the theory under I is based on scattering in all directions whereas the theory under II is based on forward scattering.[2] The radiation "reflected" back from a slab of material under the conditions of II is just the sum total of the radiation scattered back by the individual particles according to sec. 2.16, no matter what the boundary or shape of the slab is, multiple scattering being excluded.

III. *A case in common.* It would be a pity to let these formulae stand as they are without any interrelation. A case common to both may be constructed by assuming in I that the mutual distances are very large, or in II that the particles are very small.

The condition that the distances are large is not formally possible in I, but it means that there simply is no scattered field counteracting the field of the original wave; therefore the formula must still be considered correct. When in I, $\tilde{m}$ approaches 1, the right-hand member reduces to $\tilde{m}^2 - 1$, or in as good an approximation, $2(\tilde{m} - 1)$. On the other hand we may write in II for very small particles $S(0) = ik^3\alpha$; see sec. 6.12. The resulting formula arising either from I or from II is

$$\tilde{m} - 1 = 2\pi\alpha N.$$

The very special condition that the particles themselves would have the volume V and consist of a homogeneous material with the refractive index m close to 1 gives, as in sec. 6.22, the relation $2\pi\alpha = (m - 1)V$, which leads to the even simpler formula

$$\tilde{m} - 1 = (m - 1)NV.$$

[2] This difference was ignored in the paper by Urick and Ament (1949). (See reference at end of this chapter.)

This expresses the obvious result that $\tilde{m} - 1$ is proportional to the occupied fraction NV of the total volume.

There has been some confusion on the question whether to write $\tilde{m}^2 - 1$ or $2(\tilde{m} - 1)$ in the formulae for $\tilde{m}$ close to 1. This makes no difference mathematically in the formulae derived in this chapter, but, since the Lorentz-Lorenz formula contains $\tilde{m}^2$, there has been a tendency to prefer the notation with $\tilde{m}^2$. The derivation given in sec. 4.3 gives no good ground for this preference.

References

The scheme of this chapter is new. The intuitive method of relating the extinction and the forward scattering (sec. 4.2) follows:

H. C. van de Hulst, *Physica* ('s Grav.), **15**, 740 (1949).

The author did not know at that time that this extinction theorem had quite a history in quantum mechanics and sound; e.g.,

E. Feenberg, *Phys. Rev.*, **40**, 40 (1932).
W. Heisenberg, *Z. Physik*, **120**, 513 (1943).
H. Levine and J. Schwinger, *Phys. Rev.*, **74**, 958 (1948).
G. C. Wick, *Phys. Rev.*, **75**, 1459 (1949).
M. Lax, *Phys. Rev.*, **78**, 306 (1950).
P. M. Morse and H. Feshbach, *Methods of Theoretical Physics*, part II, pp. 1069 and 1547, New York, McGraw-Hill Book Co, 1953.

Nor did the writer realize that the same problem is contained in the propagation problem (sec. 4.3). This had been proposed and solved for small dielectric and metallic particles (virtually no extinction) by

Lord Rayleigh, *Phil. Mag.*, **47**, 375 (1899) (*Sci. Papers* 247).
J. C. Maxwell Garnett, *Phil. Trans.*, **203**, 385 (1904); **205**, 237 (1906).

and in a more general form by

R. Gans and H. Happel, *Ann. Physik*, **29**, 294 (1909).

while it also had attracted the attention of more recent writers:

E. Schoenberg, *Z. Physik*, **109**, 127 (1938).
R. J. Urick and W. S. Ament, *J. Acoust. Soc. Amer.*, **21**, 115 (1949).
B. H. Zimm and W. B. Dandliker, *J. Phys. Chem.*, **58**, 644 (1954).

Generalization to polarized light (sec. 4.4) is a matter of routine. The Lorentz-Lorenz formula (sec. 4.5) and related problems are discussed in many books, e.g.,

M. Born, *Optik*, secs. 73 and 74, Berlin, J. Springer, 1933.

5. POLARIZED LIGHT AND SYMMETRY RELATIONS

All scattering processes lead to polarization. Scattering problems that can be solved without explicit reference to the state of polarization of the incident and scattered light are exceptional. In the preceding chapters (e.g., sec. 2.5) we did not fully specify the dependence of the scattering process on the state of polarization of the incident light. The general formulation would have tended to obscure the main issue. We shall now remedy this omission and describe how arbitrary particles scatter light of the most general state of polarization.

5.1. The Most General State of Polarization

The characteristic difference between a scalar wave (like sound) and a vectorial wave (like light) is that the former needs one, the latter two amplitudes to characterize a wave traveling in a certain direction. The intensity of a scalar wave is the square of the amplitude,

$$I = A^2,$$

apart from a constant factor. The corresponding formulae for light waves, developed by Stokes, are not given in most textbooks. For that reason they are briefly derived in this section. The physics behind them is not deeper than behind the formula $I = A^2$, and the formalism is not as forbidding as it may seem at first sight.

5.11. Plane of Reference

Let us consider a beam of light with a certain frequency and traveling in one direction. We choose a *plane of reference* through the direction of propagation. By **r** we shall denote the unit vector along the normal of this plane (sense arbitrary) and by **l** the unit vector in this plane and perpendicular to the direction of propagation. The sense is chosen so that $\mathbf{r} \times \mathbf{l}$ is in the direction of propagation. The two letters stand for the last letters of the words perpendicular and parallel. The choice of the plane of reference is arbitrary, in principle. In scattering problems we shall always choose the plane containing the incident and scattered beams as the common plane of reference for the two beams.

5.12.   Simple Waves

By a simple wave we understand a plane-wave solution of Maxwell's equation with general, i.e., elliptic, polarization. The electric vector is transverse, so it may be represented by

$$\mathbf{E} = \mathrm{Re}\,[E_l\mathbf{l} + E_r\mathbf{r}],$$

where E_l and E_r are complex, oscillating functions. The Stokes parameters are defined by

$$I = E_l E_l{}^* + E_r E_r{}^*,$$
$$Q = E_l E_l{}^* - E_r E_r{}^*,$$
$$U = E_l E_r{}^* + E_r E_l{}^*,$$
$$V = i(E_l E_r{}^* - E_r E_l{}^*),$$

where an asterisk denotes the conjugate complex value. They are real numbers that satisfy the relation

$$I^2 = Q^2 + U^2 + V^2.$$

The first one, I, represents the intensity, i.e., the flow of energy per unit area. The other parameters have the same dimension. A constant factor common to all four parameters has for convenience been omitted. In an alternative system $I_l = E_l E_l{}^*$ and $I_r = E_r E_r{}^*$ are used instead of I and Q. The full set then is (I_l, I_r, U, V).

If the propagation constant k and the circular frequency ω are prescribed, the most general simple wave is defined by the positive amplitudes a_l, a_r, and the phases ε_1, ε_2, in

$$E_l = a_l e^{-i\varepsilon_1}e^{-ikz+i\omega t},$$
$$E_r = a_r e^{-i\varepsilon_2}e^{-ikz+i\omega t}.$$

This gives

$$I = a_l{}^2 + a_r{}^2,$$
$$Q = a_l{}^2 - a_r{}^2,$$
$$U = 2a_l a_r \cos \delta,$$
$$V = 2a_l a_r \sin \delta,$$

where $\delta = \varepsilon_1 - \varepsilon_2$.

Independently, we may give a geometric representation of a simple wave with the most general (elliptic) polarization. This is

$$\mathbf{E} = a\mathbf{p} \cos \beta \sin (\omega t - kz + \alpha) + a\mathbf{q} \sin \beta \cos (\omega t - kz + \alpha).$$

Here $\mathbf{p}$ and $\mathbf{q}$ are unit vectors along the long and short axes, a^2 is the intensity, α is a phase angle. The ellipticity is given by $\tan \beta$; if it is 0, we have linear polarization, -1 gives left-handed circular light, and $+1$ right-handed circular light. The orientation of the ellipse is given by the angle χ (Fig. 6).

The geometrical specification, by means of (a, α, β, χ), may be compared to the analytical one by means of $(a_l, a_r, \varepsilon_1, \varepsilon_2)$. A simple reduction leads to the relations

$$a_l{}^2 = a^2 \left(\cos^2 \beta \cos^2 \chi + \sin^2 \beta \sin^2 \chi\right),$$
$$a_r{}^2 = a^2 \left(\cos^2 \beta \sin^2 \chi + \sin^2 \beta \cos^2 \chi\right),$$
$$\tan (\alpha + \varepsilon_1) = - \cot \chi \cot \beta,$$
$$\tan (\alpha + \varepsilon_2) = + \tan \chi \cot \beta.$$

Hence

$$\tan \delta = \frac{\tan 2\beta}{\sin 2\chi}.$$

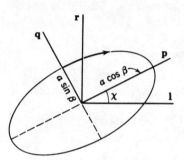

So the Stokes parameters, in the geometric notation, are

$$I = a^2,$$
$$Q = a^2 \cos 2\beta \cos 2\chi.$$
$$U = a^2 \cos 2\beta \sin 2\chi,$$
$$V = a^2 \sin 2\beta.$$

Fig. 6. Parameters defining the state of polarization of a simple (elliptically polarized) wave. The direction of propagation is into the paper.

If another plane of reference is chosen, only χ changes, so that I, $Q^2 + U^2$, and V are invariant for this choice. Taking both l and r in the opposite sense does not change the Stokes parameters.

5.13. Actual Light and Optical Equivalence

Actual light consists of many "simple waves" in very rapid succession. The duration of a coherent wave train is of the order of 10^{-8} seconds, or shorter, depending on the sharpness of the spectral line that is used as a source of monochromatic light. Measurable intensities therefore refer always to a *superposition of many millions of simple waves with independent phases.*

By the Stokes parameters of the entire beam of light we shall understand the sums

$$I = \Sigma I_i, \; Q = \Sigma Q_i, \; U = \Sigma U_i, \; V = \Sigma V_i,$$

where the index i denotes each independent simple wave and any constant factors are again omitted. We now have

$$I^2 \geqslant Q^2 + U^2 + V^2.$$

The equals sign holds only if β and χ are the same for all simple waves;

the light is then called fully polarized.   In all other cases it is called partially polarized, except when $Q = U = V = 0$, which is the definition of natural or unpolarized light.

The importance of the Stokes parameters is seen from the following argument.   Let a simple wave with arbitrary polarization pass through an arbitrary optical instrument which produces no incoherent effects so that a simple wave also emerges.   The instrument may cause scattering, reflection, or refraction, and may include birefringent crystals, Polaroid sheets, quarter-wave plates, etc.   The ingoing wave is represented by the two field components E_{l0} and E_{r0} with respect to an arbitrary plane of reference through the direction of propagation of the ingoing beam. The outgoing wave is, likewise, represented by the field components E_l and E_r with respect to a plane of reference through the outgoing beam.

In optical instruments it is not possible (as, for example, in radio waves) to introduce non-linear effects.   Since only linear processes are involved we have

$$E_l = A_2 E_{l0} + A_3 E_{r0},$$
$$E_r = A_4 E_{l0} + A_1 E_{r0}.$$

Computing the Stokes parameters of the outgoing wave (I, Q, U, V) we find that they can be expressed in terms of the Stokes parameters of the ingoing wave (I_0, Q_0, U_0, V_0), by means of another linear transformation, which reads in an abridged notation:

$$(I, Q, U, V) = \mathbf{F} \cdot (I_0, Q_0, U_0, V_0).$$

Here $\mathbf{F}$ is a matrix of 16 coefficients, each of which is a real number that is a quadratic expression of the coefficients A_1, A_2, A_3, and A_4. The explicit form is derived in sec. 5.14.

Let us now consider an actual (partially polarized) beam.   Just following the definitions, i.e., making the summations of the parameters for simple waves both for the ingoing and the outcoming beam, we find that the final transformation formula is exactly the same one as quoted above, with the same matrix $\mathbf{F}$.

The principle of optical equivalence states: It is impossible by means of any instruments to distinguish between various incoherent sums of simple waves that may together form a beam with the same Stokes parameters (I, Q, U, V).

The physical arguments for this principle are as follows: Any laboratory method for measuring the state of polarization of a beam uses an instrument which makes a linear transformation, as discussed above (Nicol prism, quarter-wave plate), and then measures an intensity.   The formulae derived above show that such instruments can yield no other data than linear combinations of the original Stokes parameters.   Different instruments can measure different combinations, so we can use (by well-known methods) a set of instruments in order to determine the four Stokes parameters separately. But *this is all*.

The principle of optical equivalence shows that the Stokes parameters are not merely *some* interesting quantities but the complete set of quantities that are needed to characterize the intensity and state of polarization of a beam of light, inasmuch as it is subject to practical analysis. Further distinctions are theoretically possible but do not correspond to measurable differences. In particular there is only one kind of natural light:

$$(I, Q, U, V) = (I, 0, 0, 0),$$

which, theoretically, may be composed of simple waves in an infinite variety of ways.

5.14. The Form of the Transformation Matrix

The remaining problem is to write the transformation matrix **F** for any given process. It is useful to start from the transformation matrix $\begin{pmatrix} A_2 & A_3 \\ A_4 & A_1 \end{pmatrix}$ for the amplitudes. We define the real numbers:

$$M_k = A_k A_k{}^* = |A_k|^2,$$
$$S_{kj} = S_{jk} = {}^1\!/_2(A_j A_k{}^* + A_k A_j{}^*),$$
$$-D_{kj} = D_{jk} = {}^i\!/_2(A_j A_k{}^* - A_k A_j{}^*).$$

Here j, $k = 1, 2, 3, 4$; asterisks denote conjugate complex numbers. Straightforward substitution then gives a transformation for the I, Q, U, V, governed by the matrix:

$$\mathbf{F} = \left[\begin{array}{cccc} \frac{1}{2}(M_2 + M_3 + M_4 + M_1), & \frac{1}{2}(M_2 - M_3 + M_4 - M_1), & S_{23} + S_{41}, & -D_{23} - D_{41} \\ \frac{1}{2}(M_2 + M_3 - M_4 - M_1), & \frac{1}{2}(M_2 - M_3 - M_4 + M_1), & S_{23} - S_{41}, & -D_{23} + D_{41} \\ S_{24} + S_{31}, & S_{24} - S_{31}, & S_{21} + S_{34}, & -D_{21} + D_{34} \\ D_{24} + D_{31}, & D_{24} - D_{31}, & D_{21} + D_{34}, & S_{21} - S_{34} \end{array} \right].$$

This matrix contains 7 independent constants, resulting from the 8 constants in the A's minus an irrelevant phase. So there must be 9 relations between the 16 coefficients. These relations have not explicitly been derived.

Somewhat simpler but less symmetric is the matrix defined by

$$(I_l, I_r, U, V) = \mathbf{F}'(I_{l0}, I_{r0}, U_0, V_0).$$

It reads:

$$\mathbf{F}' = \left\{ \begin{array}{cccc} M_2, & M_3, & S_{23}, & -D_{23} \\ M_4, & M_1, & S_{41}, & -D_{41} \\ 2S_{24}, & 2S_{31}, & S_{21} + S_{34}, & -D_{21} + D_{34} \\ 2D_{24}, & 2D_{31}, & D_{21} + D_{34}, & S_{21} - S_{34} \end{array} \right\}.$$

These most general formulae will scarcely ever be needed, for in most practical situations symmetry properties reduce the number of independent constants.

One simple case is extremely important. If $A_3 = A_4 = 0$, the transformation equations for the field are simply

$$E_l = A_2 E_{l0},$$
$$E_r = A_1 E_{r0}.$$

The matrix $\mathbf{F}'$ then assumes the very simple form

$$\mathbf{F}' = \left\{ \begin{array}{cccc} M_2 & 0 & 0 & 0 \\ 0 & M_1 & 0 & 0 \\ 0 & 0 & S_{21} & -D_{21} \\ 0 & 0 & D_{21} & S_{21} \end{array} \right\},$$

in which only 3 independent parameters occur because there is one interrelation: $S_{21}^2 + D_{21}^2 = M_2 M_1$.

The full transformation equations for the Stokes parameters are in this case

$$I_l = |A_2|^2 \cdot I_{l0},$$
$$I_r = |A_1|^2 \cdot I_{r0},$$
$$U = |A_1| \cdot |A_2| \cdot \{U_0 \cos \delta - V_0 \sin \delta\},$$
$$V = |A_1| \cdot |A_2| \cdot \{U_0 \sin \delta + V_0 \cos \delta\},$$

where δ is the phase difference between A_1 and A_2.

An illustrative example is the action of a properly oriented quarterwave plate,

$$\mathbf{A} = \left\{ \begin{array}{cc} 1 & 0 \\ 0 & i \end{array} \right\}, \quad \mathbf{F}' = \left\{ \begin{array}{cccc} 1 & 0 & 0 & 0 \\ 0 & 1 & 0 & 0 \\ 0 & 0 & 0 & -1 \\ 0 & 0 & 1 & 0 \end{array} \right\},$$

which transforms an elliptical wave,

$$(I_0, Q_0, U_0, V_0) = (1, \cos 2\beta, 0, \sin 2\beta),$$

into a linear wave,

$$(I, Q, U, V) = (1, \cos 2\beta, -\sin 2\beta, 0).$$

Scattering by an arbitrary particle (sec. 4.41) defines a transformation applied to the amplitude of the incident wave and giving the amplitudes of the scattered wave. The matrix $\mathbf{S}$ takes the place of the transformation matrix $\mathbf{A}$. The corresponding transformation of the Stokes parameters may be copied from the equations given above. Spherical particles have a scattering matrix of the simplest type, in which $S_3(\theta) = S_4(\theta) = 0$. The corresponding intensity formulae were given in sec. 4.42. The full transformation equations for the Stokes parameters, given above, admit of the following conclusions. Incident linear light with an arbitrary plane of polarization (only restriction: $V = 0$) gives, in general, elliptically polarized light. However, if the incident light is plane-polarized in one of the main planes, then it remains linearly polarized in the corresponding plane and we have to look only at the two upper transformation equations. This is what we shall do in most of this book.

5.2. Symmetry Relations for Scattering in an Arbitrary Direction

5.21. General Method

The rest of this chapter deals with the following problem: A cloud, or medium, of scattering particles contains many identical particles with different orientations in space. What simplifications of the scattering matrix result if certain assumptions about the distribution of orientations are made? At the same time we shall discuss the simplifications arising from the additional assumption that particles of one kind and also particles of a second kind, which are the mirror images of particles of the first kind, are present in equal numbers. If the particles have a plane of symmetry they are their own mirror images, and the second assumption is fulfilled automatically.

Questions of this type have been studied before in various special cases. The purpose of this chapter is not to give a detailed derivation but rather to give a complete survey of the symmetry relations for ready reference. Many different cases must be distinguished; some of them correspond to rather artificial assumptions, which are not likely to be encountered in practical problems.

The general method is the following. If the scattering matrix of a particle in a particular position and for a particular direction is known, the scattering matrix of the same particle, or its mirror image, in certain symmetrical positions is also known. This holds first for the four complex elements of the amplitude matrix $\mathbf{S}$ (defined in sec. 4.41). The corresponding relations for the sixteen real elements of the intensity matrix $\mathbf{F}$ (defined in sec. 2.5) are easily derived by means of the formula given in sec. 5.14 (where the amplitude matrix is written $\mathbf{A}$ instead of $\mathbf{S}$). The elements of the matrices $\mathbf{F}$ have to be added when different particles or differently oriented particles occur together in the same medium. This follows from the consideration that the waves scattered by these particles are essentially incoherent (secs. 1.22 and 4.22). The addition causes certain non-diagonal elements of the final matrix **F** to vanish and others to become equal, depending on the exact assumptions made.

This section and the following ones were largely inspired by a paper by F. Perrin (1942). Perrin has discussed only the case of random orientation so that his formulae are less complete. Also, he discusses only the intensity matrix and its transformation properties and does not mention the amplitude matrix. This seems a more tedious and less foolproof method than ours, in which the transformation properties of the so much simpler amplitude matrix are used. Perrin's results and ours agree in all cases that he has discussed.

In agreement with Perrin's notation we shall (only in this chapter) write the matrix **F** in the form

$$\begin{pmatrix} a_1 & b_1 & b_3 & b_5 \\ c_1 & a_2 & b_4 & b_6 \\ c_3 & c_4 & a_3 & b_2 \\ c_5 & c_6 & c_2 & a_4 \end{pmatrix}.$$

If no assumptions are made at all, the quadratic relations between the elements of the matrix **F** of a given particle in a given position are lost in the addition of the elements of differently oriented particles. They are then replaced by inequalities, which we shall ignore. Generally, therefore, there are 16 independent parameters. Under the various symmetry assumptions made below, the number of free parameters is reduced. Equal elements will be denoted by the same symbol, zero elements by 0.

5.22. Resulting Formulae

If we consider a direction for which the scattering angle is not 0° or 180°, the directions of incidence and scattering define a plane, *the scattering plane*. This is taken as the plane of reference (plane ZOY in Fig. 7). As usual, the perpendicular unit vector **r** is normal to this plane (sense arbitrary). The sense of the parallel unit vectors **l** and **l′** is then fixed by the requirement that $\mathbf{r} \times \mathbf{l}$ is in the direction of propagation of the incident light and $\mathbf{r} \times \mathbf{l′}$ is in the direction of propagation of the scattered light. The line in the scattering plane that bisects the angle $\pi\text{-}\theta$ between the incident and scattered beam will be called the *bisectrix*. The plane through the bisectrix and perpendicular to the plane of scattering will be called the *bisectrix plane*.

A given position of the particle is taken as the initial position (a). There are three other positions in which the scattering matrix may be expressed in terms of the same coefficients as in position (a). Rotation

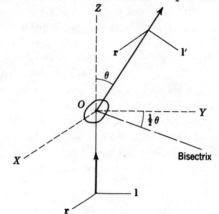

Fig. 7. Pairs of unit vectors ($\mathbf{r}$, $\mathbf{1}$) and ($\mathbf{r}$, $\mathbf{1}'$) in terms of which the states of polarization of the incident and scattered light are described. The scattering plane is the ZOY plane.

of 180° about the bisectrix puts the same particle in the position (b), which we shall call the "reciprocal position" of (a). Mirroring with respect to the plane of scattering gives a position (c) of the mirror image

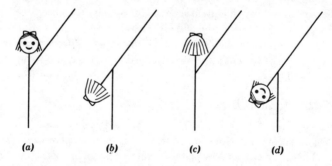

(a)   (b)   (c)   (d)

Fig. 8. Two positions of an arbitrary particle and two positions of its mirror particle that give rise to symmetry relations of the scattering functions.

of the initial particle. Mirroring with respect to the bisectrix plane gives another position (d) of the mirror particle. Any two of these three transformations in succession give the third one. For instance, (d) is the reciprocal position of (c). This is illustrated in Fig. 8, where one particle is represented by a doll with her left hand up and the mirror image by a doll with her right hand up. In all four parts of this illustration the incident light is thought to come from below and the scattering angle is 45° (as in Fig. 7). The corresponding amplitude matrices, all for scattering in the same direction, are in this order:

$$
(a) \qquad\qquad (b) \qquad\qquad (c) \qquad\qquad (d)
$$

$$
\begin{pmatrix} S_2 & S_3 \\ S_4 & S_1 \end{pmatrix} \qquad
\begin{pmatrix} S_2 & -S_4 \\ -S_3 & S_1 \end{pmatrix} \qquad
\begin{pmatrix} S_2 & -S_3 \\ -S_4 & S_1 \end{pmatrix} \qquad
\begin{pmatrix} S_2 & S_4 \\ S_3 & S_1 \end{pmatrix}.
$$

These transformation properties will be accepted without detailed proof. The proof of the transformation $(a) \to (c)$ is simple, because, apart from the change of sign of one coordinate, it is the same scattering problem that has to be solved. The transformation $(a) \to (b)$ is the reciprocity theorem for vector waves. It is a natural extension of the reciprocity theorem for the scattering of scalar waves. Blatt and Weisskopf (1952) prove this theorem by means of an inversion of the time and a resolution of the ingoing spherical wave into plane waves. Morse and Feshbach (1953) prove it from a variational principle for the scattered amplitude. The vector theorem may be proved along similar lines. An indirect proof may be based on the reciprocity theorems for antennas by considering the reciprocity of a four-terminal network consisting of two antennas communicating via the scattering particle (Silver, 1949). A direct proof was formulated by Saxon (1955). Finally, the transformation $(a) \to (d)$ follows by successive applications of the two others.

We now shall make certain assumptions about the distribution of particles among these positions. Each set of assumptions is denoted by a number from 1 to 13. Four sets of assumptions lead to certain symmetry properties of the matrix **F** for arbitrary θ. The resulting forms have been derived by the method explained in sec. 5.21, but all details of the derivation are omitted.

1. The cloud contains one kind of particles. For each particle in one position a particle in the reciprocal position is found. Adding the matrices **F** of (a) and (b) we find a matrix of the following symmetry:

$$
\begin{pmatrix}
a_1 & b_1 & b_3 & b_5 \\
b_1 & a_2 & b_4 & b_6 \\
-b_3 & -b_4 & a_3 & b_2 \\
b_5 & b_6 & -b_2 & a_4
\end{pmatrix} \text{10 parameters.}
$$

2. The cloud contains particles and their mirror particles. For any

particle in position (a) a mirror particle in position (c) is found. The resulting matrix $\mathbf{F}$ is:

$$\begin{pmatrix} a_1 & b_1 & 0 & 0 \\ c_1 & a_2 & 0 & 0 \\ 0 & 0 & a_3 & b_2 \\ 0 & 0 & c_2 & a_4 \end{pmatrix} \text{ 8 parameters.}$$

3. The cloud contains particles and their mirror particles. For any particle in position (a) a mirror particle in position (d) is found. This is a very artificial assumption. The matrix $\mathbf{F}$ becomes:

$$\begin{pmatrix} a_1 & b_1 & b_3 & b_5 \\ b_1 & a_2 & b_4 & b_6 \\ b_3 & b_4 & a_3 & b_2 \\ -b_5 & -b_6 & -b_2 & a_4 \end{pmatrix} \text{ 10 parameters.}$$

4. The cloud contains particles and their mirror particles, while any two of the preceding assumptions are made. The third one follows automatically so that there are equal numbers of particles in positions (a), (b), (c), and (d). The resulting matrix $\mathbf{F}$ has the form:

$$\begin{pmatrix} a_1 & b_1 & 0 & 0 \\ b_1 & a_2 & 0 & 0 \\ 0 & 0 & a_3 & b_2 \\ 0 & 0 & -b_2 & a_4 \end{pmatrix} \text{ 6 parameters.}$$

5. The cloud contains one kind of (asymmetric) particles with *random orientation*. The assumptions of 1 are then fulfilled and the symmetry is, as in that case, 10 parameters.

6. The cloud contains particles and their mirror images in equal numbers and in random orientation. Or: the cloud consists of particles that have a plane of symmetry and are randomly oriented. This assumption includes the set of assumptions of 4 and the symmetry is, as in that case, 6 parameters.

5.3. Symmetry Relations for $\theta = 0°$ and $180°$

The scattering angles $\theta = 0°$ and $180°$ call for a separate discussion. An angle θ close to $0°$ means scattered light propagated in nearly the same direction as the non-scattered light; this we usually call "forward scattering." The angle $\theta = 180°$ describes light being scattered back into the direction of the primary source; this we call "backward scattering." It may be noted that a different terminology is often adopted in the older literature. The supplement of θ is there defined as the scattering angle, and even the terms forward and backward are sometimes employed in the reverse sense.

The peculiarity of both $\theta = 0°$ and $\theta = 180°$ is that there is not one particular plane of scattering. Any plane through the axis (the co-incident directions of incident and scattered rays) might be considered as the plane of scattering. So, according to the convention for other angles, any plane through this axis may be chosen as the plane of reference with respect to which the amplitudes and the Stokes parameters of the incident and scattered beams are defined.

The possibility exists that for some physical reason, e.g., gravitation, or a magnetic field, one particular plane through the axis is preferred and acts as a plane with respect to which the distribution of orientation of the particles shows certain symmetry properties. We exclude this possibility in the present section, for the various formulae of the preceding section then hold without change.

5.31. Forward Scattering

Let one particular plane be chosen as the plane of reference and let one particular orientation of the particle be chosen as the initial position (a). The scattering matrix in this situation (a) is again denoted by **S**, and its elements are $\begin{pmatrix} S_2 & S_3 \\ S_4 & S_1 \end{pmatrix}$. Besides the positions (b), (c), (d), defined in sec. 5.22, there now is an infinity of positions of the particle, in which the scattering matrix may be expressed in the same elements. These positions originate from one of those by rotating the particle around the axis by an arbitrary angle φ. The same transformation of the scattering matrix results if the particle is kept fixed and the plane of reference rotated by the angle φ in the reverse direction.

Rotation of the plane of reference over an angle φ changes the diad of the amplitudes $(E_l E_r)$ of a certain beam of light in such a way as corresponds to the multiplication of this diad by the matrix

$$\mathbf{R}_\varphi = \begin{pmatrix} c & s \\ -s & c \end{pmatrix}, \quad \text{where } c = \cos\varphi, s = \sin\varphi.$$

The angle φ is called positive when the direction is clockwise as viewed in the direction of propagation of the beam. This rule has to be applied twice, because the plane of reference is rotated both for the incident and the scattered waves. The result is that the matrix **S** is replaced by the matrix $\mathbf{R}_{\varphi} \cdot \mathbf{S} \cdot \mathbf{R}_{-\varphi}$, which has the elements

$$\begin{pmatrix} c^2 S_2 + cs S_3 + cs S_4 + s^2 S_1, & -cs S_2 + c^2 S_3 - s^2 S_4 + cs S_1 \\ -cs S_2 - s^2 S_3 + c^2 S_4 + cs S_1, & s^2 S_2 - cs S_3 - cs S_4 + c^2 S_1 \end{pmatrix}.$$

Similar, but slightly more complicated, transformation properties hold for the 4×4 matrix **F**.

One assumption which gives rise to a simplified matrix is the assumption of *rotational symmetry in the cloud*. By this we shall understand the assumption that the distribution of the particles in all possible orientations does not change if the entire cloud is rotated about the axis. This means that with any particular orientation (a) a large number of other orientations (a') occur that originate from (a) by rotation about the axis over an angle φ, and that the angles φ are evenly distributed in the interval 0 to 2π. The *rotational symmetry* thus refers to a property of the distribution of orientation in the cloud, not to a property of the individual particles.

If we do not wish to assume symmetry relations with respect to a preferential plane through the axis, the positions (b) and (c) of the preceding sections are not uniquely defined if (a) is given. The position (d) is uniquely defined, however, for the "bisectrix plane" is here the plane perpendicular to the axis. So only the set of assumptions (3) of sec. 5.22 is meaningful for $\theta = 0$, and the result is the same. The assumption of rotational symmetry may be made alone (case 7) or in combination with any of the preceding sets of assumptions (cases 8 to 11). The assumption (2), for instance, is meaningful when combined with rotational symmetry because it means that equal numbers occur when mirrored with respect to *any* plane through the axis. The complete set of possibilities for $\theta = 0$ is now as follows.

3. Particles and mirror particles in equal numbers. For any particle in one position a mirror particle is found in the position that originates from reflection with respect to a plane perpendicular to the axis. This assumption is identical with 3 in sec. 5.22, and the final matrix has, as given there, 10 parameters.

7. There is one kind of (asymmetric) particles in the cloud. Only the assumption of rotational symmetry is made. The combined matrix **F** is

$$\begin{pmatrix} a_1 & 0 & 0 & b_5 \\ 0 & a_2 & b_4 & 0 \\ 0 & -b_4 & a_2 & 0 \\ c_5 & 0 & 0 & a_4 \end{pmatrix} \quad \text{6 parameters.}$$

8. There is one kind of (asymmetric) particles. The assumption of rotational symmetry is combined with the set (1). This means that equal numbers of particles occur "upside down," i.e., in the reciprocal position. Result:

$$\begin{pmatrix} a_1 & 0 & 0 & b_5 \\ 0 & a_2 & b_4 & 0 \\ 0 & -b_4 & a_2 & 0 \\ b_5 & 0 & 0 & a_4 \end{pmatrix} \quad \text{5 parameters.}$$

9. Particles and mirror particles occur in equal numbers. Rotational symmetry is combined with set 2. This means that for any particle there is a particle which is its mirror image in any plane through the axis. Result:

$$\begin{pmatrix} a_1 & 0 & 0 & 0 \\ 0 & a_2 & 0 & 0 \\ 0 & 0 & a_2 & 0 \\ 0 & 0 & 0 & a_4 \end{pmatrix} \quad \text{3 parameters.}$$

10. Particles and mirror particles occur in equal numbers. Rotational symmetry is combined with set 3, the assumption of mirror symmetry with respect to a plane perpendicular to the axis. Result:

$$\begin{pmatrix} a_1 & 0 & 0 & b_5 \\ 0 & a_2 & 0 & 0 \\ 0 & 0 & a_2 & 0 \\ -b_5 & 0 & 0 & a_4 \end{pmatrix} \quad \text{4 parameters.}$$

11. Particles and mirror particles occur in equal numbers. The assumption of rotational symmetry is combined with set 4, i.e., with all the previous assumptions. Result: The matrix $\mathbf{F}$ is not simplified beyond the form already obtained in 9. There are 3 parameters.

12. Random orientation for one kind of asymmetric particles. Result as in 8. There are 5 parameters.

13. Random orientation for particles and mirror particles in equal numbers, or for particles that are their own mirror image. Result as in 11 or 9. There are 3 parameters.

5.32   Backscatter

The situation for $\theta = 180°$ resembles the one for $\theta = 0°$ in many respects. Again the scattering plane is indeterminate; there is one "axis" along which the incident radiation travels in one direction, the scattered radiation in the reverse direction. Rotation of the plane of reference over an angle φ changes the diad (E_l, E_r) as before but requires the matrix $\mathbf{S}$ to be replaced by $\mathbf{R}_{-\varphi}\mathbf{S}\mathbf{R}_{-\varphi}$, which has the elements

$$\begin{pmatrix} c^2S_2 + csS_3 - csS_4 - s^2S_1, & -csS_2 + c^2S_3 + s^2S_4 - csS_1 \\ csS_2 + s^2S_3 + c^2S_4 + csS_1, & -s^2S_2 + csS_3 - csS_4 + c^2S_1 \end{pmatrix}.$$

Again, similar transformation properties hold for the 4×4 matrix $\mathbf{F}$.

Before we apply these formulae to various assumptions about the distribution of particles over all possible orientations we have to note one very curious theorem. Although the plane of scattering is indeterminate, the bisectrix is uniquely determined for $\theta = 180°$. This bisectrix coincides with the incident and scattered beams, i.e., with the line joining the particle and the primary light source. So a rotation of $180°$ about this bisectrix, which puts the particle in the reciprocal position, is also a rotation of $180°$ about the axis. The scattering matrix of the particle in the new position is $\begin{pmatrix} S_2 & -S_4 \\ -S_3 & S_1 \end{pmatrix}$ according to the reciprocity principle and unchanged according to the rotation formula. Since these two matrices must be the same, the theorem follows that *the scattering matrix of an arbitrary particle referred to an arbitrary plane of reference has for $\theta = 180°$ the property $S_3 + S_4 = 0$.* We shall further write the matrix for $\theta = 180°$ in the form $\begin{pmatrix} S_2 & S_3 \\ -S_3 & S_1 \end{pmatrix}$.

The enumeration of the various sets of assumptions that may bring about certain symmetry properties of the final matrix $\mathbf{F}$ without introducing a preferred plane through the axis is now fairly simple.

0a. No assumptions made. The theorem just derived gives a matrix $\mathbf{F}$ with symmetry of the type

$$\begin{pmatrix} a_1 & b_1 & b_3 & b_5 \\ b_1 & a_2 & b_4 & b_6 \\ -b_3 & -b_4 & a_3 & b_2 \\ b_5 & b_6 & -b_2 & a_4 \end{pmatrix} \quad \text{10 parameters.}$$

1a. Assumptions identical with the ones made in set 1. Change to the reciprocal position is just a rotation of $180°$ by which the matrix $\mathbf{F}$ does not change. The result is the same as in *0a*: 10 parameters.

7a. Same assumptions as 7. The cloud consists of one kind of (asymmetric) particles, and the assumption of rotational symmetry is made without any further assumptions. The resulting matrix $\mathbf{F}$ is

$$\begin{pmatrix} a_1 & 0 & 0 & b_5 \\ 0 & a_2 & 0 & 0 \\ 0 & 0 & -a_2 & 0 \\ b_5 & 0 & 0 & a_4 \end{pmatrix} \quad \text{4 parameters.}$$

8a. Same assumptions as 8. Matrix as in case 7a: 4 parameters.

11a. Same assumptions as 9 or 10 or 11, for one involves the other when $\theta = 180°$. The cloud contains particles and mirror particles in equal numbers. Rotational symmetry is combined with the assumption that any particle has its mirror particle with respect to any plane through the axis. The matrix is

$$\begin{pmatrix} a_1 & 0 & 0 & 0 \\ 0 & a_2 & 0 & 0 \\ 0 & 0 & -a_2 & 0 \\ 0 & 0 & 0 & a_4 \end{pmatrix} \quad \text{3 parameters.}$$

12a. Random orientation for one kind of asymmetric particles. Result as in 7a. There are 4 parameters.

13a. Random orientation for particles and mirror particles in equal numbers, or for particles that are their own mirror image. Result as in 11a. There are 3 parameters.

5.4. Symmetry Relations for Dispersion and Extinction

The preceding sections deal with the intensities of the scattered light. Throughout these sections we used the rule that the matrices $\mathbf{F}$ of the individual particles must be added. This was based on the rule that intensities may be added, which in turn depends on the assumption that phase relationships between the waves scattered by different particles may be neglected (sec. 4.22). Consequently, the symmetry rules derived above for $\theta = 0°$ strictly do not hold for $\theta = 0°$. They hold for angles that are small enough to put the scattering matrix, for reasons of continuity, equal to the scattering matrix for $\theta = 0°$ (this requires $\theta \ll \lambda/a$, if a is the linear size of the largest particles) and yet large enough to introduce sufficient phase shifts between particles at different locations.

5.41. Formal Results

In the strict case of $\theta = 0°$ no scattered light can be observed, but the scattered wave interferes with the original wave and gives rise to a modified wave propagation. This was explained in detail in chap. 4. In the simple case of spherical particles this means that we can determine the phase velocity from the imaginary part of $S(0)$ and the extinction from the real part of $S(0)$; see sec. 4.42. The wave propagation in the most general case is governed by a matrix of 4 coefficients (sec. 4.41):

$$\begin{pmatrix} S_2(0), & S_3(0) \\ S_4(0), & S_1(0) \end{pmatrix}.$$

The big difference with the preceding sections is that the light is scattered by all particles exactly in phase. Hence the components of the matrix $\mathbf{S}(0)$ and not those of the matrix $\mathbf{F}$ have to be added. The four complex coefficients have 8 independent parameters (also the phase is relevant). So we have:

    0. No assumptions; symmetry of the form

$$\begin{pmatrix} b & c \\ d & a \end{pmatrix} \quad \text{8 parameters.}$$

This is the most general combination of linear and circular birefringence and linear and circular dichroism.

The symmetry assumptions that can be made are exactly the ones made in the preceding sections. Addition of the matrices $\mathbf{S}$ gives the following straightforward results which are presented together with a qualitative interpretation. The scope of this book does not permit a detailed discussion. All effects can be read from the general propagation formula given in sec. 4.41.

If a certain preferred plane through the axis exists it is taken as the plane of reference and the symmetries are (compare sec. 5.22):

1:
$$\begin{pmatrix} b & c \\ -c & a \end{pmatrix} \quad \text{6 parameters.}$$

All effects, linear and circular birefringence, and linear and circular dichroism, are still present but related in a not too obvious manner.

2 or 4:
$$\begin{pmatrix} b & 0 \\ 0 & a \end{pmatrix} \quad \text{4 parameters.}$$

Only linear birefringence and dichroism, the type of symmetry is destroyed if another plane of reference is chosen.

If no preferred plane through the axis exists an arbitrary plane may be taken as the plane of reference (cf. sec. 5.31).

3:
$$\begin{pmatrix} b & c \\ c & a \end{pmatrix} \quad \text{6 parameters.}$$

Interpretation again not very obvious. The type of symmetry, but not the values of a, b, and c, is preserved if another plane of reference is chosen.

7 or 8 or 12: $\begin{pmatrix} a & c \\ -c & a \end{pmatrix}$ 4 parameters.

The type of symmetry *and* the values of the complex quantities a and c are invariant against rotation of the plane of reference.

Applying a transformation matrix of this form to the diad $(E_l, E_r) = C(1, \pm i)$, that corresponds to right- and left-handed circular polarization, we find that the circular polarization remains but that the factor C is multiplied by $a \pm ic$. This means circular birefringence and circular dichroism. If $c = S_3(0)$ is a purely real number, only circular dichroism exists. If $c = S_3(0)$ is a purely imaginary number, only circular birefringence exists; this effect is more commonly known as optical rotation of the plane of polarization.

9 or 10 or 11 or 13: $\begin{pmatrix} a & 0 \\ 0 & a \end{pmatrix}$ 2 parameters.

In this most simple case there is no birefringence or dichroism of any kind. We may call it *scalar propagation*. The two parameters are the real and imaginary parts of $S(0)$, which according to secs. 4.3 or 4.42 define the extinction coefficient and the refractive index of the medium that is composed of the scattering particles. These two parameters are the same for any state of polarization of the incident light, and this state of polarization is preserved while the wave travels through the medium. Full formulae for *a medium consisting of spherical particles* are given in sec. 4.42.

5.42. Summary

A medium of odd particles not randomly oriented can produce any effects of linear and circular birefringence and dichroism mixed. Since the particle as a whole is characterized by the scattering matrix **S** it does not matter whether any asymmetry in **S** is caused by the *shape* or the composition of these particles; e.g., linear birefringence is produced by elongated particles of a homogeneous material as well as by spheres of a substance that is itself birefringent.

In order to have linear birefringence we need at least case 2, i.e., the existence of mirror symmetry with respect to a preferred plane. Then circular birefringence is excluded.

In order to have circular birefringence (optical rotation) we need at least case 7, i.e., asymmetric particles (we may also say: optically active particles of one kind) and rotational symmetry in the distribution. Then linear birefringence is excluded.

Random orientation leads to circular birefringence if the particles are optically active and if there are not equal numbers of both kinds (case 12). Random orientation gives scalar propagation if the particles are not optically active or if equal numbers of both kinds occur together (case 13).

In general, linear dichroism occurs with linear birefringence and circular dichroism with circular birefringence. This is because in general the numbers $S_1(0)$, $S_2(0)$, $S_3(0)$, and $S_4(0)$ may have any complex values. One further remark should be made, however. Most of the applications are either in molecular physics or in the field of artificial composite media for microwaves. In both cases the particles are likely to be small compared to the wavelength. In this domain of the scattering theory the matrix $\mathbf{S}(0)$ has a special form that does not admit all the possibilities noted above. Linear birefringence is quite easy to produce; linear dichroism does not occur unless the particles are absorbing (they need not be dichroic). Circular birefringence or dichroism does not arise in the first approximation (Rayleigh scattering) and, therefore, is small as long as the particles are small compared to the wavelength.

References

The Stokes parameters and the principle of optical equivalence have been introduced by

G. C. Stokes, *Trans. Cambr. Phil. Soc.*, **9**, 399 (1852).

Modern presentations are given by

F. Perrin, *J. Chem. Phys.*, **10**, 415 (1942),
R. Clark Jones, *J. Opt. Soc. Amer.*, **37**, 107 and 110 (1947),

the latter with a formula for experimental determination of the parameters. An application to Compton scattering is made by

V. Fano, *J. Opt. Soc. Amer.*, **39**, 859 (1949).

Several applications to multiple scattering are made by

S. Chandrasekhar, *Radiative Transfer*, Oxford, Oxford Univ. Press, 1950.

An interesting extension to include spectral analysis of the wave was proposed by

E. Wolf, *Nuovo Cimento*, **12**, 884 (1954).

The explicit form of the transformation matrix in sec. 5.14 was newly derived. The reciprocity principle (sec. 5.22) in its simplest form was expressed by

Lord Rayleigh, *Theory of Sound*, Vol. I, secs. 107-111, London, Macmillan and Co., 1877.

Reference is made to the proofs of the reciprocity theorem for the scattering of scalar waves by

J. M. Blatt and V. F. Weisskopf, *Theoretical Nuclear Physics*, pp. 336-339, 528-530, New York, John Wiley & Sons, 1952.

P. M. Morse and H. Feshbach, *Methods of Theoretical Physics*, part II, pp. 1130-1131 New York, McGraw-Hill Book Co., 1953.

Reciprocity theorems in antenna theory are discussed by

S. Silver (ed.), *Microwave Antenna Theory and Design*, chap. 2, M.I.T. Radiation Laboratory Series **12**, 1949.

The reciprocity theorem for vector waves was proved by

D. S. Saxon, *Lectures on the Scattering of Light* (Notes prepared by R. S. Fraser), pp. 78-100, U.C.L.A. Dept. of Meteorology, 1955.

D. S. Saxon, *Phys. Rev.*, **100**, 1771 (1955).

The sections on symmetry relations were inspired by Perrin's paper cited above; a new method is used and the number of cases is greatly extended.

PART II

Special Types of Particles

The basic theory of part I permits us to solve any scattering or propagation problem if the four components of the scattering matrix

$$\begin{pmatrix} S_2(\varphi, \theta) & S_3(\varphi, \theta) \\ S_4(\varphi, \theta) & S_1(\varphi, \theta) \end{pmatrix}$$

are known for each individual particle for the particular way it is oriented in space.

The theory of part II permits us to calculate these scattering functions for certain special types of particles. This part of the book has three distinct divisions. The particles in chaps. 6 to 8 are special by their size (or size combined with refractive index) and may have an arbitrary shape. The particles in chaps. 9 to 14 are spheres and may have an arbitrary size. The particles in chaps. 15 to 17 have other geometrical forms.

PART III

Special Types of Parabolas

6. PARTICLES SMALL COMPARED TO THE WAVELENGTH

The particles treated in this chapter may have arbitrary forms and they are small compared to the wave length both outside and inside the particle, as is specified more precisely in sec. 6.4.

6.1. Polarizability and Rayleigh Scattering
6.11. If the Polarizability Is a Tensor

The simplification introduced by the small size is that the particle may be considered to be placed in a *homogeneous* electric field $\mathbf{E}_0$, which we may call the "applied" field. The particle's own field, caused by the electric polarization of the particle, modifies this field both inside and near the particle. The combined field will be denoted by $\mathbf{E}$. Let $\mathbf{p}$ be the induced dipole moment; then the electrostatic formula,

$$\mathbf{p} = \alpha\mathbf{E}_0,$$

is applicable. This relation defines α, the polarizability of the particle. Since the dimension of $\mathbf{E}$ is charge per area and the dimension of $\mathbf{p}$ is charge times length, α has the dimension of a volume. In the case of homogeneous bodies with volume V, we shall also introduce α', the average volume polarizability, by $\alpha = \alpha'V$. The quantity α' is dimensionless.

In general, α is a tensor. This means that the directions of $\mathbf{p}$ and $\mathbf{E}_0$ coincide only if the field is applied in one of three mutually perpendicular directions. Let these directions by denoted by the unit vectors $\mathbf{n}_1$, $\mathbf{n}_2$, and $\mathbf{n}_3$. The particle is then characterized by the three tensor components α_1, α_2, and α_3 in such a way that any external field

$$\mathbf{E}_0 = E_1\mathbf{n}_1 + E_2\mathbf{n}_2 + E_3\mathbf{n}_3$$

gives the dipole moment

$$\mathbf{p} = \alpha_1 E_1\mathbf{n}_1 + \alpha_2 E_2\mathbf{n}_2 + \alpha_3 E_3\mathbf{n}_3.$$

These formulae, known from electrostatic theory, hold without change if the applied field is the periodic field of an incident plane-polarized wave:

$$\mathbf{E}_0 e^{i\omega t};$$

the induced dipole moment then is

$$\mathbf{p}e^{i\omega t}.$$

The components of the polarizability tensor may be complex and may depend on ω. We may now repeat some well-known facts from electromagnetic radiation theory. The oscillating dipole radiates in all directions. This is the type of scattering called *Rayleigh scattering*. Let a point P be at a distance $r \gg \lambda$ from the particle and in a direction that makes an angle γ with $\mathbf{p}$ (Fig. 9). The electric field of the scattered wave is

$$E = \frac{k^2 p \sin \gamma}{r} e^{-ikr}$$

times the unit vector directed as the component of $\mathbf{p}$ perpendicular to the radius vector.

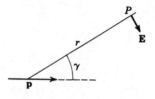

Fig. 9. Electric dipole scattering.

The corresponding intensities of the incident and scattered radiation are (for gaussian units, time average of the Poynting vector):

$$I_0 = \frac{c}{8\pi} |E_0|^2, \qquad I = \frac{c}{8\pi} |E|^2.$$

Integrating I over a big sphere we find that the total energy scattered in all directions per unit time is

$$W = \tfrac{1}{3} k^4 c \, |p|^2,$$

and dividing this by I_0 we obtain the scattering cross section

$$C_{sca} = \frac{8}{3}\pi k^4 |\alpha|^2,$$

where $|\alpha|^2$ is defined by

$$|\alpha|^2 = l^2 |\alpha_1|^2 + m^2 |\alpha_2|^2 + n^2 |\alpha_3|^2$$

and l, m, n are the direction cosines of $\mathbf{E}_0$ with respect to the three main axes of the polarizability tensor.

The directions appearing in this problem must be clearly distinguished. The value of $|\alpha|$ is determined by the orientation of $\mathbf{E}_0$ with respect to the particle; the direction of propagation of the incident light is irrelevant. On the other hand, the angular distribution of scattered light is determined by the angle with $\mathbf{p}$. The scattered intensity is 0 in the direction of $\mathbf{p}$ but may have a non-zero value in the direction of $\mathbf{E}_0$. What is expressed here in some sentences can also be expressed in formulae. The intensity formulae for the most general case, in which α_1, α_2, and α_3 are not equal and in which the incident light is elliptically polarized or natural light, are implicit in the derivation for particles with random orientation in sec. 6.5.

6.12. If the Polarizability Is Isotropic

The situation becomes much simpler in the very common case that

$$\alpha_1 = \alpha_2 = \alpha_3 (\equiv \alpha).$$

In that circumstance the directions of $\mathbf{p}$ and $\mathbf{E}_0$ always coincide, and in the preceding formulae α may be considered a scalar. The angle γ then is the angle of a direction of scattering with $\mathbf{E}_0$. If θ is the scattering angle as before, the perpendicular component (r-component) has $\gamma = 90°$, the parallel component (l-component) has $\gamma = 90° - \theta$. So the scattered field just derived is obtained by the scattering tensor (sec. 4.41):

$$\begin{pmatrix} S_2 & S_3 \\ S_4 & S_1 \end{pmatrix} = ik^3\alpha \begin{pmatrix} \cos\theta & 0 \\ 0 & 1 \end{pmatrix}.$$

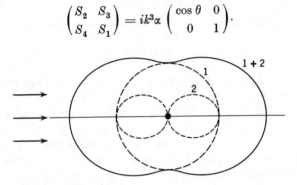

Fig. 10. Rayleigh scattering: polar diagram of scattered intensity if incident radiation is unpolarized, $1 =$ polarized with electric vector $\perp$ plane of drawing, $2 =$ polarized with electric vector in plane of drawing, $1 + 2 =$ total.

We may also go directly to intensities and find, for instance, for incident natural light with the intensity I_0 the scattered intensity

$$I = \frac{(1 + \cos^2\theta)k^4 |\alpha|^2}{2r^2} I_0.$$

The term 1 comes from $S_1(\theta)$ and corresponds to the r-component (electric vector perpendicular to the plane of scattering), and the term $\cos^2\theta$ comes from $S_2(\theta)$ and corresponds to the l-component of the scattered light (electric vector parallel).

Figure 10 illustrates this formula by the well-known scattering diagram. The solid line denotes the total intensity, the dotted lines the intensities of the polarized components; the light scattered by 90° is fully polarized in the r-direction. Although this diagram is commonly referred to as *the* diagram for Rayleigh scattering, it may be noted that it holds only if the incident light is natural and the particles are isotropic (α a scalar). For anisotropic particles the 90° scattering is not fully polarized (sec. 6.52). The formulae for incident light of arbitrary (partially elliptical) polarization may be read from sec. 4.42.

6.13. Absorbing Particles

Absorbing particles are characterized by complex values of the polarizability. In the simple case that $\alpha_1 = \alpha_2 = \alpha_3 = \alpha$, the amplitude function for $\theta = 0$ is the scalar

$$S(0) = ik^3\alpha,$$

so that by the general relation of sec. 4.21 we find

$$C_{abs} = 4\pi k \operatorname{Re} (i\alpha).$$

A small correction is needed here. The general relation just used should give the extinction cross section, which includes scattering as well as absorption. The fact that the result is 0 if we substitute a real value for α shows that scattering is not yet included. This is due to the fact that we have neglected radiation reaction on the oscillating dipole. This reaction actually causes a small phase lag of p with respect to E_0 even if the particle is non-absorbing. The simplest way to compute the scattering cross section is to integrate the scattered intensity over all directions, as was done in sec. 6.11. The total extinction cross section is

$$C_{ext} = C_{sca} + C_{abs},$$

where C_{sca} follows from sec. 6.11 and C_{abs} from the formula given above. If we should like to obtain C_{ext} at once from the general extinction formula, we should use the more refined formula for $S(0)$ that includes the radiation reaction:

$$S(0) = ik^3\alpha + \frac{2}{3} k^6\alpha^2.$$

A fuller inquiry into the higher-order terms may be made by means of the rigorous theory for spherical particles for which series expansions are given in secs. 10.3 and 14.21. This shows that the second term of $S(0)$ given above is indeed the first real term if α is real but imaginary terms of the order of k^5 are also present.

It is important to observe that, if α is not dependent on λ, the scattering is proportional to $\lambda^4 V^2$, absorption to $\lambda^{-1} V$. For very small particles ($V \to 0$) the absorption, if present at all, tends to be the stronger effect.

The more general case in which α is a tensor involves no fundamental difficulties. Generally $S(0)$ is replaced by a tensor of four components, and the medium of such particles may be dichroic. Full details may be taken from sec. 4.41.

6.2. Some Simple Applications

6.21. Determination of Size and Number

The emphasis in sec. 6.13 was on the extinction. If we consider a medium consisting of small scattering and absorbing particles, and if, according to the theory of sec. 4.3, we write the complex refractive index of the medium as

$$\tilde{m} = \tilde{n} - i\tilde{n}',$$

the extinction effect is expressed by $\tilde{n}'$. The real part of the refractive index $\tilde{n}$ (which is the same as $\tilde{m}$ for non-absorbing particles) is just as important. In fact when before 1850 Thomson, Mosotti, and others started to devise models for the refractive index of a medium (like a gas or a fluid) they had exactly this picture in mind: a medium consisting of small scattering particles. The relation of our formulae to the classical formulae of molecular optics has been explained in sec. 4.5. Under the very simplest condition, that the particles are small and far apart, the refractive index is given by the formula (III) that may be derived from either theory. If, in addition, α is supposed to be real we have

$$\tilde{n} = 1 + 2\pi\alpha N.$$

Here an important comment must be made. The scattering diagram is similar for all particles with sizes $\ll \lambda$. So it is impossible to infer the size of the particles from the form of the scattering diagram. The intensity of the scattered light is proportional to the (unknown) number of particles per cm³, N, so that it is in itself also insufficient for a determination of size.

Fortunately it is possible to combine two effects. Measurements of the scattering (under 90°, or integrated over all angles) give the quantity $N\alpha^2$. Measurements of the refractive index of the medium give the quantity $N\alpha$. Thus we can find N and α separately. The problem is then solved, for knowing N and the density of the medium we find at once the mass of one particle. Or knowing α and the composition of each particle we find at once the volume, to which α is, in general, proportional.

The principle outlined above is the basis of a good number of practical applications. It has been applied for a long time for determining the number of Avogadro from the light scattered by an ideal gas (e.g., air) of known refractive index (sec. 6.53). It has more recently been revived as a powerful method (the Debye method) for determining the molecular weight of high polymers and simpler molecules (sec. 19.12). The detailed procedures and formulae must be omitted here.

It is important to note in the present context that similar methods can be applied with exactly the same ease and accuracy in any medium of small solid particles or droplets, provided that the wavelength is sufficiently large compared to their size. Since α is proportional to the volume V of one particle, the refractive index n depends (in this approximation) only on the volume concentration NV, independently of the size of the particles. But the scattering for a given volume concentration is more effective the coarser the particles are.

6.22. Particles with Refractive Index Close to 1

The dipole moment $\mathbf{p}$ of a solid particle in an electric field is equal to

$$\mathbf{p} = \int \mathbf{P} \, dV,$$

where dV is a volume element of the particle and $\mathbf{P}$ the polarization per unit volume, which at any point inside the particle is given by

$$\mathbf{P} = (m^2 - 1)(\mathbf{E}/4\pi),$$

where $\mathbf{E}$ is the electric field and m the complex refractive index at that point. This is a direct consequence of Maxwell's equations. The complication of the problem is that $\mathbf{E}$ in turn depends on the polarization of the particle. This complication does not occur if m is close to 1 throughout the particle. The actual field $\mathbf{E}$ can then be put equal to the applied field $\mathbf{E}_0$, and we have

$$\alpha = \frac{1}{4\pi} \int (m^2 - 1) \, dV.$$

If in addition the particle is homogeneous,

$$\alpha = (m^2 - 1)(V/4\pi).$$

Both results are independent of direction (α is isotropic) and hold for particles of arbitrary form. The resulting cross sections according to secs. 6.11 and 6.13 are

$$C_{sca} = \frac{k^4 V^2}{6\pi} \left| m^2 - 1 \right|^2$$

$$C_{abs} = -kV \, \text{Im} \, (m^2 - 1).$$

The formula $m^2 = \varepsilon - 4\pi i\sigma/\omega$ (secs. 9.12 and 14.1), where $\omega = kc$, $\varepsilon \approx 1$, and $\gamma = 4\pi\sigma/c$ under the conditions of this section, reduces the latter expression to $C_{abs} = 4\pi\sigma V/c = \gamma V$. This result follows also directly from the Joule heat developed in the particle.

In all the formulae for m close to 1 we may replace $m^2 - 1$ with $2(m - 1)$ to the same degree of approximation. In particular we have for the entire medium

$$\tilde{m} = 1 + (m - 1)NV$$

(compare secs. 4.5 and 6.21).

6.23. Free Electrons

Another very simple application is the action of free electrons. A free electron (charge $-e$, mass m) exposed to an applied oscillating field $\mathbf{E}_0$, circular frequency ω, radiates as a dipole $\mathbf{p} = \alpha\mathbf{E}_0$, where

$$\alpha = \frac{-e^2}{m\omega^2}.$$

With $\omega = kc$ this gives

$$C_{sca} = \frac{8\pi}{3} \cdot \frac{e^4}{m^2c^4} = 0.67 \times 10^{-24} \text{ cm}^2$$

and

$$\tilde{n} = 1 - \frac{2\pi e^2 N}{m\omega^2}.$$

The first result is the rigorous expression for the Thomson scattering coefficient; it is independent of wavelength because the dependence of α on ω cancels the k^4 in the general formula. The second result is the approximation (for n close to 1) of the more general Lorentz formula for a gas of free electrons, which is

$$n^2 = 1 - \frac{4\pi e^2 N}{m\omega^2}.$$

Two remarks may be made:

(a). This is not the only example in which n is smaller than 1 and the phase velocity is larger than c. It occurs quite often for media consisting of particles with the sizes of the order of λ, viz., whenever the imaginary part of $S(0)$ is negative.

(b). Bound electrons scatter light as "virtually free" electrons when the frequency is very large. This occurs in X-ray scattering. The result is that most media have a refractive index for X-ray scattering which is close to 1 but very, very slightly below it. Consequently, if we want to compute the X-ray scattering by a solid particle, the relevant formula are virtually always those of chap. 7.

6.3. Spheres and Ellipsoids

The only non-trivial case in which the polarizability of a particle can be computed by elementary means is that of homogeneous spheres and ellipsoids. For such particles the counter field due to the polarization, and, consequently, also the total field $\mathbf{E}$ are constant inside the particle. We refer to other books for the full derivation and simply quote the results and some numerical data.

6.31. Spheres

Lorentz had derived that spheres with radius a, volume V, have

$$\alpha = \frac{3(m^2 - 1)}{4\pi(m^2 + 2)} \qquad V = \frac{m^2 - 1}{m^2 + 2}\, a^3,$$

independently of direction. Direct substitution into the formulae of secs. 6.11 and 6.13 gives C_{sca} and C_{abs} and, dividing by πa^2, we find the efficiency factors ($x = ka$):

$$Q_{sca} = \frac{8}{3}\, x^4 \left| \frac{m^2 - 1}{m^2 + 2} \right|^2 , \qquad Q_{abs} = -4x\, \text{Im} \left\{ \frac{m^2 - 1}{m^2 + 2} \right\} .$$

These are the first terms of series expansions derived in secs. 10.3 and 14.21. The special case of very large m gives

$$\alpha = a^3, \quad C = \frac{8}{3}\, \pi a^6 k^4, \quad Q = \frac{8}{3}\, x^4.$$

See sec. 10.61, however, for totally reflecting spheres ($m = \infty$). The special case of m close to 1 gives

$$Q_{sca} = \frac{32}{27}\, x^4 \,|\, m - 1\,|^2, \qquad Q_{abs} = -\frac{8x}{3}\, \text{Im}\, (m - 1).$$

6.32. Ellipsoids

If $\mathbf{E}_0$ is applied along one of the main axes ($j = 1, 2, 3$) of the ellipsoid, $\mathbf{E}$ is given at any point inside the ellipsoid by

$$\mathbf{E} = \mathbf{E}_0 - L_j \cdot 4\pi \mathbf{P},$$

where the L_j are three factors depending on the ratios of the axes. Combining this equation with the second one of sec. 6.22, we can eliminate $\mathbf{E}$ and find

$$\mathbf{p} = \mathbf{P} V = \alpha_j \mathbf{E}_0.$$

This results in the equation:

$$\frac{V}{4\pi\alpha_j} = L_j + \frac{1}{m^2 - 1} ,$$

which gives the three main values α_1, α_2, α_3, of the polarizability tensor.

For an arbitrary ratio of the semiaxes, a, b, and c, we have

$$L_1 = \int_0^\infty \frac{abc\, ds}{2(s + a^2)^{3/2}\, (s + b^2)^{1/2}\, (s + c^2)^{1/2}}$$

and the same formula, with cyclical changes, for L_2 and L_3. The relation

$$L_1 + L_2 + L_3 = 1$$

always holds. Spheres have $L = \frac{1}{3}$, independently of direction, from which we find the formulae of sec. 6.31. Further special formulae for the values of L are collected below.

Spheroids ($b = c$)

prolate ($a > b$):

$$e^2 = 1 - (b^2/a^2), \quad L_1 = \frac{1 - e^2}{e^2}\left(-1 + \frac{1}{2e}\ln\frac{1 + e}{1 - e}\right).$$

oblate ($a < b$):

$$f^2 = (b^2/a^2) - 1, \quad L_1 = \frac{1 + f^2}{f^2}\left(1 - \frac{1}{f}\arctan f\right).$$

a close to b, oblate or prolate:

$$L_1 = \frac{1}{3} + \frac{4}{15}\frac{b - a}{a}.$$

Numerical values are found in Table 1.

Flat elliptical disk (both b and $c \gg a$):

$$L_1 = 1, \quad L_2 = L_3 = 0.$$

Long elliptical cylinder (both b and $c \ll a$):

$$L_1 = 0, \quad L_2 = \frac{c}{b + c}, \quad L_3 = \frac{b}{b + c}.$$

For ellipsoids not belonging to these special classes L_1, L_2, and L_3 may be found by numerical integration or may be read from graphs prepared by Osborn (1945). Sometimes, however, interpolation is easier and just as accurate. Table 2 is a self-evident example of such an interpolation. A rule, sometimes useful, is that approximate answers (l_1, l_2, l_3) are obtained from

$$\begin{cases} l_1 + l_2 + l_3 = 1, \\ l_1 : l_2 : l_3 = 1/a : 1/b : 1/c. \end{cases}$$

The errors $|l - L|$ are < 0.036 in all cases.

Table 1. Values of L for spheroids, $b = c$

| | Prolate | | | Oblate | |
b/a	L_1	$L_2 = L_3$	a/b	L_1	$L_2 = L_3$
0 (rods)	0.000	0.500	0 (disks)	1.000	0.000
0.2	0.056	0.472	0.2	0.750	0.125
0.4	0.134	0.433	0.4	0.588	0.206
0.6	0.210	0.395	0.6	0.478	0.261
0.8	0.276	0.362	0.8	0.396	0.302
1 (spheres)	0.333	0.333	1 (spheres)	0.333	0.333

Table 2. Example of interpolation

	a	b	c	L_1	L_2	L_3
Elliptical disk	1	3	0	0.000	0.000	1.000
Prolate spheriod	1	3	1	0.445	0.110	0.445
Interpolated	1	3	2	0.579	0.157	0.264
Oblate spheriod	1	3	3	0.636	0.182	0.182
Elliptical cylinder	1	3	∞	0.750	0.250	0.000

The scattering and absorption cross sections are, finally, found by computing α_1, α_2, and α_3 from the equation in the beginning of this section and by substituting these values into any of the equations of sec. 6.11 or 6.13. The formulae hold for real and complex values of m alike.

Table 3 shows the formulae of the polarizability for some values of L and their numerical values for some m. These numbers show how strongly the polarizability and, therefore, the scattering efficiency of a small particle depend both on its refractive index and on its form. Strongly elongated spheroids (needles) have $L = 0$ along their length and $L = \frac{1}{2}$ across; flat elliptical disks have $L = 1$ for a field applied perpendicular to their plane and $L = 0$ for a field in their plane.

Table 3. Representative values of $4\pi\alpha/V$

	$m = 2.0$	$m = 1.5$	$m = 1.25$	m arbitrary
$L = 0$	3.00	1.250	0.562	$m^2 - 1$
$L = \frac{1}{3}$	1.50	0.882	0.474	$\dfrac{3(m^2 - 1)}{m^2 + 2}$
$L = \frac{1}{2}$	1.20	0.769	0.439	$\dfrac{2(m^2 - 1)}{m^2 + 1}$
$L = 1$	0.75	0.556	0.360	$\dfrac{(m^2 - 1)}{m^2}$

A numerical example may illustrate the use of these formulae. Let plane-polarized radiation fall perpendicularly on a needle, of which both length and width are small compared to λ. The scattering cross section will then be strongest if the electric field is along the needle and weakest if it is perpendicular to the needle. For a very elongated needle the ratio of the scattering cross sections is as $(3/1.2)^2 = 6.25$ for $m = 2$ and as $(0.562/0.439)^2 = 1.64$ for $m = 1.25$. If, however, the needle is metallic, the refractive index is complex, e.g. $m = 1.27 - 1.37i$. Then $m^2 - 1 = -1.27 - 3.48i$ and $4\pi\alpha/V = -1.27 - 3.48i$ for a field along the axis ($L = 0$) and $4\pi\alpha/V = 1.77 - 1.10i$ for a field across the axis ($L = \frac{1}{2}$). The ratio of the absorption cross sections then is as the ratio of the imaginary parts, i.e., as $3.48/1.10 = 3.16$.

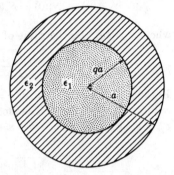

Fig. 11.   Sphere covered with spherical shell of different material.

6.33.   Ellipsoids of Anisotropic Material

Clark Jones (1945) has discussed the extension of these results to ellipsoids of anisotropic material. Two tensors appear in the equations, viz., L (with main axes coinciding with the main axes of the ellipsoid and with principal values L_j as defined above) and ε, corresponding to our m^2, whose orientation depends on the axes of the material (e.g., a crystal) from which the ellipsoid has been cut. The fundamental equations are

$$\mathbf{E} = \mathbf{E}_0 - L \cdot 4\pi\mathbf{P},$$
$$4\pi\mathbf{P} = (\varepsilon - 1)\mathbf{E}.$$

As for isotropic materials, the $\mathbf{P}$ appears to be a constant so that $\mathbf{p} = \mathbf{P}V = \alpha\mathbf{E}_0$, where α is again a tensor. Elimination of $\mathbf{E}$ gives an explicit expression for α, which is complicated if the orientations of the ellipsoid and of ε are not related. Elimination of $\mathbf{P}$ gives

$$\mathbf{E}_0 = \{1 + L(\varepsilon - 1)\}\mathbf{E}.$$

6.34.   Spherical Shells

Recently the problem of a sphere covered with a concentric spherical shell of different material has received some attention. Let (Fig. 11)

$$\varepsilon = \varepsilon_1 \quad \text{for} \quad 0 < r < qa,$$
$$\varepsilon = \varepsilon_2 \quad \text{for} \quad qa < r < a,$$
$$\varepsilon = 1 \quad \text{for} \quad r > a.$$

The polarizability α of such a body is not readily found in the literature, although it is very likely that somebody has solved this problem before. The solution is implicit, however, in the full solution for the scattering of electromagnetic waves of arbitrary wavelength by a body of this type (see sec. 16.1). From Güttler (1952) the expression of the coefficient a_1 may be copied. By a comparison of the formulae of secs. 6.12 and 10.3 we have in the case of a size small compared to the wavelength:

$$S_1(0) = S_2(0) = ik^3 \, \alpha = \frac{3}{2} \, a_1 x^3,$$

where $x = ka$. So the value of α is found. The result is:

$$\alpha = a^3 \frac{(\varepsilon_2 - 1)(\varepsilon_1 + 2\varepsilon_2) + q^3(2\varepsilon_2 + 1)(\varepsilon_1 - \varepsilon_2)}{(\varepsilon_2 + 2)(\varepsilon_1 + 2\varepsilon_2) + q^3(2\varepsilon_2 - 2)(\varepsilon_1 - \varepsilon_2)} \, .$$

In the special cases, $q = 0$, or $q = 1$, or $\varepsilon_1 = \varepsilon_2$, or $\varepsilon_2 = 1$, this expression reduces to the one for a homogeneous sphere. In the case of $\varepsilon_1 = 1$, $\varepsilon_2 = \varepsilon$, i.e., a *hollow spherical shell*, it may be rearranged to

$$\alpha = a^3 \frac{(1 - q^3)(\varepsilon - 1)(2\varepsilon + 1)}{(1 - q^3)(\varepsilon + 2)(2\varepsilon + 1) + 9q^3\varepsilon} \, .$$

Table 4 gives some numerical values when $\varepsilon = 81$ (water at very low frequencies). The table shows that a hollow water shell with a thickness one-fifth the outer radius has a polarizability that is only 5 per cent lower than the polarizability of a solid water sphere with the same outer radius. The polarizability is half that of a solid sphere when the shell thickness is 2 per cent of the radius.

Table 4. Polarizability of a hollow shell with $\varepsilon = 81$

q	q^3	α/a^3
0	0	0.964
0.464	0.10	0.959
0.585	0.20	0.950
0.670	0.30	0.942
0.737	0.40	0.930
0.794	0.50	0.914
0.843	0.60	0.894
0.888	0.70	0.851
0.929	0.80	0.794
0.965	0.90	0.648
0.980	0.94	0.521
0.990	0.97	0.352

6.4. Conditions for Rayleigh Scattering; Small Particles with $m = \infty$

So far we have made the obvious condition that the external field could be considered as a homogeneous field:

$$(a) \quad \text{size} \ll \lambda/2\pi \quad [\text{sphere: } x \ll 1].$$

However, a second condition is needed for Rayleigh scattering. The applied field should penetrate so fast into the particle that the static polarization is established in a time short compared to the period. The velocity inside the particle is c/m. So the further condition is:

$$(b) \quad |m| \cdot \text{size} \ll \lambda/2\pi \quad [\text{sphere: } |mx| \ll 1].$$

Since the wavelength inside the particle is λ/m we may express this condition also in this form: The size should be small compared to the wavelength inside the particle. If (a) is fulfilled and (b) is not, we are in the "resonance region." Here the inner field is not in phase with the external field. Waves penetrate slowly into the particle and may give rise to various systems of standing waves. Besides the electric dipole radiation we have magnetic dipole radiation and quadrupole radiation, etc.; each of them comes into resonance at well-defined values of the ratio of size to wavelength. This resonance is connected with the proper modes of vibration of the particle. No general theory for ellipsoids in this case has been developed. The resonance effects for spheres are discussed in secs. 10.5 and 14.31.

If $|m|$ is very large, the opposite of (b) may be true while assumption (a) still remains true. This means that the field hardly penetrates at all into the particle. The reason is most often a conductivity so large that the "skin depth" (sec. 14.41) is small compared to the radius.

In the limit of $m = \infty$ the body is a perfect conductor and the internal field is 0. This condition has often been chosen as a practicing example for its simplicity. Such "small, perfectly conducting" particles, however, are not representative for the small particles treated in the other sections of this chapter. They are not included in the assumptions made, as the particle size is not small compared to the *internal* wavelength. A separate theory is, therefore, needed. A somewhat detailed derivation for spheres is given in sec. 10.61. The scattering consists of electric dipole and magnetic dipole radiation. The analogous problem for homogeneous ellipsoids (including spheres) may be solved in a formal manner, as follows.

Let both the electric and the magnetic field of the incident wave be oriented along a main axis of the ellipsoid and let the corresponding numbers L (sec. 6.32) be L_e and L_m. The electric polarizability is, by the substitution of $m^2 = \infty$ into the formula of sec. 6.32,

$$\alpha = \frac{V}{4\pi L_e}.$$

Using the rule in the footnote on p. (sec. 9.11) 115 we have, with $\mu = 0$, a similar equation for the magnetic polarizability:

$$\alpha' = \frac{V}{4\pi(L_m - 1)}.$$

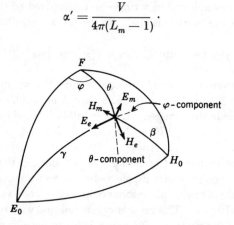

Fig. 12. Orientation of vectors in combined electric and magnetic dipole scattering by small, conducting ellipsoid.

Figure 12 shows part of the surface of a sphere centered around the scattering particle. Here E_0, H_0, and F represent the directions of the electric field, the magnetic field, and of propagation of the incident wave. An arbitrary direction described by the angles θ, φ may also be characterized by the angles made with the electric and magnetic dipoles, γ and β, where

$$\cos\gamma = \sin\theta\cos\varphi, \quad \cos\beta = \sin\theta\sin\varphi,$$

The fields radiated by the electric dipole and the magnetic dipole, respectively, are

$$E_e = H_e = C\alpha\sin\gamma$$
$$E_m = H_m = C\alpha'\sin\beta,$$

with $C = E_0 k^2 r^{-1} \exp(-ikr)$. With omission of the phase factor and taking account of the fact that α' is negative, they are oriented as shown: E_e points towards E_0 and H_m points away from H_0. Decomposing and adding the θ- and φ-components (in the sense shown by the figure) we obtain

$$E_\theta = C(\alpha \cos \theta + \alpha') \cos \varphi$$
$$E_\varphi = -C(\alpha + \alpha' \cos \theta) \sin \varphi,$$

by which the intensity and state of polarization of the scattered radiation is determined. The intensity is

$$I = \frac{I_0 k^4}{r^2} [\alpha^2 (1 - \sin^2 \theta \cos^2 \varphi) + \alpha'^2 (1 - \sin^2 \theta \sin^2 \varphi) + 2\alpha\alpha' \cos \theta].$$

Upon integration we find the scattering cross section

$$C_{sca} = \frac{8\pi}{3} k^4 (\alpha^2 + \alpha'^2).$$

The formulae for spheres (sec. 10.61) may be derived from these equations by putting $\alpha = a^3$, $\alpha' = -\tfrac{1}{2}a^3$. Those for a very flat spheroid (disk) with normally incident radiation (sec. 16.22) are found by putting $\alpha = 4a^3/3\pi$, $\alpha' = -ca^2/3 = $ negligible, where $a = $ major semiaxis and $c = $ minor semiaxis.

It is interesting to note that any perfectly conducting spheroid illuminated by linearly polarized radiation with $\mathbf{H} \parallel$ axis has the same scattering diagram as a perfectly conducting sphere, since such a spheroid always has $\alpha' = -\tfrac{1}{2}\alpha$.

6.5. Small Particles with Random Orientation

This section deals with the scattering by a cloud consisting of many small, identical particles that are oriented at random. Not much explanation is needed. The calculation for one particle with a given orientation follows from the theory of sec. 6.11; the combined effect of many particles is found along the lines of chap. 5.

6.51. Scattering Matrix for One Particle

A small particle is characterized by the three main components of its polarizability tensor, α_1, α_2, and α_3. Its orientation in space is characterized by the three mutually perpendicular unit vectors $\mathbf{n}_1$, $\mathbf{n}_2$, and $\mathbf{n}_3$ that denote the directions of the three axes corresponding to these values of the polarizability.

Let the direction of propagation of the incident light be in the z-direction and let the direction in which we wish to find the scattered intensity be in the ZOY plane (see Fig. 7, sec. 5.22). We denote by $\mathbf{n}_x$, $\mathbf{n}_y$, $\mathbf{n}_z$ the unit vectors in the direction of the axes and by θ the usual scattering angle.

The vector pairs, with respect to which the state of polarization is described (sec. 5.22), are

$$(\mathbf{l}_0, \mathbf{r}_0) = (\mathbf{n}_y, \mathbf{n}_x)$$

for the incident wave, and

$$(\mathbf{l}, \mathbf{r}) = (\mathbf{n}_y \cos \theta - \mathbf{n}_z \sin \theta, \mathbf{n}_x)$$

for the scattered wave.

The electric vector of the incident light is written in the most general manner as

$$\mathbf{E}_0 = E_{0l}\mathbf{n}_y + E_{0r}\mathbf{n}_x$$

By means of the linear relations,

$$\mathbf{n}_x = c_{11}\mathbf{n}_1 + c_{12}\mathbf{n}_2 + c_{13}\mathbf{n}_3$$
$$\mathbf{n}_y = c_{21}\mathbf{n}_1 + c_{22}\mathbf{n}_2 + c_{23}\mathbf{n}_3$$
$$\mathbf{n}_z = c_{31}\mathbf{n}_1 + c_{32}\mathbf{n}_2 + c_{33}\mathbf{n}_3,$$

this can be written in a form suitable for applying the theory of sec. 6.11. The induced dipole moment $\mathbf{p}$ is then found and, finally, its rectangular components by multiplying $\mathbf{p}$ by $\mathbf{n}_x$, $\mathbf{n}_y$, and $\mathbf{n}_z$. The resulting values are

$$p_x = E_{0r}P_{11} + E_{0l}P_{12}$$
$$p_y = E_{0r}P_{21} + E_{0l}P_{22}$$
$$p_z = E_{0r}P_{31} + E_{0l}P_{32},$$

where the constants P stand for

$$P_{ik} = P_{ki} = c_{i1}c_{k1}\alpha_1 + c_{i2}c_{k2}\alpha_2 + c_{i3}c_{k3}\alpha_3.$$

The electric field $\mathbf{E}$ of the scattered wave is the component of $\mathbf{p}$ perpendicular to the radius vector, multiplied by the form

$$k^2 e^{-ikr}/r.$$

By our general convention $\mathbf{E}$ has to be written in the form

$$\mathbf{E} = E_l(\mathbf{n}_y \cos \theta - \mathbf{n}_z \sin \theta) + E_r\mathbf{n}_x.$$

Consequently:

$$E_l = (p_y \cos \theta - p_z \sin \theta)\, k^2 e^{-ikr}/r$$
$$E_r = p_x k^2 e^{-ikr}/r.$$

This completes the formulae by which E_l and E_r are expressed as linear combinations of E_{l0} and E_{r0}.   The transformation matrix in the notation of sec. 4.41 is

$$\begin{pmatrix} S_2 & S_3 \\ S_4 & S_1 \end{pmatrix} = ik^3 \begin{pmatrix} P_{22}\cos\theta - P_{23}\sin\theta, & P_{12}\cos\theta - P_{13}\sin\theta \\ P_{12}, & P_{11} \end{pmatrix}.$$

This matrix expresses in a quite general way the scattering of any kind of polarized light (including elliptical polarization) by a small particle in any orientation.

6.52.   Scattered Light and Depolarization Factors

In order to find the combined effect of all particles, we have to compute the transformation matrix of the Stokes parameters, following the formula of sec. 5.14).   Its elements contain products of the elements shown above. These products are quadratic in P, i.e., of the fourth degree in the c_{ik} and of the second degree in the α's.   Of these products we have to take the averages for all possible orientations of the particles.

It is clear for symmetry reasons that many of the products give a zero average.   The only types of expression of the fourth degree that do not vanish are

c_{11}^4 (fourth power of a direction cosine) av. $= 1/5$

$c_{11}^2 c_{12}^2$ (squares of cosines in same row or column) av. $= 1/15$

$c_{11}^2 c_{22}^2$ (squares of cosines neither in same row nor in same column)

av. $= 2/15$

$c_{11}c_{12}c_{21}c_{22}$ (common elements to two rows and two columns)

av. $= -1/30$

The derivation of these averages by means of the Eulerian angles may be omitted.   It is also clear for symmetry reasons that only two expressions of the second degree in α are left.   We shall denote them by

$$15\,A = \alpha_1\alpha_1^* + \alpha_2\alpha_2^* + \alpha_3\alpha_3^*,$$
$$15\,B = \tfrac{1}{2}\{\alpha_1\alpha_2^* + \alpha_2\alpha_3^* + \alpha_3\alpha_1^*$$
$$+ \alpha_2\alpha_1^* + \alpha_3\alpha_2^* + \alpha_1\alpha_3^*\},$$

where the possibility that the α's may be complex is taken into account (asterisks denote conjugate complex values). A and B are real.

It follows that we retain only the following products of factors P with non-zero average values:

$$P_{11}P_{11}^*, \ P_{22}P_{22}^*, \ P_{33}P_{33}^*, \qquad \text{av. value} = 3A + 2B.$$

$$P_{12}P_{12}^*, \ P_{23}P_{23}^*, \ P_{31}P_{31}^*, \qquad \text{av. value} = A - B.$$

$$\left.\begin{array}{l} P_{11}P_{22}^*, \ P_{22}P_{33}^*, \ P_{33}P_{11}^*, \\ P_{22}P_{11}^*, \ P_{33}P_{22}^*, \ P_{11}P_{33}^*, \end{array}\right\} \qquad \text{av. value} = A + 4B.$$

Straightforward substitution in the transformation matrix (sec. 5.14) now gives the scattered intensity per particle in the cloud:

$$\begin{pmatrix} I \\ Q \\ U \\ V \end{pmatrix} = \frac{k^4}{r^2} \begin{pmatrix} 4A + B - \tfrac{1}{2}(2A + 3B)\sin^2\theta, & -\tfrac{1}{2}(2A + 3B)\sin^2\theta, & 0 & 0 \\ -\tfrac{1}{2}(2A + 3B)\sin^2\theta, & (2A + 3B)(1 - \tfrac{1}{2}\sin^2\theta), & 0 & 0 \\ 0 & 0 & (2A + 3B)\cos\theta, & 0 \\ 0 & 0 & 0 & 5B\cos\theta \end{pmatrix} \begin{pmatrix} I_0 \\ Q_0 \\ U_0 \\ V_0 \end{pmatrix}$$

The matrix is the matrix **F** defined in sec. 5.13 and used throughout the further sections of chap. 5. The symmetry relations are seen to be fulfilled. The form found above has the symmetry

of case 6, sec. 5.22, for arbitrary θ,

of case 13, sec. 5.31, for $\theta = 0$,

of case 13a, sec. 5.32, for $\theta = 180°$.

It has been customary, however, in this case to use the matrix **F'**, defined in sec. 5.14. The relation then is

$$\begin{pmatrix} I_l \\ I_r \\ U \\ V \end{pmatrix} = \frac{k^4}{r^2} \begin{pmatrix} (2A + 3B)\cos^2\theta + A - B, & A - B, & 0 & 0 \\ A - B, & 3A + 2B, & 0 & 0 \\ 0 & 0 & (2A + 3B)\cos\theta, & 0 \\ 0 & 0 & 0 & 5B\cos\theta \end{pmatrix} \begin{pmatrix} I_{l0} \\ I_{r0} \\ U_0 \\ V_0 \end{pmatrix}$$

This formula contains the complete information on the intensity and polarization of the scattered light when the incident light has an arbitrary state of polarization. Two simple cases are (both in the I_l, I_r, U, V representation):

1. Incident light is *natural* light ($\tfrac{1}{2}I_0, \tfrac{1}{2}I_0, 0, 0,$). The light scattered at $\theta = 90°$ has the parameters

$$\tfrac{1}{2}I_0 k^4 r^{-2} \ (2A - 2B, \ 4A + B, \ 0, \ 0),$$

i.e., partially plane-polarized with the ratio of the intensities I_l/I_r equal to

$$\Delta = \frac{2A - 2B}{4A + B}.$$

2. Incident light is *plane-polarized* in the r-direction, so that its

parameters are $(0, I_0, 0, 0)$. The light scattered at $\theta = 90°$ is represented by

$$I_0 k^4 r^{-2} (A - B, 3A + 2B, 0, 0)$$

and the ratio of intensities is

$$\Delta' = \frac{A - B}{3A + 2B}.$$

The quantities Δ and Δ' are the *depolarization factors*;[1] they are related by

$$\Delta = \frac{2\Delta'}{1 + \Delta'}, \qquad \Delta' = \frac{\Delta}{2 - \Delta}$$

Both are 0 if the polarizability is isotropic and they have the maximum values $\Delta = 1/2$, $\Delta' = 1/3$, if the particle can be polarized in only one direction.

These formulae are most frequently used for scattering by single molecules; values of Δ for many molecules are, e.g., given by Stuart (1936). Scattering measurements permit the determination of only one relation between the three polarizability values, α_1, α_2, and α_3. A quite different experiment (Kerr effect) is needed to determine a second relation. The third combination is given by the measured value of the refractive index (see below).

6.53. Transmitted Light, Refractive Index, and Extinction

The plane wave transmitted by the medium suffers a change in intensity (extinction) and phase (refractive index) according to the general formulae of chap. 4. The scattering matrix for $\theta = 0$ is

$$\begin{pmatrix} S_2(0) & S_3(0) \\ S_4(0) & S_1(0) \end{pmatrix} = ik^3 \begin{pmatrix} P_{22}, & P_{12} \\ P_{12}, & P_{11} \end{pmatrix}.$$

This matrix determines the propagation if the orientation of the particles is not at random. In random orientation we have to average the separate elements; the non-diagonal elements vanish, and the diagonal elements are equal because

$$(P_{11})_{av} = (P_{22})_{av} = \tfrac{1}{3}(\alpha_1 + \alpha_2 + \alpha_3) = \alpha.$$

We take this as the definition of α. There is a relation

$$3 |\alpha|^2 = 5A + 10B.$$

---

[1] This somewhat confusing term springs from the fact that before 1930 it was customary to denote the ratio I_l/I_r of a partially plane-polarized beam as its depolarization, a term that had no direct relation to scattering theories.

The result is that we are left with the "scalar" propagation case discussed in secs. 4.3 and 4.42, with

$$S(0) = ik^3\alpha.$$

The resulting value for $\tilde{m}$ is

$$\tilde{m} = 1 + 2\pi N\alpha,$$

which comprises the effects of a real refractive index,

$$\tilde{n} = 1 + 2\pi N \, \mathrm{Re} \, (\alpha),$$

and of an extinction caused by absorption. The average cross section of absorption per particle is

$$C_{abs} = -4\pi k \, \mathrm{Im} \, (\alpha),$$

as in sec. 6.13.

However, as with isotropic particles, α does not yet include the effect of the radiative reaction on the dipole, so that the magnitude of the total scattering has to be computed separately. (This is the only contribution to the extinction if α is real.) The total scattering is most easily computed for incident natural light. Integration of

$$I = k^4 r^{-2}\{4A + B - \tfrac{1}{2}(2A + 3B) \sin^2 \theta\}I_0$$

over the surface of a sphere with radius r gives the result $C_{sca}I_0$. We find

$$C_{sca} = \frac{8\pi k^4}{3} \cdot 5A,$$

and the extinction coefficient of the medium (in the absence of absorption) is $\gamma = NC_{sca}$.

Combining the result for C_{sca} and for $\tilde{n} - 1$ (in the absence of absorption) we obtain

$$C_{sca} = \frac{8\pi k^4 \alpha^2}{3} f$$

and

$$\gamma = \frac{2k^4}{3\pi} \frac{(\tilde{n} - 1)^2}{N} f,$$

where

$$f = \frac{3A}{A + 2B} = \frac{3(2 + \Delta)}{6 - 7\Delta},$$

and Δ is the depolarization factor defined at the end of sec. 6.52.

It is more usual, but in the given approximation not more correct, to write $\frac{1}{4}(\tilde{n}^2 - 1)^2$ instead of $(\tilde{n} - 1)^2$. Replacing k by $2\pi/\lambda$ we have the result

$$\gamma = \frac{8\pi^3}{3\lambda^4} \frac{(\tilde{n}^2 - 1)^2}{N} \cdot f$$

in a form closely similar to that in which it was first derived by Cabannes, and in which it has been given in the study by de Vaucouleurs (1951).

The last formula connects two directly observable quantities, $\tilde{n} =$ the observed refractive index of the gas, or medium consisting of small particles, and $\gamma =$ its attenuation coefficient per unit length. It has been extensively applied in the theory of atmospheric extinction. The experimental study of de Vaucouleurs has shown that earlier measurements of the depolarization factors have been systematically in error. For example, the values for atmospheric air are

As adopted previously: $\Delta = 0.0415$, $f = 1.073$;

As measured by de Vaucouleurs: $\Delta = 0.031$, $f = 1.054$

The new values give exactly the correct value for N.

References

Most of the topics of this chapter are found in standard textbooks and handbooks:

J. A. Stratton, *Electromagnetic Theory*, New York, McGraw-Hill Book Co., 1941.
M. Born, *Optik,* Berlin, J. Springer, 1933.
H. A. Stuart, *Hand und Jahrb. Chem. Physik*, 8II, 1 (1936).

The first papers on Rayleigh scattering are:

Lord Rayleigh, *Phil. Mag.*, **41**, 107, 274, and 447 (1871) (*Sci. Papers* 8 and 9).

The solution for ellipsoids (sec. 6.32) is found in

Lord Rayleigh, *Phil. Mag.*, **44**, 28 (1897) (*Sci. Papers* 230),
R. Gans, *Ann. Physik*, **37**, 881 (1912),

and its extension to anisotropic materials (sec. 6.33) in

R. Clark Jones, *Phys. Rev.*, **68**, 93 and 213 (1945).

Graphs for a rapid determination of the factors L_i were given by

J. A. Osborn, *Phys. Rev.*, **67**, 351 (1945).

The results in secs. 6.34 and 6.4 are largely new. The problem of anisotropic particles in random orientation (sec. 6.5) was not solved until

Lord Rayleigh, *Phil. Mag.*, **35**, 373 (1918) (*Sci. Papers* 430).

The solution is repeated in the textbooks quoted and in

J. Cabannes, *La diffusion moléculaire de la lumière*, Paris, Les Presses universitaires de France, 1929.

The most recent determination of Avogadro's number by the method outlined in sec. 6.53 is by

G. de Vaucouleurs, *Ann. phys.*, **6**, 211 (1951).

The intensity and polarization of the blue daylight sky can be found from the theory of multiple scattering by Rayleigh's law. This theory which starts from a formulation by means of the Stokes parameters (sec. 6.52) was discussed by

H. C. van de Hulst, *The Atmospheres of the Earth and Planets*, chap. 3 (2nd ed.), G. P. Kuiper, ed., Chicago, Univ. of Chicago Press, 1952,

and the full solution presented by

S. Chandrasekhar, *Radiative Transfer*, Oxford, Oxford Univ. Press, 1950.
S. Chandrasekhar and D. D. Elbert, *Trans. Am. Phil. Soc.*, **44**, 643 (1954).

7. RAYLEIGH-GANS SCATTERING

The assumptions made in this chapter are:

(a) The refractive index (which may be complex and is measured relative to the surrounding medium) is close to 1:

$$|m - 1| \ll 1.$$

(b) The "phase shift" is small:

$$2ka \, |m - 1| \ll 1.$$

Here $k = 2\pi/\lambda$ and a is a length of the order of the size of the particle. For spheres a will denote the radius.

A restriction is therefore placed on the size of the particles, yet this restriction is not that the size should be $\ll \lambda$, as in chap. 6, but includes the wider range of possibilities that the size is $\ll \lambda/|m - 1|$.

(c) It will be shown that, as a consequence, the efficiency factor is small:

$$Q_{ext} \ll 1.$$

Also the set of conditions (a) and (c) is sufficient to characterize the domain of validity of the theory in this chapter. The resulting formulae are so simple that they have been applied very often, sometimes beyond the range of their validity. Some correct applications are: small-angle X-ray scattering and scattering by most opal glasses.

7.1. The General Formulae

7.11. Particles with Positive $m - 1$

The basis of the theory of Rayleigh-Gans scattering is ordinary Rayleigh scattering. We have found in sec. 6.12 that any small particle with volume dV has the scattering functions

$$\left. \begin{array}{c} S_1(\theta) \\ S_2(\theta) \end{array} \right\} = ik^3\alpha \left\{ \begin{array}{c} 1 \\ \cos\theta \end{array} \right. ,$$

where from sec. 6.22 we have under the condition (a):

$$\alpha = \frac{(m^2 - 1)\,dV}{4\pi} = \frac{m - 1}{2\pi}\,dV.$$

The non-diagonal scattering functions $S_3(0)$ and $S_4(0)$ are zero.

This formula is applied to each separate volume element of the scattering particle. By reason of condition (b) the "applied" field, to which each volume element is exposed, differs neither in phase nor in amplitude appreciably from the original wave. The wave goes through the particle nearly as if there were no particle at all. Similarly, the (very weak) wave scattered by a volume element in a certain direction can leave the particle without being modified or distorted by the presence of the other volume elements.

An attempt to prove these assertions in a formal manner on the basis of condition (b) will not be made. Their correctness is easily seen for large particles ($ka \gg 1$) for which ray-optics holds.

The physical basis of Rayleigh-Gans scattering is, therefore, very simple: each volume element gives Rayleigh scattering and does so independently of the other volume elements. The waves scattered in a given direction by all these elements interfere because of the different positions of the volume elements in space. In order to calculate the interference effects we have to refer the phases of all scattered waves to a common origin of coordinates and then add the complex amplitudes. This means that a phase factor $e^{i\delta}$ is added to the expression given above. Each element now gives

$$\left. \begin{array}{c} S_1(\theta) \\ S_2(\theta) \end{array} \right\} = \frac{ik^3(m-1)}{2\pi} \, e^{i\delta} \, dV \left\{ \begin{array}{c} 1 \\ \cos\theta \end{array} \right.$$

and the entire particle gives

$$\left. \begin{array}{c} S_1(\theta) \\ S_2(\theta) \end{array} \right\} = \frac{ik^3(m-1) \, V}{2\pi} \, R(\theta, \varphi) \left\{ \begin{array}{c} 1 \\ \cos\theta \end{array} \right. ,$$

where

$$R(\theta, \varphi) = \frac{1}{V} \int e^{i\delta} \, dV.$$

The phases δ depend on θ and on the position of P. They may be computed as follows. Let $\mathbf{n}$ be the unit vector in the direction of the incident beam and $\mathbf{m}$ the same for the scattered wave. Let O be an arbitrary origin and let the vector OP be $\mathbf{r}$. The path from the (infinitely distant) source to P is longer by $\mathbf{rn}$ than the path to O and the path from P to the (infinitely distant) observer is shorter by $\mathbf{rm}$ than the path from O. The phase lag of a ray scattered at P over a ray scattered at O is found by adding the two effects and multiplying by k:

$$\delta = k\mathbf{r}(\mathbf{m} - \mathbf{n}).$$

The vector $\mathbf{m} - \mathbf{n}$ has the length $2 \sin \frac{1}{2}\theta$ along the bisectrix of the directions of **m** and —**n**. This can also be seen geometrically. In Fig. 13 the bisectrix of ∠AOB is OQ. The plane brought through P perpendicular to this bisectrix cuts it at Q so that $OQ = b$. All points in this plane have the same phase shift

$$\delta = kb \cdot 2 \sin \tfrac{1}{2}\theta.$$

This suggests an integration by means of "slices" perpendicular to the bisectrix, each slice having an area B and a thickness db. Then

$$R(\theta, \varphi) = \frac{1}{V} \int_{-\infty}^{\omega} B e^{ikb2 \sin \frac{1}{2}\theta} \, db.$$

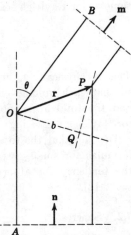

We may conclude that the scattered amplitude is simply the amplitude found from Rayleigh scattering multiplied by a function $R(\theta, \varphi)$ that is independent of polarization. So the polarization of the scattered light is exactly as it is for Rayleigh scattering. In particular, natural incident light gives fully polarized light at $\theta = 90°$.

The intensity is found by multiplying the intensity for Rayleigh scattering by $|R(\theta, \varphi)|^2$. Natural incident light of intensity I_0 gives (secs. 6.12 and 6.22) the scattered intensity

Fig. 13. Geometry of Rayleigh-Gans scattering: **n** = direction of incidence and **m** = direction of scattering; for convenience **r** is drawn in same plane as **m** and **n**.

$$I = \frac{1 + \cos^2 \theta}{2} \frac{k^4 V^2}{r^2} \left(\frac{m-1}{2\pi}\right)^2 |R(\theta, \varphi)|^2 I_0.$$

The factor $R(\theta, \varphi)$ can be found by a simple integration for most particles which have a simple geometrical form. For $\theta = 0$ we have $R = 1$; so the unmodified Rayleigh scattering holds in the forward direction. In other directions $|R| < 1$. The overall result is a pattern with strongest brightness in the small angles.

7.12. Absorbing Particles and Particles with $m < 1$

All formulae derived above hold for a complex value of m near 1, with the exception that in the intensity formula $(m-1)^2$ should be replaced by $|m-1|^2$.

A real value for m smaller than 1 can exist if the outer medium is not a vacuum; a practical example, with $m = 0.90$ to 0.95, is found in opal glass. It is also common in X-ray scattering. It is interesting that the refractive indices $1 + \varepsilon$ and $1 - \varepsilon$, where ε is small, give exactly the same scattering pattern. This also holds true for the wider problem discussed in chap. 11. However, no such rule exists if ε is not very small. Separate numerical computations are then needed for $m < 1$ and $m > 1$.

Particles with complex m have absorption in addition to scattering. Since the amplitude of forward scattered light is exactly the amplitude computed from Rayleigh scattering, the absorption formula is also the same as for Rayleigh scattering (sec. 6.31):

$$Q_{abs} = -\frac{4x}{3} \operatorname{Im} (m^2 - 1) = -\frac{8x}{3} \operatorname{Im} (m - 1).$$

This is physically clear: the incident wave is virtually undisturbed, so that each volume element absorbs just as if it were all by itself, and the total absorption is proportional to the total volume just as for Rayleigh scattering.

The fact that the total absorption is not reduced, with respect to the formulae for small particles, whereas the total scattering is, enhances the tendency for absorption to be the preponderant effect (compare sec. 6.13).

7.2. Spheres

7.21. Scattering Diagram

The scattering by a sphere under the conditions of this chapter was first discussed by Rayleigh (1881) and much later independently by Gans (1925).

Let a be the radius and let us make the substitutions

$$x = ka, \quad b = za, \quad u = 2x \sin \tfrac{1}{2}\theta, \quad \delta = zu;$$

then the "slice" has a thickness $a\,dz$ and a radius $a\sqrt{(1 - z^2)}$ so that we can integrate at once:

$$R(\theta, \varphi) = \frac{1}{V} \int_{-1}^{+1} e^{izu}\pi a^2(1 - z^2)a \, dz = G(u),$$

where

$$G(u) = \frac{3}{2} \int_0^1 \cos zu \cdot (1 - z^2) \, dz$$

$$= \frac{3}{u^3} (\sin u - u \cos u) = \left(\frac{9\pi}{2u^3}\right)^{1/2} J_{3/2}(u).$$

Numerical values of $G(u)$ and its zeros are found in sec. 7.4.

The full amplitude functions for a sphere under these conditions become

$$\left.\begin{array}{c} S_1(\theta) \\ S_2(\theta) \end{array}\right\} = ik^3a^3(m-1)\left(\frac{2\pi}{u^3}\right)^{1/2} J_{3/2}(u) \begin{cases} 1 \\ \cos\theta \end{cases},$$

and the scattered intensity for incident natural light is

$$I = \frac{k^4 V^2 \left| m-1 \right|^2}{8\pi^2 r^2} I_0 \cdot G^2 \left(2x \sin \tfrac{1}{2}\theta\right) \cdot (1 + \cos^2\theta),$$

which can be given equivalent forms by any of the substitutions

$$2(m-1) = m^2 - 1, \quad k = 2\pi/\lambda, \quad V = \tfrac{4}{3}\pi a^3, \text{ etc.}$$

According to the general rule explained in sec. 7.11 the forward intensity is exactly that of Rayleigh scattering, for $G(0) = 1$. The backscattering is somewhat smaller than the forward scattering if x is small; with increasing x this asymmetry becomes more and more pronounced. For $x = 2.25$ a minimum appears at $\theta = 180°$, corresponding to the first zero of $G(u)$. This minimum develops with further increasing x into a dark ring which shifts gradually to smaller angles. A new minimum appears at $\theta = 180°$ for $x = 3.87$, and so on. For large values of x the concentration of light towards $\theta = 0°$ has become so large and so many minima have come to lie in the region of small θ that the scattering pattern resembles (but is not identical to) a Fraunhofer diffraction pattern; see sec. 11.33 for the transition between both.

Looking at a fixed θ while the particle size grows, we see a train of maxima and minima shifting from the back side to the front side of the pattern. By counting the minima we may, in principle, estimate the particle size.

7.22. Total Scattering

The next question is to find the total scattering. Integrating I over a large sphere and dividing the result by πa^2 we find the efficiency factor Q_{sca} (which equals Q_{ext} since we deal with dielectric spheres). By means of direct integration we find

$$Q_{sca} = \left| m-1 \right|^2 \varphi(x),$$

where

$$\varphi(x) = \frac{4}{9} x^4 \int_0^\pi G^2(2x \sin \tfrac{1}{2}\theta) \cdot (1 + \cos^2\theta) \sin\theta \, d\theta.$$

The function $\varphi(x)$ was found by Rayleigh to have the form

$$\varphi(x) = \frac{5}{2} + 2x^2 - \frac{\sin 4x}{4x} - \frac{7}{16x^2}(1 - \cos 4x)$$

$$+ \left(\frac{1}{2x^2} - 2\right)\{\gamma + \log 4x - \text{Ci}(4x)\},$$

where γ = Euler's constant = 0.577 and Ci is the cosine integral defined by

$$\text{Ci}(x) = -\int_x^\infty \frac{\cos u}{u}\,du.$$

Separate integrals for the front half, $0 < \theta < \pi/2$, and the back half, $\pi/2 < \theta < \pi$, have been computed by Ryde and Cooper. A table is more useful than the formulae. (See Table 5.)

Table 5. Total scattering by a sphere in the Rayleigh-Gans domain

x	$Q(x)$	Front Half	Back Half	$\overline{\cos\theta}$
0.2	0.00187	0.00094	0.00093	0.00
0.4	0.0285	0.0147	0.0138	0.02
0.6	0.133	0.072	0.061	0.04
0.8	0.379	0.218	0.161	0.09
1	0.813	0.500	0.313	0.16
1.2	1.44	0.96	0.48	0.24
1.4	2.24	1.64	0.60	0.32
1.6	3.20	2.54	0.66	0.39
1.8	4.28	3.68	0.60	0.46
2	5.50	5.03	0.47	0.53
3	14.49	14.26	0.23	0.79
4	27.84	27.49	0.35	0.91
6	67.0	66.7	0.33	0.98
8	122.5	122.1	0.36	1.00
10	194.0	193.7	0.33	1.00

The last column gives rough estimates of $\overline{\cos\theta}$ which may be useful in computations of radiation pressure (see sec. 2.3).

For $x \ll 1$ we have equal parts in the front and back halves and return to Rayleigh scattering with

$$\varphi(x) = \frac{32}{27}x^4,$$

which gives

$$Q_{sca} = \frac{32}{27}(m-1)^2 x^4$$

in agreement with sec. 6.31.

The limiting case of $x \gg 1$ gives $\varphi(x) = 2x^2$, all from the front half; the integral over the back half converges to $1 - \log 2 = 0.31$. Thus for large x:

$$Q_{sca} = 2(m - 1)^2 x^2,$$

a result which will be derived in a quite different way in sec. 11.22. This result holds as long as condition (b) holds, so that in *all correct* applications of the Rayleigh-Gans scattering we have $Q_{sca} \ll 1$, which agrees with statement (c) at the beginning of this chapter.

7.23. Spherical Objects with an Inhomogeneous Density Distribution

The general theory of secs. 7.11. and 7.12 is by no means confined to homogeneous particles. A simple example of inhomogeneous particles is that of a collection of discrete scattering centres discussed in sec. 7.5. Another simple case, to be discussed now, is a particle in which the volume polarizability α' has a spherically symmetric distribution.

Let us first consider a spherical shell, radius a, thickness da, volume polarizability α', volume $4\pi a^2\, da$. With the same substitutions as in sec. 7.21 we obtain a "slice" in the form of an onion ring with

$$B = 2\pi a\, da, \quad b = a\, dz,$$

so that the integration gives

$$R(\theta, \varphi) = \frac{\sin (2ka \sin \tfrac{1}{2}\theta)}{2ka \sin \tfrac{1}{2}\theta} = E(2ka \sin \tfrac{1}{2}\theta).$$

This same function appears in sec. 7.32.

Next we take a collection of such shells, forming a body in which α' is a function of the distance from the center r only. Then the polarizability of the entire particle is

$$\alpha = \int_0^\infty 4\pi r^2 \alpha'(r)\, dr,$$

and a simple integration gives

$$R(\theta, \varphi) = \frac{1}{\alpha} \int_0^\infty 4\pi r^2 \alpha'(r)\, \frac{\sin (2kr \sin \tfrac{1}{2}\theta)}{2\, kr \sin \tfrac{1}{2}\theta}\, dr.$$

This formula was first derived for X-ray scattering by Ehrenberg and Schäfer (1932).

As the simplest example we may mention a particle in which α' is strongest at the center and has a gaussian distribution:

$$\alpha'(r) = \alpha_0' e^{-(r/a)^2}.$$

We shall assume that α_0' is real. The polarizability of the entire particle, which occurs in the Rayleigh formula, is

$$\alpha = \int \alpha' \, dV = \pi^{3/2} a^3 \alpha_0'.$$

The additional amplitude factor $R(\theta, \varphi)$, obtained from the formula above, or from direct "slicing," is

$$R(\theta, \varphi) = e^{-(ka \sin \frac{1}{2}\theta)^2}$$

The scattered intensity for incident unpolarized light of intensity I_0 is then in any direction:

$$I = \frac{I_0}{r^2} \, k^4 \alpha^2 \cdot \tfrac{1}{2}(1 + \cos^2 \theta) \, e^{-2(ka \sin \frac{1}{2}\theta)^2},$$

from which we find the scattering cross section by integration (sec. 2.2):

$$C_{sca} = 2\pi k^4 \alpha^2 \left\{ \left(\frac{1}{p} + \frac{1}{p^3} \right)(1 - e^{-2p}) - \frac{1}{p^2} (1 + e^{-2p}) \right\},$$

where $p = k^2 a^2$. When this result is expanded in powers of a, the dominant term is

$$C_{sca} = \frac{8\pi}{3} \, k^4 \alpha^2,$$

in agreement with the result for Rayleigh scattering (sec. 6.11).

This model was advanced by Peterlin (1951) as a simple representation of points of disturbance in a crystal, and Hart and Montroll (1951) mention it as a model for certain high polymers. The latter authors give an exponent which is too small by the factor 4π. Peterlin remarks that a (or an effective value of a) may be determined from the slope of the graph obtained by plotting the logarithm of $I(\theta)/(1 + \cos^2 \theta)$ against $2 \sin \frac{1}{2}\theta$. The ratio $I(135°)/I(45)°$ decreases from 0.87 for $2a/\lambda = 0.1$ to 0.11 for $2a/\lambda = 0.4$. In all formulae $\lambda = 2\pi/k$ means the wavelength in the surrounding medium, or λ_{vac}/n_0, if n_0 is the refractive index of the medium.

The computations on this model may be extended to still larger values of ka, for which the phase shift is not negligible. We then come to the realm of chap. 11 (see sec. 11.1 for the precise distinctions). The resulting cross section is (by the method of sec. 11.21):

$$C_{ext} = 2\pi a^2 \int_0^1 (1 - \cos qz) \frac{dz}{z} = 2\pi a^2(\gamma + \log q - \text{Ci } q),$$

where $q = 2k\alpha/a^2$, $\gamma = 0.57721$, and Ci q means the cosine integral. The physical meaning of q is the phase shift suffered by a ray passing through the center of the body. The same result has been independently derived by Glauber.

The "intermediate case" in the sense of sec. 11.1 is found from this formula by making q very small and from the preceding formula by making ka very large. The result obtained in both ways is

$$C_{ext} = C_{sca} = \frac{2\pi k^2 \alpha^2}{a^2} = 2\pi^4 a^4 k^2 \alpha_0'^2.$$

7.3. Ellipsoids and Cylinders

The calculation of $R(\theta, \varphi)$ by means of the formula in sec. 7.11 can readily be made for a variety of forms. We give here the results for a few simple forms, but the list may be extended at will.

7.31. Ellipsoids

Consider an ellipsoid with arbitrary axes, with light incident from an arbitrary direction, and the scattered intensity is sought in another arbitrary direction. Let the two planes that are tangent to the ellipsoid and perpendicular to the bisectrix b cut this bisectrix at the points C and D. Unlike the situation for spheres, the points C and D are not identical with the points at which these planes touch the ellipsoid. Slices cut by planes perpendicular to b and cutting b at Q all are elliptical slices of the same form. Their volumes decrease with increasing value of OQ in the same manner as for spheres. Only the maximum value of OQ is different: it is not the radius, as for spheres, but the value of $OC = OD$. The integration results in the same function,

$$R(\theta, \varphi) = G(2k \cdot OC \cdot \sin \tfrac{1}{2}\theta),$$

in which only the argument is different from that for spheres, for OC now depends on the directions of incidence and of scattering, i.e., on θ and φ. It is not of interest to write a formula, for the geometrical definition is the clearest one; the computation of OC in terms of the three semiaxes and the direction cosines of incident and scattered light is a matter of geometry.

The total scattering is not computed as easily as for spheres.

7.32. Circular Cylinders of Finite Length

Another form for which the Rayleigh-Gans scattering can be easily and rigorously computed is the circular cylinder. Let its length be l and its diameter $2a$, and let the phase shift be small for a ray traversing the cylinder in any direction. The orientation of the cylinder with respect to the incident wave is arbitrary.

The volume integration (sec. 7.11) cannot easily be performed by slices perpendicular to the bisectrix, but it is easily done by circular slices perpendicular to the cylinder axis. Let β be the angle between the cylinder axis and the bisectrix. Integration over one circular disk of radius a gives

$$R(\theta, \varphi) = \frac{1}{\pi a^2} \int_{-a}^{a} e^{2ik \sin \frac{1}{2}\theta \cdot y \sin \beta} 2\sqrt{(a^2 - y^2)}\, dy$$

$$= F(2ka \sin \tfrac{1}{2}\theta \sin \beta),$$

where $F(u)$ is defined by

$$F(u) = \frac{4}{\pi} \int_0^1 \cos tu \cdot (1 - t^2)^{1/2}\, dt = \frac{2}{u} J_1(u).$$

Numerical values and zeros of this function are given in sec. 7.4.

The phase in this integration has been referred to the center of the disk. In integrating over all disks that make up the cylinder of length l we refer the phases to the center of the cylinder, thus introducing an additional phase shift

$$2k \sin \tfrac{1}{2}\theta \cdot z \cos \beta,$$

which gives an additional factor in $R(\theta, \varphi)$:

$$\frac{1}{l} \int_{-l/2}^{+l/2} e^{2ik \sin \frac{1}{2}\theta \cdot z \cos \beta}\, dz = E(kl \sin \tfrac{1}{2}\theta \cos \beta),$$

where $E(u)$ is defined by

$$E(u) = \int_0^1 \cos tu \cdot dt = \frac{\sin u}{u} = \left(\frac{\pi}{2u}\right)^{1/2} J_{1/2}(u).$$

Numerical values of this function are also found in sec. 7.4. The final factor $R(\theta, \varphi)$, which has to be multiplied with the amplitude in the Rayleigh formula, is

$$R(\theta, \varphi) = F(2ka \sin \tfrac{1}{2}\theta \sin \beta) \cdot E(kl \sin \tfrac{1}{2}\theta \cos \beta).$$

Both F and E are 1 if their arguments are $\ll 1$. So we find at once the special results:

Thin disk $\quad (kl \ll 1)$: $\quad R(\theta, \varphi) = F(2ka \sin \tfrac{1}{2}\theta \sin \beta)$,

Thin rod $\quad (ka \ll 1)$: $\quad R(\theta, \varphi) = E(kl \sin \tfrac{1}{2}\theta \cos \beta)$.

The ratio of the arguments is $\tan \beta / \tan \beta_0$, where β_0 is defined by $\tan \beta_0 = l/2a$. Even when the disks are not very thin or the cylinders not very thin, a simplified expression can be used whenever β differs greatly from β_0. If $\tan \beta \ll \tan \beta_0$, we may apply the expression for a thin rod, for F remains 1 until E has become very small. If $\tan \beta \gg \tan \beta_0$, we may apply the expression for a thin disk, for E remains 1 until F has become very small.

The dependence of β on θ and φ is made explicit when we define the direction of the cylinder axis (n) with respect to the direction of propagation of the incident light (i) and of the scattered light (s) by means of the angles α and φ (Fig. 14). Then the bisectrix (b) has a direction as shown and

$$\cos \beta = -\cos \alpha \sin \tfrac{1}{2}\theta + \sin \alpha \cos \tfrac{1}{2}\theta \cos \varphi.$$

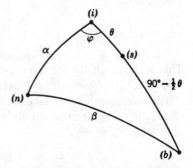

Fig. 14. Spherical triangle showing position of cylinder axis (n) with respect to directions of incidence (i) and scattering (s). The bisectrix has the direction (b).

The total scattering for an arbitrary orientation and arbitrary dimensions may be found by integrating the intensity (sec. 7.11) over all directions of scattering. We shall merely discuss one special case which serves as a check on the calculations in secs. 15.32 and 15.41.

Let the cylinder be very long ($l \gg 2a$, β_0 near $90°$), and let the cylinder axis be perpendicular to the incident radiation ($\alpha = 90°$). Then the scattered light is concentrated near the directions perpendicular to the axis, because of the E factor. Denoting the angle between (n) and (s) by $90° - \varepsilon$ (Fig. 14) we have $\cos \varphi = \sin \varepsilon / \sin \theta$, so that $\cos \beta = \sin \varepsilon / 2 \sin \tfrac{1}{2}\theta$. For small ε we, therefore, have

$$R(\theta,\varphi) = F(2ka \sin \tfrac{1}{2}\theta) \, E(\tfrac{1}{2}kl \sin \varepsilon).$$

The corresponding intensity of Rayleigh-Gans scattering for incident plane-polarized light is

$$I = \frac{x^4 l^2}{4r^2} \, (m - 1)^2 \, I_0 \, R^2(\theta, \varphi) \begin{cases} 1 \\ \cos^2 \theta \end{cases},$$

where $x = ka$. The upper value holds for case I (electric field//axis) and the lower value for case II (magnetic field//axis). The integration has to include only a narrow zone about the meridional plane. So the element of solid angle is simply $d\theta \, d\varepsilon$. The integral divided by the geometrical cross section $2al$ gives the efficiency factor:

$$Q_{sca} = \tfrac{1}{2}\pi(m-1)^2\varphi(x),$$

where

$$\varphi(x) = x^3 \int_0^\pi F^2(2x \sin \tfrac{1}{2}\theta) \, d\theta \qquad \text{(in case I),}$$

$$\varphi(x) = x^3 \int_0^\pi F^2(2x \sin \tfrac{1}{2}\theta) \cos^2 \theta \, d\theta \qquad \text{(in case II).}$$

The integrations have not been performed, but the asymptotic formulae are simple. For small x the factor F^2 may be omitted, and we find the formulae for very thin cylinders (sec. 15.41). For large x both cases give the same result:

$$\varphi(x) = 16x^2/3\pi,$$

and therefore

$$Q_{sca} = \tfrac{2}{3}\{2x(m-1)\}^2.$$

This is the result for the intermediate case (in the sense of sec. 11.1) for cylinders. It agrees with the expression for small ρ found in sec. 15.32.

The same results may be obtained by integrating the intensity over a closed surface around the particle in the near zone (zone 2 in the terminology of sec. 15.22), where

$$I = \frac{\pi x^4}{2rk} \, (m-1)^2 I_0 F^2(2x \sin \tfrac{1}{2}\theta) \begin{cases} 1 \\ \cos \theta \end{cases};$$

the derivation may be left to the reader.

7.33. Diffraction by a Partially Transparent Circular Disk

Let the cylinder be a circular disk ($l \ll a$), and let the disk be perpendicular to the incident light. In the formulae of sec. 7.32 we then have $\beta = 90° - \tfrac{1}{2}\theta$, and the scattering pattern consists of two strong lobe systems, one forward and one backward, in which the intensity in the exact forward and backward direction is determined by

$$R(0°) = 1, \; R(180°) = E(kl).$$

Barring the directions close to $180°$ (or assuming $kl \ll 1$) we have exactly

$$R(\theta) = F(ka \sin \theta).$$

It is not accidental that the same function with the same argument occurs in the formula for diffraction by an opaque circular disk (and perpendicular incidence). We may establish the connection as follows.

1. If $x = ka$, $\rho = kl(m-1)$, the result just derived holds for $a \gg l$, $m-1 \ll 1$, and $\rho \ll 1$; the values of x and kl are arbitrary. We now make the additional assumption that $kl \gg 1$, so also $x \gg 1$. The intensity

in the front lobe of the scattering pattern, which is then by far the largest of the two lobes, is (with $V = \pi a^2 l$)

$$I = \frac{x^4(m-1)^2 l^2}{8r^2} I_0 F^2(x \sin \theta) \cdot (1 + \cos^2 \theta).$$

2. A quite different discipline, similar to the one applied in chap. 11 gives the formula

$$S(\theta, \varphi) = \tfrac{1}{2}x^2(1 - e^{-i\rho}) \cdot F(x \sin \theta),$$

from which

$$I = \frac{x^4 I_0}{4k^2r^2} \left| 1 - e^{-i\rho} \right|^2 F^2(x \sin \theta).$$

These formulae hold for small angles only and under the conditions $m - 1 \ll 1$ and $x \gg 1$. They give the well-known diffraction by an *opaque disk*, (sec. 8.31), if the additional assumption is made that ρ has a large imaginary part. Instead, we now supplement these conditions with $\rho \ll 1$ (ρ is real) and thus find the set of conditions under which this theory may be applied equally well as the Rayleigh-Gans scattering derived above:

$$I = \frac{x^4 \rho^2}{4k^2 r^2} I_0 F^2(x \sin \theta).$$

The corresponding efficiency factor is found alike from $S(0)$ and from integration over all directions:

$$Q = \rho^2.$$

This is the analogue of the laws $Q = \tfrac{1}{2}\rho^2$ for spheres (secs. 7.22 and 11.22) and $Q = \tfrac{2}{3}\rho^2$ for cylinders (secs. 7.32 and 15.32) in the corresponding intermediate case.

7.34. Randomly Oriented Rods and Disks

The factor by which the intensity of the Rayleigh formula has to be multiplied in order to give the average pattern of the randomly oriented particles in the Rayleigh-Gans domain is

$$\overline{R^2} = \frac{1}{4\pi} \int R^2(\theta, \varphi)\, d\omega,$$

where the values of θ (and φ) are considered fixed and the integration over the solid angles refers to the orientations that the particles can have.[1]

---

[1] Particles that do not have an axis of symmetry require an integration over the three angles of Euler.

Starting from the formula for *thin rods* of length l we find

$$\overline{R^2} = \int_0^1 E^2(z \cos \beta) \, d(\cos \beta) = \frac{1}{z} \int_0^{2z} \frac{\sin w}{w} \, dw - \left(\frac{\sin z}{z} \right)^2.$$

Here $z = kl \sin \frac{1}{2} \theta$ and the integral function (sine integral) can be looked up in Jahnke-Emde or another table. The lobes and sharp minima in the pattern have been washed out by the integration. But the characteristic dissymmetry, which gives strong preference to forward angles, if kl is larger than 1, remains. It provides a rapid means for estimates of l in research on viruses.

A similar result may be written for randomly oriented *disks*:

$$\overline{R^2} = \int_0^1 F^2(z \sin \beta) \, d(\cos \beta) = \frac{2}{z^2} \{1 - F(2z)\}.$$

Here $z = 2 \, ka \sin \frac{1}{2} \theta$. The same results are found by employing a formula equivalent to that of Debye (sec. 7.5) for randomly oriented bodies consisting of discrete scattering centers (Kratky and Porod, 1949).

7.4. Some Properties of the Functions $E(u)$, $F(u)$, $G(u)$

The functions $E(u)$, $F(u)$, and $G(u)$ defined in this chapter form a set of similar functions. They are the Bessel functions of order $\frac{1}{2}$, 1, and $\frac{3}{2}$, divided by the first term of the series expansions of these same functions. Each function is 1 for $u = 0$, and the approximate expressions are

For small u	*For large u*
$E(u) = 1 - (u^2/6) + \ldots,$	$E(u) = (1/u) \sin u.$
$F(u) = 1 - (u^2/8) + \ldots,$	$F(u) = (8/\pi u^3)^{1/2} \sin (u - \pi/4).$
$G(u) = 1 - (u^2/10) + \ldots,$	$G(u) = -(3/u^2) \cos u.$

These functions appear in this book in two different contexts. In the present chapter they result from a *volume* integration over the scattering body; the arguments have the form $u = 2kb \sin \frac{1}{2}\theta$, and the formulae are valid both for small and for large values of kb. In secs. 8.31 and 8.32, $E(u)$ and $F(u)$ result from a *surface integration* over the screened-off part of a wave front and define, respectively, the diffraction pattern of an opaque strip (or cylinder) and of a circular disk (or sphere). The argument is $x \sin \theta$, and the formulae are derived only for large x and small θ. A connection between these apparently unrelated results is pointed out for disks in sec. 7.33.

Numerical values of these functions are given in Table 6. The squares of these values appear in the scattered intensity. Values of the zeros, which are needed to compute the angles at which dark rings occur, are also given.

For completeness some often needed integrals are given. They occur in the expressions for the total scattering when kb, or x, is very large.

$$\int_0^\infty E^2(u)\, du = \frac{\pi}{2}, \qquad \int_0^\infty F^2(u)\, du = \frac{16}{3\pi},$$

$$\int_0^\infty F^2(u)u\, du = 2, \qquad \int_0^\infty G^2(u)\, du = \frac{3\pi}{5},$$

$$\int_0^\infty G^2(u)u\, du = \frac{9}{4}, \qquad \int_0^\infty G^2(u)u^2\, du = \frac{3\pi}{2}.$$

Table 6. Functions Defining the Most Common Diffraction Patterns

u	$E(u)$	$F(u)$	$G(u)$	u	$E(u)$	$F(u)$	$G(u)$
0	1.000	1.000	1.000	5.2	−0.170	−0.132	−0.071
0.2	0.993	0.995	0.996	5.4	−0.143	−0.128	−0.080
0.4	0.974	0.980	0.984	5.6	−0.113	−0.119	−0.085
0.6	0.941	0.956	0.965	5.8	−0.080	−0.107	−0.086
0.8	0.897	0.922	0.938	6.0	−0.047	−0.092	−0.084
1.0	0.842	0.880	0.904				
				7	0.094	−0.001	−0.040
1.2	0.777	0.830	0.863	8	0.124	0.059	0.013
1.4	0.704	0.774	0.817	9	0.046	0.054	0.035
1.6	0.625	0.712	0.766	10	−0.054	0.009	0.024
1.8	0.541	0.646	0.711	11	−0.091	−0.032	−0.002
2.0	0.455	0.577	0.653	12	−0.045	−0.037	−0.018
2.2	0.368	0.505	0.593				
2.4	0.281	0.433	0.531				
2.6	0.198	0.362	0.468				
2.8	0.120	0.293	0.406			*Zeros*	
3.0	0.047	0.226	0.346		$E = 0$	$F = 0$	$G = 0$
				$u =$	3.142	3.832	4.49
3.2	−0.018	0.163	0.287		6.283	7.016	7.73
3.4	−0.075	0.105	0.231		9.425	10.173	10.90
3.6	−0.123	0.053	0.179		12.566	13.324	14.08
3.8	−0.161	0.007	0.131				
4.0	−0.189	−0.033	0.087				
4.2	−0.208	−0.066	0.048				
4.4	−0.216	−0.092	0.014			*Half-Intensity*	
4.6	−0.216	−0.112	−0.015		$E^2 = \frac{1}{2}$	$F^2 = \frac{1}{2}$	$G^2 = \frac{1}{2}$
4.8	−0.208	−0.124	−0.038	$u =$	1.39	1.61	1.81
5.0	−0.192	−0.131	−0.057				

7.5.   Discrete Scattering Centers; Form Factor

X-rays tend to go through a particle virtually unhampered and, therefore, form a typical example in which the conditions of this chapter hold. Although a complete discussion of X-ray scattering is beyond the scope of this book, some rules for applying the results of this chapter to X-ray scattering may be found helpful.

The fundamental fact is that each loosely bound electron in an atom, molecule, or larger particle scatters light according to the Thomson scattering formula (sec. 6.23). Let us call this particle an atom, and let there be Z scattering electrons. The amplitudes of the light scattered by each electron in a certain direction have to be added. Precisely the same argument as that of sec. 7.11 now gives the result that, starting from the scattering by one electron, we have to multiply

the amplitude by $\quad ZR(\theta, \varphi) \quad = f,$

the intensity by $\quad Z^2 |R(\theta, \varphi)|^2 = |f|^2 = F.$

The factors f and F are called the form factors. The intensity scattered by the atom for incident natural light of intensity I_0 becomes (compare sec. 6.23):

$$I = \frac{1 + \cos^2 \theta}{2} \frac{e^4}{r^2 m_0^2 c^4} F I_0.$$

The computation of f and F follows the same pattern as in the preceding sections. Two cases may be distinguished.

1. The distribution of the electrons is considered to be continuous. This is so for the density-probability in an atom and may also be assumed for a larger particle that has a spherical shape. Here any of the results in the preceding sections applies directly. In particular the results for a spherically symmetrical distribution, derived in sec. 7.23, have proved useful.

2. The electrons are considered as discrete scattering centers. The integrals of sec. 7.11 are then replaced by sums over the Z electrons. The evident analogue of the formula of sec. 7.11 is

$$R = f/Z,$$

where $f = \sum_j e^{i \, \delta_j}$ and δ_j is the phase shift for the jth electron. Multiplication with the complex conjugate gives

$$F = |f|^2 = \sum_j \sum_k \cos (\delta_j - \delta_k).$$

The double sum includes all combinations in which $j \neq k$ twice and has Z^2 terms in total.

If the distribution of the electron stays fixed in the atom but the atom assumes all random orientations, it is possible to derive the average value of F for all orientations and for a given θ. The result is

$$\bar{F} = \sum_j \sum_k \frac{\sin{(2\,ka_{jk} \sin{\tfrac{1}{2}\theta})}}{2\,ka_{jk} \sin{\tfrac{1}{2}\theta}} \,,$$

where a_{jk} is the distance between the jth and kth electron. This formula was first derived by Debye in 1915.

References

The term Rayleigh-Gans scattering is not very specific, for both authors have worked a great deal in scattering theories of various types. The term has been used in the literature for the specific topic of this chapter, and there is no reason to replace it with another one.

The papers referred to are:

Lord Rayleigh, *Phil. Mag.*, **12**, 81 (1881) (*Sci. Papers* 74).
R. Gans, *Ann. Physik*, **76**, 29 (1925).

The full formulae for spheres (sec. 7.22) were worked out by

Lord Rayleigh, *Proc. Roy. Soc.*, **A90**, 219 (1914) (*Sci. Papers* 381).
J. W. Ryde, *Proc. Roy. Soc.*, **A131**, 451 (1931).
J. W. Ryde and B. S. Cooper, *Proc. Roy. Soc.*, **A131**, 464 (1931).
R. W. Hart and E. W. Montroll, *J. Appl. Phys.*, **22**, 376 (1951).

Those for spherical shells (sec. 7.23) by

Lord Rayleigh, *Proc. Roy. Soc.*, **A94**, 296 (1918) (*Sci. Papers* 427).

Those for a sphere with gaussian distribution of polarizability (sec. 7.23) by

A. Peterlin, *Kolloid-Z.*, **120**, 75 (1951),

and in the paper by Hart and Montroll; the corresponding formula for large sizes by Glauber, as communicated by

E. W. Montroll and J. M. Greenberg, *Proc. Symposia Applied Math.*, Am. Math. Soc., **5**, 103 (1954).

The solution for straight circular cylinders (sec. 7.32) has been derived independently, though it is not certain that the result is new. The result for stiff rods with random orientation (sec. 7.34) was derived (in a different way) by

T. Neugebauer, *Ann. Physik*, **42**, 509 (1943).
B. H. Zimm, R. S. Stein, and P. Doty, *Polymer Bull.*, **1**, 90 (1945).

The result for thin disks with random orientation was first derived by
O. Kratky and G. Porod, *J. Colloid Sci.*, **4,** 35 (1949).

Table 6 has been newly computed. The integrals can be derived from
a more general one in

G. N. Watson, *Theory of Bessel Functions*, p. 403, Cambridge, Cambridge Univ.
Press, 1922.

Analogous or identical results in the theory for small-angle X-ray
scattering are found for almost all topics discussed in this chapter. The
form factor for an arbitrary particle with random orientation (sec. 7.5)
was derived by

P. Debye, *Ann. Physik*, **46,** 809 (1915).

and for spherically symmetric particles (sec. 7.23) by

W. Ehrenberg and K. Schäfer, *Physik. Z.*, **33,** 97 (1932).

General expositions and a discussion of special cases in the X-ray
domain are given by

H. G. Trieschmann, *Hand und Jahrb. Chem. Physik*, **8 II,** 105 (1936).
A. Guinier, *Ann. Phys.* **12,** 161 (1939).
G. Fournet and A. Guinier, *J. phys. radium*, **11,** 516 (1950).
A. Guinier and G. Fournet, *Small-Angle Scattering of X-Rays*, New York, John
Wiley & Sons, 1955.

A special case of multiple scattering is discussed by

D. L. Dexter and W. W. Beeman, *Phys. Rev.*, **76,** 1782 (1949).

8. PARTICLES VERY LARGE COMPARED TO THE WAVELENGTH

8.1. Fundamental Separation of Diffraction, Reflection, and Refraction

Large particles call for a way of looking at scattering problems that is quite different from the way of treating small particles. The fundamental fact is that the incident beam of light, which forms a plane wave front of infinite extent, may be thought to consist of separate rays of light which pursue their own path. We have seen in sec. 3.13 that a small part of a wide wave front can be looked upon as defining a ray which has for some length its own existence independent of the entire wave front. A length l requires a width of the order of $\sqrt{(l\lambda)}$, and the requirement for any separate existence at all is a width $> \lambda$. For a particle with a size of 20 or more times the wave length it is possible to distinguish fairly sharply between the rays incident on the particle and the rays passing along the particle. Among the former it is possible to distinguish rays hitting various parts of the particle's surface. Such rays may be said to be localized.

The experimental counterpart of the notion of localization is that it is possible to place a screen in the incident beam of light with a hole in such a position that a particular part of the particle is exposed to this light. This is the same principle as the Hartmann test for lenses or mirrors. A Hartmann test is possible only if the particle is large compared to the wavelength. A telescope lens is indeed a fine example of such a particle. The formal confirmation of this localization principle in the exact solution of the scattering problem for a sphere or cylinder is found in their asymptotic forms (sec. 12.3).

The rays hitting the particle and passing along it give rise to two distinct phenomena (I and II, below), both of which are in the terminology of this book included in the term scattering.

I. *Reflection and refraction.* Rays hitting the particle's surface are partially reflected and partially refracted. Refracted light may emerge after another refraction, possibly after several internal reflections. The light so emerging and the light directly reflected from the outer surface both contribute to the total *scattering* by the particle. The energy that does not emerge is lost by *absorption* inside the particle. Evidently, the amounts of energy absorbed or scattered and the angular distribution and polarization of the scattered light depend greatly on the form and composition of the particle and on the condition of its surface. The formulae for smooth spheres are worked out in sec. 12.21.

The reflection may be specular or diffuse. The theory of diffuse reflection will not be discussed in this book. It is sufficient to recall that any minute portion of a diffusely reflecting surface has to consist of some system following the ordinary laws of wave optics, for instance, a corrugated surface with plane faces of many orientations. The statistical effect of regular reflection against these faces gives rise to the phenomenon observed as diffuse reflection. Only a surface area that greatly exceeds the wave length can exhibit diffuse reflection. For that reason this concept appears exclusively in the present chapter (sec. 8.42).

II. *Diffraction.* The rays passing along the particle form a plane wave front from which a part, in the form and size of the geometrical shadow of the particle, is missing. This incomplete wave front gives rise by Huygens' principle to a certain angular distribution of intensity (at very large distances), the Fraunhofer diffraction pattern. Although the term diffraction is often used for the entire scattering process, we shall reserve it specifically for the Fraunhofer diffraction (see sec. 3.3). The intensity distribution in this diffraction pattern depends on the form and size of the particle but is independent of its composition or of the nature of its surface. For example, a black (fully absorbing) body, a white (diffusely reflecting) body, a totally reflecting body, or a glass body of the same shape, all have identical diffraction patterns. Diffracted light (defined in *this* sense!) has the same state of polarization as the incident light, and its pattern is independent of this polarization.

The two parts just described are distinct not only in their dependence on the nature of the particle but also in the angular distribution of scattered light. Let the size of the particle be fixed and the wavelength gradually decreasing. The scattering pattern due to reflection and refraction follows from rays that can be more and more sharply localized and will, eventually, approach the pattern that follows from the theory of geometrical optics. At the same time, the diffraction pattern will be more and more compressed into a narrow but very intense lobe around the forward direction, $\theta = 0$. In fact, the separation described in this section is strictly possible only if the particle is very large, i.e., if this lobe is very narrow. The entire pattern in this limit consists of two parts: a very narrow and very intense central lobe due to diffraction (II) and a less intense radiation into all directions dependent on the optical properties of the particle (I).

If, inversely, we make λ increase, we shall finally reach a stage where the two patterns are comparable in intensity and comparable in angular extent. This means generally that the theory of large particles breaks down and that more rigorous methods for solving the problem are needed. An exception is formed by particles with a refractive index close to 1. They have a range of sizes where the two patterns exist side by side with comparable intensity and combine by optical interference (= addition of amplitudes) (sec. 11.3). In the following sections we shall exclude this exceptional case and assume that the diffraction pattern is much more intense and much narrower than the reflection and refraction pattern.

8.2. General Theory of the Diffracted Light

8.21. Babinet's Principle

For very large particles the diffracted light is separable from the remaining scattering pattern, because it is confined to very small angles θ. A general formula may be derived on the basis of two imaginary experiments. Experiment 1: We replace the particle by a black disk in the form and size of its geometric shadow area. The incomplete wave front passing along this disk is identical to that passing along the particle. Hence the diffracted light is the same for particle and disk, both in amplitude and in phase. Experiment 2: The entire wave front is covered by a black screen, except for a hole in the form of the geometric shadow of the particle.

In this experiment as well as in experiment 1 a partially screened plane wave front is given, and the "disturbance" at any point beyond this wave front may be derived from Huygens' principle in Fresnel's formulation (sec. 3.12). If the two parts of the wave front are combined it is intact and the ordinary plane wave results. Therefore, the disturbance at any point in experiment 1 is the disturbance in the plane wave *minus* the disturbance found in experiment 2. Disregarding the plane wave, we may state that the disturbances in the two experiments are *equal in amount and opposite in sign*. Their squares, determining the intensity of the diffracted light, are equal. So the two experiments give identical scattering patterns. This result is known as Babinet's principle. If "diffraction" is used in the more general sense of sec. 3.3d, then also Babinet's principle requires a more general formulation (sec. 16.21).

Let the z-axis be in the direction of propagation of the incident light; let the orgin O be in the shadow area of the wave front and let the disturbance be sought at a distant point P (Fig. 15) which has the polar coordinates (r_0, θ, φ).

The distance from P to a point $(x, y, 0)$ of the shadow area is

$$r = r_0 - (x \cos \varphi + y \sin \varphi) \sin \theta.$$

This supposes that the quadratic terms in x and y are $\ll \lambda$, which is true if $r_0 \gg$ (largest dimension of G)$^2/\lambda$. Physically this means that, seen from P, the shadow is well within the central Fresnel zone. The disturbance at P in experiment 1, i.e., the disturbance caused by diffraction about the particle, is

$$u_p = u_0 e^{-ikz} - \frac{iu_0}{r\lambda} e^{-ikr_0} \iint\limits_{G} e^{-ik(x\cos\varphi + y\sin\varphi)\sin\theta} \, dx\, dy.$$

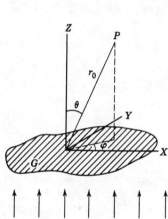

Here u_0 is the disturbance in the original wave on the plane wave front through O. The first term represents the plane wave; from it we have subtracted the disturbance that would be found in experiment 2, according to sec. 3.1. We shall denote the double integral appearing in this formula by

$$G \cdot D(\theta, \varphi)$$

where $G = \iint dx\, dy$ is the geometric area of the shadow and

$$D(\theta, \varphi)$$
$$= \frac{1}{G} \iint e^{-ik(x\cos\varphi + y\sin\varphi)\sin\theta} \, dx\, dy$$

Fig. 15. Diffraction by large body with geometrical shadow area G.

is a complex function that is 1 for $\theta = 0$. This function essentially describes the amplitude in the diffraction pattern. Comparing the results just derived with the general definition of the amplitude functions $S(\theta, \varphi)$ in sec. 4.41, we find that to the approximation to which this diffraction theory holds $S_3(\theta, \varphi)$ and $S_4(\theta, \varphi)$ are 0 and $S_1(\theta, \varphi)$ and $S_2(\theta, \varphi)$ are equal and given by

$$S(\theta, \varphi) = \frac{k}{\lambda} G D(\theta, \varphi) = \frac{k^2}{2\pi} G D(\theta, \varphi).$$

Consequently, we may apply the formulae of sec. 4.42 and find the scattered intensity:

$$I = \frac{G^2}{\lambda^2 r^2} I_0 \left| D(\theta, \varphi) \right|^2.$$

In the special case of $\theta = 0$, $D = 1$, this checks with Fresnel's formula derived in sec. 3.12.

The fact that the total energy scattered in the entire diffraction pattern is GI_0 (see next section) requires that

$$\int_0^\infty \int_0^{2\pi} |D(\theta, \varphi)|^2 \, \theta \, d\theta \, d\varphi = \frac{\lambda^2}{G}.$$

This formula can serve as a check on computations of $D(\theta, \varphi)$.

Some special examples of the function $D(\theta, \varphi)$ for particles of simple forms are mentioned in sec. 8.3.

8.22.   The Extinction Paradox

The total extinction by a particle that is very large compared to the wave length may be found in two ways.

First we may see what happens to the energy. All energy falling on the particle is scattered or absorbed, i.e., in any event removed from the proceeding wave. This gives an effective cross section equal to G, the geometric area. Besides that we have the diffraction forming an angular pattern that is identical with the diffraction through a hole of area G by Babinet's principle. This gives diffraction (= small-angle scattering) again with a cross section G. The total energy removed from the proceeding wave corresponds to a cross section

$$C_{ext} = 2G,$$

i.e. to an efficiency factor $Q_{ext} = 2$.

The second method is a direct application of the general extinction formula derived in sec. 4.21. Substituting

$$S(0) = \frac{k^2}{2\pi} \, G,$$

we find at once that $C_{ext} = 2G$.

This noteworthy paradox, that a large particle removes from the incident beam exactly *twice* the amount of light it can intercept, has attracted attention from various sides. Its paradoxical character is removed if we recall the exact assumptions that have been made in its derivation. They are (a) that *all* scattered light, including that at small angles, is counted as removed from the beam, and (b) that the observation is made at a very great distance, i.e., far beyond the zone where a shadow can be distinguished. A flower pot in a window prevents only the sunlight falling on it from entering the room, and not twice this amount, but a meteorite of the same size somewhere in interstellar space between a star and one of our big telescopes will screen twice this light. If many such rocks filled interstellar space, the light scattered by reflection and refraction would be seen as a faint luminosity all over the sky, and the diffracted light would show as narrow halos of light around each star. If the brightness of the star without this halo is measured, we have to use $Q = 2$. If, however, the method of measurement included the entire diffraction halo with the star's image (and this is certainly true if the telescope objective is smaller than the piece of rock), then it would be legitimate to use $Q = 1$. Exactly the same reasoning holds for colloidal solutions, and the effective value of Q, depending on the angular field of the measuring equipment, can be worked out numerically (e.g., Gumprecht and Sliepcevich, 1953).

8.3. Diffraction by Large Spheres and Thick Cylinders

8.31. Spheres

A *sphere* has a circular shadow of radius a and area $G = \pi a^2$. The function $D(\theta, \varphi)$ is found by direct integration and has the form

$$D(\theta, \varphi) = F(x \sin \theta) = \frac{2J_1 (x \sin \theta)}{x \sin \theta},$$

where $x = ka$. The same function appears in a different context in sec. 7.32; a table of its values and its zeros is given in sec. 7.4. Its square represents the intensity distribution. This distribution is the same as in the diffraction pattern of a circular aperture with central maximum and dark and bright rings, which is well known to every telescope user (Fig. 16). The first dark ring is at

$$u = \frac{2\pi a \sin \theta}{\lambda} = 3.83.$$

For later reference (secs. 11.31 and 12.32) we wish to have the complete expression of the amplitude functions with the proper phase. Turning to sec. 8.21 we find that

$$S_1(\theta) = S_2(\theta) = x^2 \frac{J_1(x \sin \theta)}{x \sin \theta}.$$

8.32. Cylinders

A *long cylinder* has, in almost any position, a shadow area which is virtually a rectangle with sides b and c. Choosing the x-axis along the side b and the y-axis along the side c we obtain by direct integration:

$$D(\theta, \varphi) = E(\tfrac{1}{2}kb \sin \theta \cos \varphi) \cdot E(\tfrac{1}{2}kc \sin \theta \sin \varphi),$$

where the function $E(u) = (\sin u)/u$ is again tabulated in sec. 7.4. It corresponds to the well-known diffraction pattern of a straight slit. The first minimum of $E(u)$ is at $u = 3.14$.

Long cylinders with random orientation present a diffraction problem which is not as simple as it seems. It is needed in the theory of diffraction by ice needles in cirrus clouds; a wrong solution has been given in the standard work of Pernter and Exner and has for a long time persisted in the meteorological literature. Meyer (1950) has drawn attention to the error.

Let the cylinders have length l and diameter $2a$, with $l \gg 2a$, while both are $\gg \lambda$. In a given direction (θ, φ) we have to find the sum of the intensities diffracted by each cylinder. It is easiest to divide by the total number of cylinders in the cloud so that we obtain the average intensity diffracted by one cylinder. By the formula found in sec. 8.21 this is

$$I = \frac{k^2 I_0}{4\pi^2 r^2} \overline{G^2 D^2(\theta, \varphi)}.$$

Let us first average over all rods that are inclined by an angle γ with respect to the direction of propagation. They have $G = 2al \sin \gamma =$ constant, so that only the average of

$$D^2 = E^2(ka\theta \cos \psi)\, E^2(\tfrac{1}{2}kl\theta \sin \gamma \sin \psi)$$

over all azimuth angles ψ has to be found. A rigorous average has not been derived, but when we confine ourselves to angles for which $kl\theta \sin \gamma \gg 1$, only small angles ψ contribute to the average, which becomes after some calculation

$$\overline{D^2} = \frac{2}{kl\theta \sin \gamma} E^2(ka\theta).$$

The average of $G^2 D^2$ now contains

$$\overline{\sin \gamma} = \tfrac{1}{2} \int_0^\pi \sin^2 \gamma \, d\gamma = \frac{\pi}{4},$$

so that, collecting these results, we obtain

$$I = \frac{I_0}{r^2} \cdot \frac{kla^2}{2\pi\theta} E^2(ka\theta),$$

for

$$\theta \gg \frac{1}{kl}.$$

The important feature is the factor θ in the denominator, in addition to the θ-dependence expressed by the ordinary diffraction pattern of a straight rod. This factor was omitted by Pernter and Exner. With this factor added, I in the bright rings is proportional to θ^{-3} just as it is for spheres, and no *conspicuous* difference between the diffraction by water drops and by randomly oriented ice needles is left. The two patterns are compared in Fig. 16.

As a check we may multiply by $2\pi r^2\theta\,d\theta$ and integrate in order to find the total amount of scattered light. The fact that the formula is wrong for very small θ does not matter. The result is

$$\int_0^\infty I 2\pi r^2\theta\,d\theta = I_0 \cdot \frac{\pi l a}{2} = I_0, \overline{G}$$

as it should be.

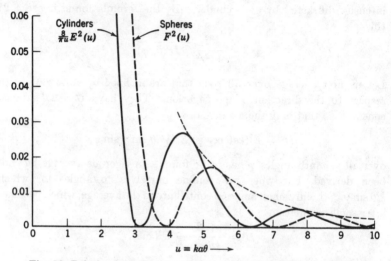

Fig. 16. Relative intensities across first and second diffraction rings in the diffraction patterns of spheres and of randomly oriented cylinders, all with radius a.

8.4. Large Convex Particles with Random Orientation

This section deals with two theorems holding for large, convex particles that are oriented at random. Such particles are, e.g., ellipsoids, rods disks, pieces of rock, crystals.

8.41. Average Geometrical Cross Section

The average geometrical cross section of a convex particle with random orientation is one-fourth its surface area.

The proof is simple. A convex particle, when illuminated from any direction, has a light side and a dark side separated by a closed curve on the surface. Any small surface area dS is on the dark side when its outward normal makes an angle $< 90°$ with the direction of propagation of the incident light; it is on the light side when this angle is $> 90°$. If many identical particles are randomly oriented, the normals of corresponding surface areas on all particles also have random positions. An area dS is dark in half the positions and obliquely illuminated in the other half. By a simple integration we find that it intercepts on the average the radiation $\frac{1}{4}I_0\,dS$. Integrating over the entire surface we find

$$\overline{G} = \tfrac{1}{4}S,$$

which proves the theorem. A sphere, with $G = \pi a^2$, $S = 4\pi a^2$, is the simplest example. This, apparently, is one of those theorems which gets discovered and rediscovered many times. The writer has not traced the literature about it. A simple extension to include non-convex particles is not possible.

8.42. Pattern of Reflected Radiation

The scattering pattern caused by reflection on very large convex particles with random orientation is identical with the scattering pattern by reflection on a very large sphere of the same material and surface condition.

This theorem follows from the same reasoning. The normals of any one surface element that assumes random positions are distributed in the same manner as the normals of all surface elements on a sphere. So also the scattering pattern by reflection must be identical. The other parts of the scattering pattern, viz., the refracted light and the diffracted light (see sec. 8.1), do not follow this theorem.

A few examples may be given.

Polished, totally reflecting particles. The reflection is specular and total. Section 12.44 shows for spheres that the scattering by reflection is isotropic and that natural incident light gives natural scattered light. By virtue of the theorem the same holds for convex particles of other forms. The scattered intensity at a distance r per particle is

$$I = I_0\overline{G}/4\pi r^2.$$

Polished metallic particles. The reflection is specular but partial. The refracted light is absorbed, so that it does not spoil the scattering pattern. Section 12.44 shows for spheres that the reflected light has a flat maximum in the forward direction, $\theta = 0$ (the diffracted light has a very sharp peak in the same direction). Polarization effects occur. All results are directly applicable to convex particles of other forms.

White particles. The reflection is total but diffuse. The concept "white," like the concept of diffuse reflection, makes sense only for surface areas with dimensions $\gg \lambda$ (sec. 8.1). The exact law of diffuse reflection depends on the nature of the surface or the cause of diffuse reflection, a discussion of which is beyond the scope of this book. Most white surfaces obey fairly closely Lambert's law. This law states that the surface brightness of a white surface is the same in all directions, independently of the direction from which it is illuminated. It is consistent with this law to postulate that the reflected light is unpolarized, irrespective of the polarization of the incident light. If a surface element dS receives the flux $F\ dS$, it reflects the flux $(\cos \alpha)F\ dS/\pi$ per unit solid angle in a direction that makes an angle α with the normal.

The scattering pattern of a sphere whose surface elements follow Lambert's law has been computed and tabulated by Schoenberg (1929). Compared to isotropic scattering (polished particles) we have an additional factor ("gain"):

$$f(\theta) = \frac{8}{3\pi}\ (\sin \theta - \theta \cos \theta),$$

so that

$$I = \frac{I_0 \overline{G} f(\theta)}{4\pi r^2}.$$

Table 7 gives some values of $f(\theta)$. The maximum light is found in the direction $\theta = 180°$, i.e., in the direction back to the light source. This is qualitatively illustrated by the moon, although it does not follow Lambert's law! The full moon ($\theta = 180°$) is brightest, the new moon ($\theta = 0°$) is dark. A polished moon, or a planet covered by a reflecting ocean, would show just the reverse effect. It would have a bright reflection spot at its center when full but an even brighter reflection spot near its edge when nearly new. The rest of the disk would look black.

Table 7. Scattering by a White, Diffusely Reflecting Sphere following Lambert's Law

θ	$f(\theta)$	θ	$f(\theta)$	θ	$f(\theta)$
0°	0	70°	0.443	140°	2.134
10°	0.0015	80°	0.630	150°	2.349
20°	0.0119	90°	0.849	160°	2.518
30°	0.0395	100°	1.093	170°	2.628
40°	0.0917	110°	1.355	180°	2.667
50°	0.1742	120°	1.624		
60°	0.2907	130°	1.888		

References

The theory of sec. 8.1 is classical and was worked out by Fresnel, Poisson, and others around 1820. A useful book is

A. Gray and G. B. Matthews, *A Treatise on Bessel Functions and Their Applications to Physics*, London, Macmillan & Co. (2nd. ed.), 1922.

Babinet's principle dates from 1837.   A few papers that discuss the extinction paradox (sec. 8.22) in optics are:

O. Struve, *Ann. astrophys.*, **1**, 143 (1938).

J. Bricard, *J. phys.*, **4**, 57 (1943); *Compt. rend.*, **223**, 1164 (1946).

H. C. van de Hulst, "Thesis Utrecht," *Recherches astron. Obs. d'Utrecht*, **11**, part 1, p. 10 (1946).

D. Sinclair, *J. Opt. Soc. Amer.*, **37**, 475 (1947).

L. Brillouin, *J. Appl. Phys.*, **20**, 1108 (1949).

Similar explanations are found in acoustics and wave-mechanics, e.g.,

H. Wergeland, *Avhandl. Norske Videnskaps-Akad. Oslo. I. Mat. Naturvid. Kl.*, No. 9 (1945).

J. M. Blatt and V. F. Weisskopf, *Theoretical Nuclear Physics*, p. 324, New York, John Wiley & Sons, 1952.

P. M. Morse and H. Feshbach, *Methods of Theoretical Physics*, part II, pp. 1381 and 1555, New York, McGraw-Hill Book Co., 1953.

These books refer to the diffracted wave as the shadow-forming wave. Although this is mathematically correct as, in the near field, this wave annihilates the original wave behind the particle, the term might convey the incorrect idea that anything more subtle than simple screening forms the shadow.

The dependence on the angular aperture of the measuring instruments was computed by

R. O. Gumprecht and C. M. Sliepcevich, *J. Phys. Chem.*, **57**, 90 (1953).

For the problem of randomly oriented needles (sec. 8.32) reference is made to

J. M. Pernter and F. M. Exner, *Meteorologische Optik, Vienna*, W. Braumüller, 1910.

R. Meyer, *Ber. deut. Wetterdienstes U.S. Zone*, Bad Kissingen, **12**, 182 (1950).

The phase function for white spheres (sec. 8.42) is discussed by

E. Schoenberg, *Handb. Astrophysik*, **2**, 255 (1929).

who tabulates ^{10}log $\{f(\theta)/f(180°)\}$ for several assumptions about the law of diffuse reflection.

9. RIGOROUS SCATTERING THEORY FOR SPHERES OF ARBITRARY SIZE (MIE THEORY)

All problems in theoretical optics are problems in Maxwell's theory and should be treated as such when a full, formal solution is required. Frequently a short route towards physical insight in a problem must be preferred to a deduction of its formal solution from a given set of equations. This is the reason why Maxwell's equations have not appeared in this book until the present chapter. The scattering of light by a homogeneous sphere cannot be treated in a general way, other than by the formal solution of Maxwell's equations with the appropriate boundary conditions. Readers not interested in the mathematics of this solution may turn at once to sec. 9.3 for final results and to chaps. 10–15 for special cases and numerical results.

9.1. Maxwell Equations

9.11. General Equations

A thorough treatment of Maxwell's theory is beyond the scope of this book. There are excellent modern textbooks to which the reader is referred. The author's references are to Stratton's *Electromagnetic Theory* (1941), although both the units and the notations adopted in this book differ from Stratton's.

The following points may be briefly noted:

1. Gaussian units, which have been used in all classical papers on the subject, are employed in this book. They are non-rationalized c.g.s. units electrostatic for E, D, I, σ, ρ, and electromagnetic for H. By this choice elegant, symmetric formulae result. The disadvantage, namely, that the conversion from fields into currents and charges is more elaborate, is immaterial in calculations of the kind presented here.

2. The author uses the time factor $\exp(i\omega t)$, where Stratton uses $\exp(-i\omega t)$. This choice, giving a refractive index with a negative imaginary part, is the classical one and has also been made in some modern tables.

3. Choice of the angle θ as the angle between the directions of propagation of the incident wave and the scattered wave (see Figs. 1, 13, 15, 19) agrees with that of Stratton. The classical choice of the supplement of θ as the scattering angle was an unfortunate one.

The two equations of Maxwell are

$$\operatorname{curl} \mathbf{H} = \frac{4\pi \mathbf{I}}{c} + \frac{1}{c}\frac{d\mathbf{D}}{dt}, \tag{1}$$

where $$D = \varepsilon E \quad \text{and} \quad I = \sigma E,$$

and $$\text{curl } E = -\frac{1}{c}\frac{dH}{dt}. \tag{2}$$

The meaning of the symbols is: t = time, c = velocity of light, H = magnetic field strength, E = electric field strength, D = dielectric displacement, I = current density, ε = dielectric constant, σ = conductivity. The magnetic permeability (μ) has been put 1, because this is true for all applications we shall consider.[1] If μ is not 1, it appears in all formulae as a second parameter beside the refractive index m, defined below.

The third independent equation expresses the conservation of charge:

$$\text{div } I + \frac{d\rho}{dt} = 0, \tag{3}$$

where ρ is the density of charge.

Three derived equations will be found useful.   By taking the divergence of both sides of equation 1 we obtain:

$$4\pi \text{ div } I + \frac{d}{dt} \text{ div } D = 0. \tag{4}$$

When combined with equation 3 this gives

$$\text{div } D = 4\pi\rho, \tag{5}$$

if this is supposed to be correct at a particular time.   In the same manner the divergence of equation 2 gives

$$\text{div } H = 0. \tag{6}$$

In the following subsections we shall find the forms assumed by these equations in the particular conditions of interest to our problems.   E.g., equation 1 assumes the forms of equations 1a, 1b, and 1c.

9.12.   Periodic Fields

From now on we consider periodic phenomena with a circular frequency ω.   It is advantageous to write all quantities as complex functions of the time in the form

$$A = (\alpha + i\beta)e^{i\omega t},$$

[1] One advantage of keeping μ in the formulae would be that the boundary conditions for the outer fields at a perfectly conducting surface (sec. 9.14, equations 2c and 6c) are the same as the boundary conditions at the interface of vacuum with an arbitrary medium (sec. 9.13) if in the latter we let $\varepsilon \to \infty$, $\mu \to 0$. This theorem is used in sec. 10.4.

where it is understood that the physical quantity represented by A shall equal the real value Re (A). The symbol Re means that the real part is taken. With this periodicity Maxwell's equations assume the much simpler forms

$$\text{curl } \mathbf{H} = ikm^2\mathbf{E}, \tag{1a}$$

$$\text{curl } \mathbf{E} = -ik\mathbf{H}, \tag{2a}$$

where

$$k = \frac{\omega}{c} = \frac{2\pi}{\lambda}$$

and

$$m^2 = \varepsilon - \frac{4\pi i\sigma}{\omega}.$$

Both k and m are extremely important parameters: k is *the propagation constant* (= *wave number*) *in vacuum*. The wave length in vacuum, λ, follows from it by $\lambda = 2\pi/k$; it often is a matter of taste whether λ or k is used in a formula. The parameter m is *the complex refractive index of the medium* at the frequency ω. It should be noted that m cannot generally be determined from the static values of ε and σ but should be determined by measurements at the circular frequency ω.

The divergence of equation 1a gives, in correspondence with equation 4,

$$\text{div } (m^2\mathbf{E}) = 0. \tag{4a}$$

In a homogeneous medium, where m = constant, the divergence of $\mathbf{E}$ itself vanishes, so that by equation 5 the charge density ρ is zero. A further result for a homogeneous medium, found from equations 1a and 2a, is that any rectangular component of $\mathbf{E}$ or $\mathbf{H}$ satisfies the scalar wave equation

$$\Delta\psi = -k^2m^2\psi.$$

The simplest type of solution corresponds to a plane wave. A plane wave traveling in the positive z-direction has the form

$$\psi = e^{ikmz+i\omega t}.$$

This shows that km is *the propagation constant in the medium with refractive index* m. The wave is damped if m has a negative imaginary part and is undamped if m is real. In the latter case λ/m is the wavelength inside the medium.

9.13. Boundary Conditions

We consider a sharp boundary between homogeneous media. Let medium 1 have the values ε_1, σ_1, m_1, and medium 2 the values ε_2, σ_2, m_2.

Both m_1 and m_2 are assumed to be finite; for infinite m see sec. 9.14. Let **n** be the normal to the boundary surface, directed into medium 2. Equations 1a and 2a then change by a well-known limiting process into the boundary conditions for the tangential components:

$$\mathbf{n} \times (\mathbf{H}_2 - \mathbf{H}_1) = 0, \tag{1b}$$

$$\mathbf{n} \times (\mathbf{E}_2 - \mathbf{E}_1) = 0. \tag{2b}$$

Corresponding to equations 4 and 6 we have the boundary conditions for the normal components:

$$\mathbf{n} \cdot (m_2^2\mathbf{E}_2 - m_1^2\mathbf{E}_1) = 0, \tag{4b}$$

$$\mathbf{n} \cdot (\mathbf{H}_2 - \mathbf{H}_1) = 0. \tag{6b}$$

Like equations 4 and 6, these equations are not independent. They may be derived from equations 1a and 2a valid in the medium at either side combined with 1b and 2b valid at the boundary.

The set of equations may be made complete by introducing the surface charge density δ. Surface charges are supplied and carried away by normal components of the current densities in either medium.

As the boundary form of equation 3, expressing the conservation of charge, we find

$$\mathbf{n} \cdot (\mathbf{I}_2 - \mathbf{I}_1) + \frac{d\delta}{dt} = 0. \tag{3b}$$

Combined with equation 4b, this equation gives, in perfect correspondence with equation 5,

$$\mathbf{n} \cdot (\varepsilon_2\mathbf{E}_2 - \varepsilon_1\mathbf{E}_1) = 4\pi\delta. \tag{5b}$$

There are no surface currents here; only if m_1 or m_2 are infinite do surface currents exist. The surface charges are zero if m_1 and m_2 are real, i.e., at the boundary between two dielectrics. Since we consider *only* periodic phenomena, this statement refers to a periodically changing charge; there always may be a static surface charge.

9.14. Skin Conditions; Perfect Conductors

A different set of conditions becomes appropriate at the surface of a highly conducting medium, like metallic conductors at infrared and radio frequencies. If the fields are supplied by outside sources, they penetrate only into a thin skin of the conductor.

We assume in this section that medium 2 is a vacuum and medium 1 a metal with skin depth small compared to the vacuum wavelength. This is true if $4\pi\sigma/\omega$ (denoted by η in sec. 14.41) is large compared to 1. Here and in the further formulae all material constants refer to the metal.

These assumptions are further specifications of assumptions made in sec. 9.13. So the *boundary conditions* 1b to 6b, derived there, are still correct here, but in addition another set of equations, the *skin conditions*, may be derived. The distinction is illustrated by Fig. 17. The popular derivation of the boundary conditions for the tangential components uses a loop formed by line 2 just outside and line 1 just inside the medium. For the normal components the loop is replaced by a flat box with bottom 1 and top 2. In either case the vertical sides can be made truly infinitesimal. Therefore, the boundary conditions 1b to 6b are *exact* and refer to the fields at the top of the skin.

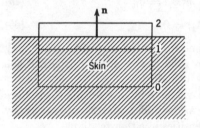

Fig. 17. Skin conditions at the boundary of a good conductor.

However, since the skin is *thin* it seems promising to consider another loop. The bottom of this loop may be placed at any level well below the skin, where the fields are virtually zero. Top and bottom of the loop are then formed by lines 2 and 0 in the figure. The vertical sides of this loop are small but not infinitesimal. So the skin conditions derived from it are approximate, not exact.

Let the z-axis be in the direction of $\mathbf{n}$, the y-axis along the horizontal sides of the loop, and the x-axis perpendicular to it. Integration of 1a over the area of the loop gives

$$-H_{2y} = ikm^2 \int E_x \, dz.$$

Only the contribution of the top side of the loop has been retained. The bottom side gives zero, because it is below the skin. The vertical sides give small contributions.[2] Assuming that ε is a small part of m^2, we may write

$$m^2 = -\frac{4\pi i\sigma}{\omega},$$

by which the equation is converted into

$$-H_{2y} = \frac{4\pi}{c} \int I_x \, dz.$$

Completed and written in vector notation, we have

$$\mathbf{n} \times \mathbf{H}_2 = \frac{4\pi}{c} \mathbf{j}, \tag{1c}$$

[2] They are smaller than the main term by the order of $|m|^{-2}$, but the detailed derivation of this result is cumbersome.

where

$$\mathbf{j} = \int \mathbf{I}_{tang} \, dz.$$

The tangential components of the volume currents in the skin when integrated over the depth of the skin form a quasi-surface current, $\mathbf{j}$. This $\mathbf{j}$ determines by formula 1c the tangential component of the external magnetic field. At the same time normal components of the volume currents exist, as before, and give rise to actual surface charges. The new equation for the conservation of charge, found upon combining equation 3 in the skin and 3b at the boundary, is

$$-\frac{d\delta}{dt} = \text{Div } \mathbf{j} \left[\equiv \frac{\delta j_x}{\delta x} + \frac{\delta j_y}{\delta y} \right]. \tag{3c}$$

In a manner similar to the derivation of formula 1c we find the approximate relation:

$$E_{2y} = ik \int H_x \, dz,$$

but its right-hand member is smaller than that of 1c by the order of magnitude $|m|^{-1}$, so that for most purposes we may neglect it and write

$$\mathbf{n} \times \mathbf{E}_2 = 0. \tag{2c}$$

The dependent equations are, again in individual correspondence with equations 4 to 6,

$$i\omega \mathbf{n} \cdot \mathbf{E}_2 = -4\pi \text{ Div } \mathbf{j}, \tag{4c}$$

$$\mathbf{n} \cdot \mathbf{E}_2 = 4\pi\delta, \tag{5c}$$

and

$$\mathbf{n} \cdot \mathbf{H}_2 = 0. \tag{6c}$$

They follow rigorously from equations 1c to 3c and the Maxwell equations in vacuum. Hence the whole set of equations 1c to 6c may be used to obtain a consistent solution of Maxwell's equations in a vacuum (medium 2) bounded by the conductor.

Perfect conductor. If σ goes to ∞, then $m \to \infty$ and the metal becomes a perfect conductor. The equations 1c to 6c are then rigorously correct and $\mathbf{j}$ is an actual surface current. It is important to note that the boundary conditions of a perfect reflector are the idealization of the *skin conditions* 1c to 6c of an imperfect conductor and not the direct consequence of its boundary conditions 1b to 6b.

9.2. Mie's Formal Solution

9.21. Solution of Vector Wave Equation

Spherical coordinates of a point P will be denoted by (r, θ, φ) with the usual meaning of the symbols. The vector OP, which has the rectangular components (x, y, z), or

$$r \cos \varphi \sin \theta, \; r \sin \varphi \sin \theta, \; r \cos \theta,$$

is denoted by $\mathbf{r}$.

The scalar wave equation

$$\Delta \psi + k^2 m^2 \psi = 0$$

is separable in these coordinates and has elementary solutions of the following type:

$$\psi_{ln} = \left. \begin{matrix} \cos l\varphi \\ \sin l\varphi \end{matrix} \right\} P_n^l (\cos \theta) \, z_n(mkr).$$

Here n and l are integers:

$$n \geq l \geq 0;$$

the first factor may be either a cosine or a sine; the second factor is an associated Legendre polynomial; the third factor may be any spherical Bessel function, defined by

$$z_n(\rho) = \sqrt{\frac{\pi}{2\rho}} Z_{n+1/2}(\rho)$$

in terms of ordinary Bessel functions. The general solution of the scalar wave equation is a linear combination of such elementary solutions.

By virtue of formulae 1a and 2a the field vectors $\mathbf{E}$ and $\mathbf{H}$ in a homogeneous medium satisfy the vector wave equation

$$\Delta \mathbf{A} + k^2 m^2 \mathbf{A} = 0.$$

Elementary solutions of this equation may be found from the following theorem, which is stated without proof. If ψ satisfies the scalar wave equation, the vectors $\mathbf{M}_\psi$ and $\mathbf{N}_\psi$ defined by

$$\mathbf{M}_\psi = \operatorname{curl} (\mathbf{r}\psi), \tag{7}$$

$$mk\,\mathbf{N}_\psi = \operatorname{curl} \mathbf{M}_\psi, \tag{8}$$

satisfy the vector wave equation and are, moreover, related by

$$mk\,\mathbf{M}_\psi = \operatorname{curl} \mathbf{N}_\psi. \tag{9}$$

This is just what we need. A simple substitution shows that, if u and v are two solutions of the scalar wave equation and $\mathbf{M}_u$, $\mathbf{N}_u$, $\mathbf{M}_v$, $\mathbf{N}_v$ are the derived vector fields, the Maxwell equations 1a and 2a are satisfied by

$$\left. \begin{matrix} \mathbf{E} = \mathbf{M}_v + i\mathbf{N}_u \\ \mathbf{H} = m(-\mathbf{M}_u + i\mathbf{N}_v) \end{matrix} \right\}. \tag{10}$$

The full components of $\mathbf{M}_\varphi$ and $\mathbf{N}_\varphi$ are

$$M_r = 0, \qquad mkN_r = \frac{\partial^2(r\psi)}{\partial r^2} + m^2 k^2 r\psi,$$

$$M_\theta = \frac{1}{r \sin \theta} \frac{\partial(r\psi)}{\partial \varphi}, \; mkN_\theta = \frac{1}{r} \frac{\partial^2(r\psi)}{\partial r \, \partial \theta},$$

$$M_\varphi = -\frac{1}{r} \frac{\partial(r\psi)}{\partial \theta}, \qquad mkN_\varphi = \frac{1}{r \sin \theta} \frac{\partial^2(r\psi)}{\partial r \, \partial \varphi}.$$

The components of $\mathbf{E}$ and $\mathbf{H}$ may, therefore, be written in terms of the scalar solutions u and v and their first and second derivatives, but these full formulae will not be needed.

9.22.   Solution of Coefficients from Boundary Conditions

We now turn to the Mie problem: scattering of a plane wave by a homogeneous sphere. In order to simplify the notation we assume that the outside medium is vacuum ($m_2 = 1$), and the material of the sphere has an arbitrary refractive index m. We assume that the incident radiation is linearly polarized. The origin is taken at the center of the sphere, the positive z-axis along the direction of propagation of the incident wave, and the x-axis in the plane of electric vibration of the incident wave. The incident wave (of amplitude 1) then is described by

$$\mathbf{E} = \mathbf{a}_x e^{-ikz + i\omega t},$$

$$\mathbf{H} = \mathbf{a}_y e^{-ikz + i\omega t},$$

where $\mathbf{a}_x$ and $\mathbf{a}_y$ are unit vectors along the x- and y-axes.

It can be proved (but we omit the derivation) that the same fields are written in the form postulated in sec. 9.21 by choosing for u and v the following functions.

Outside, incident wave:

$$u = e^{i\omega t} \cos \varphi \sum_{n=1}^{\infty} (-i)^n \frac{2n+1}{n(n+1)} P_n^1(\cos \theta) j_n(kr),$$

$$v = e^{i\omega t} \sin \varphi \sum_{n=1}^{\infty} (-i)^n \frac{2n+1}{n(n+1)} P_n^1(\cos \theta) j_n(kr),$$

where j_n is the spherical Bessel function derived from the Bessel function of the first kind, $J_{n+1/2}$.

This form for the incident wave sets the form also for the complete solution. The field outside the sphere consists of the incident wave plus the scattered wave. From a consideration of the boundary conditions and the conditions to be satisfied at infinity it is found that the following assumption is sufficiently general.

Outside, scattered wave:

$$u = e^{i\omega t} \cos \varphi \sum_{n=1}^{\infty} -a_n(-i)^n \frac{2n+1}{n(n+1)} P_n^1(\cos \theta) h_n^{(2)}(kr),$$

$$v = e^{i\omega t} \sin \varphi \sum_{n=1}^{\infty} -b_n(-i)^n \frac{2n+1}{n(n+1)} P_n^1(\cos \theta) h_n^{(2)}(kr).$$

These series contain elementary solutions with $l = 1$ only, like the series for the incident wave; a_n and b_n are coefficients to be determined presently. The spherical Bessel function $h_n^{(2)}(kr)$ derives from the Bessel function of the second kind $H_{n+1/2}^{(2)}(kr)$ and has been chosen because its asymptotic behavior,

$$h_n^{(2)}(kr) \sim \frac{i^{n+1}}{kr} e^{-ikr},$$

when combined with the factor $\exp (i\omega t)$, represents an outgoing spherical wave, as is required for the scattered wave.

Likewise, the field inside the sphere can be represented by

Inside wave:

$$u = e^{i\omega t} \cos \varphi \sum_{n=1}^{\infty} mc_n(-i)^n \frac{2n+1}{n(n+1)} P_n^1(\cos \theta) j_n(mkr),$$

$$v = e^{i\omega t} \sin \varphi \sum_{n=1}^{\infty} md_n(-i)^n \frac{2n+1}{n(n+1)} P_n^1(\cos \theta) j_n(mkr).$$

Here c_n and d_n are another pair of undetermined coefficients, and the choice of $j_n(mkr)$ is based on the facts that the refractive index is m and the fields are finite at the origin.

We use the boundary conditions (sec. 9.13, 1b and 2b) in order to find the undetermined coefficients. Apart from factors (and differentiations with respect to θ and φ) which are the same for the waves inside and outside the sphere, the field components E_θ and E_φ both contain the expressions

$$v \qquad \text{and} \qquad \frac{1}{m} \frac{\partial(ru)}{\partial r}.$$

The components H_θ and H_φ contain

$$mu \qquad \text{and} \qquad \frac{\partial(rv)}{\partial r}.$$

These four expressions have to have equal values at either side of the boundary surface, $r = a$, where a is the radius of the sphere.

The notations are simplified by introducing a new set of functions which differs from spherical Bessel functions by an additional factor z:

$$\psi_n(z) = zj_n(z) = (\pi z/2)^{1/2} J_{n+1/2}(z) = S_n(z).$$

$$\chi_n(z) = -zn_n(z) = -(\pi z/2)^{1/2} N_{n+1/2}(z) = C_n(z),$$

$$\zeta_n(z) = zh_n^{(2)}(z) = (\pi z/2)^{1/2} H_{n+1/2}^{(2)}(z).$$

These are the Riccati-Bessel functions. The notation S_n and C_n is most common at present. We use ψ_n, χ_n and ζ_n, the notations introduced in 1909 by Debye. By virtue of

$$H_n^{(2)}(z) = J_n(z) - iN_n(z)$$

we have

$$\zeta_n(z) = \psi_n(z) + i\chi_n(z).$$

The derivatives of these functions will be denoted by primes.

The arguments are

$$x = ka = \frac{2\pi a}{\lambda}, \qquad y = mka.$$

The parameter x, which equals the ratio of the circumference of the sphere to the wavelength, is the most important parameter here and in the following chapters.

With these notations the boundary conditions, expressed by the continuity of the four functions placed in brackets, assume the forms:

$$[mu]: \qquad \psi_n(x) - a_n\zeta_n(x) = mc_n\psi_n(y),$$

$$\left[\frac{1}{m}\frac{\partial(ru)}{\partial r}\right]: \quad \psi_n'(x) - a_n\zeta_n'(x) = c_n\psi_n'(y),$$

$$[v]: \qquad \psi_n(x) - b_n\zeta_n(x) = d_n\psi_n(y),$$

$$\left[\frac{\partial(rv)}{\partial r}\right]: \qquad \psi_n'(x) - b_n\zeta_n'(x) = md_n\psi_n'(y).$$

On eliminating c_n from the first pair and d_n from the second pair of equations we obtain the solutions:

$$a_n = \frac{\psi_n'(y)\psi_n(x) - m\psi_n(y)\psi_n'(x)}{\psi_n'(y)\zeta_n(x) - m\psi_n(y)\zeta_n'(x)},$$

$$b_n = \frac{m\psi_n'(y)\psi_n(x) - \psi_n(y)\psi_n'(x)}{m\psi_n'(y)\zeta_n(x) - \psi_n(y)\zeta_n'(x)}.$$

For c_n and d_n we find fractions with the same respective denominators and

$$\psi_n'(x)\,\zeta_n(x) - \psi_n(x)\,\zeta_n'(x) = i$$

as common numerator. This completes the solution of the problem. The field at any point inside or outside the sphere is now expressed in known functions.

9.3. Resulting Amplitude Functions and Efficiency Factors

9.31. Amplitude Functions

The Mie solution sketched in the preceding sections gives the fields at any point inside and outside the particle. The problem posed in chap. 4 (sec. 4.1 and 4.42) was restricted to the scattered field at very large distances from the sphere. By substituting for $h_n^{(2)}(kr)$ the asymptotic expression, already given, we find for the scattered wave

$$u = -\frac{i}{kr}\,e^{-ikr+i\omega t}\cos\varphi \sum_{n=1}^{\infty} a_n\,\frac{2n+1}{n(n+1)}\,P_n^1(\cos\theta),$$

$$v = -\frac{i}{kr}\,e^{-ikr+i\omega t}\sin\varphi \sum_{n=1}^{\infty} b_n\,\frac{2n+1}{n(n+1)}\,P_n^1(\cos\theta).$$

On deriving the tangential field components by means of equations 7 to 10 in 9.21, the following functions of the scattering angle appear:

$$\pi_n(\cos\theta) = \frac{1}{\sin\theta}\,P_n^1(\cos\theta),$$

$$\tau_n(\cos\theta) = \frac{d}{d\theta}\,P_n^1(\cos\theta).$$

Figure 18 shows the values of these functions for $n = 1$ to 6.

Alternative expressions are

$$\pi_n(\cos\theta) = \frac{dP_n(\cos\theta)}{d\cos\theta},$$

$$\tau_n(\cos\theta) = \cos\theta \cdot \pi_n(\cos\theta) - \sin^2\theta\,\frac{d\pi_n(\cos\theta)}{d\cos\theta}.$$

The resulting field components can be written at once in the form:

$$E_\theta = H_\varphi = -\frac{i}{kr}\,e^{-ikr+i\omega t}\cos\varphi\,S_2(\theta),$$

$$-E_\varphi = H_\theta = -\frac{i}{kr}\,e^{-ikr+i\omega t}\sin\varphi\,S_1(\theta),$$

where $\qquad S_1(\theta) = \sum_{n=1}^{\infty} \frac{2n+1}{n(n+1)} \{a_n \pi_n(\cos \theta) + b_n \tau_n(\cos \theta)\},$

$$S_2(\theta) = \sum_{n=1}^{\infty} \frac{2n+1}{n(n+1)} \{b_n \pi_n(\cos \theta) + a_n \tau_n(\cos \theta)\}.$$

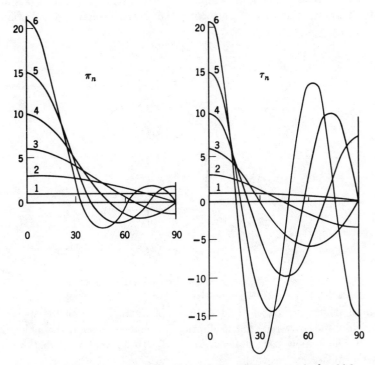

Fig. 18. The two sets of functions of the scattering angle θ, which occur in the Mie formulae, for $n = 1$ to 6.

These fields represent an outgoing spherical wave with amplitude and state of polarization dependent on direction. The radial components E_r and H_r may also be derived from Mie's solution but tend to zero with a higher power of $1/r$. It remains to be proved that the functions $S_1(\theta)$ and $S_2(\theta)$ are the *amplitude functions* in the sense defined in sec. 4.42. Observing the previous sign conventions we obtain a situation as illustrated in Fig. 19. The plane of reference is taken through the directions of propagation of the incident and scattered waves. The perpendicular and parallel components of the electric field of the incident wave are

$$E_{0r} = \sin \varphi, \quad E_{0l} = \cos \varphi.$$

Those of the scattered wave are

$$E_r = -E_\varphi, \quad E_l = E_\theta.$$

It is thus seen that the functions denoted here as $S_1(\theta)$ and $S_2(\theta)$ agree in all details with the amplitude functions defined in sec. 4.42. The matrix components $S_3(\theta)$ and $S_4(\theta)$ are zero. By means of the formulae summarized in sec. 4.42 it is possible to write at once the intensity and state of polarization of the scattered wave in any direction if radiation

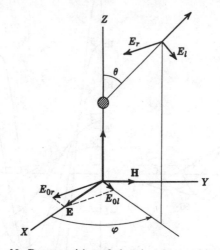

Fig. 19. Decomposition of electric vectors of incident
and scattered waves.

with an arbitrary intensity and state of polarization falls on the sphere. Generally, the scattered light is elliptically polarized, even if the incident light has linear polarization, for $S_1(\theta)$ and $S_2(\theta)$ are complex numbers with different phase.

Quite generally the computational problem involved in the Mie problem is to compute the numbers

$$i_1 = |S_1(\theta)|^2 \quad \text{and} \quad i_2 = |S_2(\theta)|^2$$

and the phase difference δ (sec. 4.42) as functions of the scattering angle θ. Most often only i_1 and i_2 are tabulated, which suffices for the scattering of unpolarized light.

If the phase function (sec. 2.1) is required, it may be computed after the tabulation of i_1, i_2, and Q_{ext}. Alternatively, it may be written as a series of Legendre functions with coefficients expressed directly in terms of the Mie coefficients by means of expressions derived by Chu and Churchill (1955).

The formulae most often needed are compiled in sec. 9.4.

9.32.   Resulting Efficiency Factors

These factors, defined for general particles in sec. 2.4, may be computed at once from the amplitude functions.   The efficiency factor for extinction may be determined from the amplitude function for $\theta = 0$.   Both $S_1(\theta)$ and $S_2(\theta)$ have the value

$$S(0) = \tfrac{1}{2} \sum_{n=1}^{\infty} (2n + 1)(a_n + b_n)$$

for $\theta = 0$; use has been made of the relations

$$\pi_n(1) = \tau_n(1) = \tfrac{1}{2}n(n + 1).$$

The fact that there is only one $S(0)$ is in agreement with the case of highest symmetry discussed in sec. 5.41 and implies that the extinction is independent of the state of polarization of the incident light.   The value of Q_{ext} follows from (secs. 2.4 and 4.21)

$$C_{ext} = \pi a^2 Q_{ext} = \frac{4\pi}{k^2} \operatorname{Re} \{S(0)\},$$

so that

$$Q_{ext} = \frac{4}{x^2} \operatorname{Re} \{S(0)\},$$

which, expressed in the Mie coefficients, reads

$$Q_{ext} = \frac{2}{x^2} \sum_{n=1}^{\infty} (2n + 1) \operatorname{Re} (a_n + b_n).$$

The further cross sections follow by integration over the entire scattering pattern, according to the formulation of secs. 2.2 and 2.4.   The form of the function $F(\theta, \varphi)$ defining the *intensity* of the scattered light in an arbitrary direction, according to sec. 2.1, can be written down at once for the special case of incident linearly polarized light, discussed in the previous sections (see Fig. 19).   We find

$$F(\theta, \varphi) = i_2(\theta) \cos^2 \varphi + i_1(\theta) \sin^2 \varphi,$$

Integration over all directions, with

$$d\omega = \sin \theta \, d\theta \, d\varphi,$$

now gives, by means of the equations of secs. 2.2 and 2.3

$$Q_{sca} = \frac{C_{sca}}{\pi a^2} = \frac{1}{x^2} \int_0^\pi \{i_1(\theta) + i_2(\theta)\} \sin \theta \, d\theta$$

and, likewise

$$\overline{\cos \theta} \cdot Q_{sca} = \frac{1}{x^2} \int_0^\pi \{i_1(\theta) + i_2(\theta)\} \cos \theta \sin \theta \, d\theta.$$

The same total cross sections are obtained if the incident light has arbitrary (e.g., partly elliptical) polarization; the proof may be left to the reader. Once these values have been found, the cross sections for absorption and radiation pressure follow by subtraction:

$$Q_{abs} = Q_{ext} - Q_{sca},$$
$$Q_{pr} = Q_{ext} - \overline{\cos \theta} \cdot Q_{sca}$$

The conversion of the above results into expressions containing the Mie coefficients is not very simple. Since S_1 and S_2 are in the form of infinite series, their squares i_1 and i_2 are in the form of doubly infinite series. However, upon integration over θ most terms in the double series give zero because of orthogonality relations of π_n and τ_n. These have been investigated by Debye, to whom we refer for further details. The results are

$$Q_{sca} = \frac{2}{x^2} \sum_{n=1}^\infty (2n + 1)\{|a_n|^2 + |b_n|^2\},$$

$$\overline{\cos \theta} \cdot Q_{sca} = \frac{4}{x^2} \sum_{n=1}^\infty \frac{n(n + 2)}{n + 1} \text{Re} \, (a_n a_{n+1}^* + b_n b_{n+1}^*)$$

$$+ \frac{4}{x^2} \sum_{n=1}^\infty \frac{2n + 1}{n(n + 1)} \text{Re} \, (a_n b_n^*).$$

A further reduction, which is helpful only for a real refractive index, will be given in sec. 10.21.

9.4. Formulae for Practical Use

Although the preceding formulae are general and strict, it is helpful to have the equations for some very simple conditions ready for practical use. They are contained in the formulations given above, and no further comments on their derivation will be given.

One particle, extinction. From radius a and wavelength λ, find $x = 2\pi a/\lambda$. If the sphere is not in vacuum but submerged in a homogeneous medium, the wavelength in the external medium, which is λ_{vac}/m_{medium}, should be used in this formula. Tables of Mie functions usually give the efficiency factor $Q_{ext}(x, m)$. If I_0 is the intensity of the incident light (watt/m²), the sphere will intercept $Q_{ext} \cdot \pi a^2 \cdot I_0$ watt from the incident beam, independently of the state of polarization of the latter. The part absorbed and the part scattered in all directions are found by replacing Q_{ext} by Q_{abs} and Q_{sca}, respectively.

One particle, scattered intensity. Tables usually give i_1 and i_2 as functions of x, m, and θ. Let r be the distance from the center of the sphere and $k = 2\pi/\lambda$. If *natural* light of intensity I_0 (watt/m²) is incident on the sphere, the scattered light in any direction has partial linear polarization. Its intensity (watt/m²) is

$$I = \frac{I_0(i_1 + i_2)}{2k^2r^2},$$

in which the terms i_1 and i_2 refer, respectively, to the intensity of light vibrating perpendicularly and parallel to the plane through the directions of propagation of the incident and scattered beams. By vibration we understand vibration of the electric vector. The degree of polarization is $(i_1 - i_2)/(i_1 + i_2)$.

Full formulae for incident polarized light are in sec. 4.42.

Medium containing N particles per unit volume. The particles are first supposed to be identical spheres in vacuum. The intensity scattered per unit volume in a given direction is simply N times the intensities mentioned above. It depends on the state of polarization of the incident light.

The intensity of the proceeding beam decreases in a distance l by the fraction $e^{-\gamma l}$, where the extinction coefficient γ is computed from

$$\gamma = N\pi a^2 Q_{ext},$$

independently of the state of polarization of the incident light. At the same time the proceeding wave is retarded . Weakening and retardation are described together as the effect of a complex refractive index of the medium:

$$\tilde{m} = 1 - iS(0) \cdot 2\pi N k^{-3}.$$

If the particles have different radii, with $N(a)\, da$ particles with radii between a and $a + da$ per unit volume, then

$$\gamma = \int_0^\infty Q_{ext} N(a)\pi a^2 \, da,$$

where Q_{ext} depends on a because of the argument $x = 2\pi a/\lambda$.

Particles suspended in a medium with refractive index different from 1. In order to keep notations simple we have assumed throughout this book that the suspending medium is vacuum, but the generalization is simple. Let the outside medium have the refractive index m_2 (real) and the spheres have the refractive index m_1 (real or complex). The m used throughout this book is

$$m = m_1/m_2;$$

the λ used throughout this book is

$$\lambda = \lambda_{vac}/m_2$$

and, consequently,

$$x = \frac{2\pi a}{\lambda} = \frac{2\pi a m_2}{\lambda_{vac}}.$$

All functions of x and m remain the same, except the complex refractive index of the medium with suspended particles, which now is $\tilde{m} \cdot m_2$.

General proofs of these relations can be given by keeping m_2 as a separate parameter in all formulae from the beginning.

References

The treatment of Maxwell's equations follows to some extent:

J. A. Stratton, *Electromagnetic Theory*, New York, McGraw-Hill Book Co. (1941).

The discussion of boundary and skin conditions in correspondence with the ordinary equations (secs. 9.12–9.14) is a novelty.

Mie's solution (secs. 9.2 and 9.3) was given by

G. Mie, *Ann. Physik*, **25,** 377 (1908).

Equivalent versions of it may be found in Stratton's book and in

P. Debye, *Ann. Physik*, **30,** 59 (1909).

H. Bateman, *Electrical and Optical Wave Motion*, Cambridge, Cambridge Univ. Press (1915).

M. Born, *Optik*, Berlin, J. Springer (1933).

A multitude of transformations has been suggested in order to make the numerical work more convenient.   Among these we mention the method of phase angles (sec. 10.4) and the method of logarithmic derivative functions, used by

L. Infeld, *Quart. Appl. Math.*, **5,** 113 (1947).

A. L. Aden, *J. Appl. Phys.*, **22,** 601 (1951).

The coefficients needed to express the phase function (sec. 2.1) as a series of Legendre functions may be computed from formulae given by

C. M. Chu and S. W. Churchill, *J. Opt. Soc. Amer.*, **45,** 958 (1955).

For references to numerical results see chaps. 10 and 14.

10. NON-ABSORBING SPHERES

This chapter deals mainly with spheres of a non-absorbing (dielectric) material. This means that the electrical conductivity σ is zero and that the refractive index m of the material is a real constant.

This assumption in itself does not introduce a drastic simplification. It allows a convenient survey, however, of the simplifications that arise in the limits if m and x have large or small values. This survey is given in the first section. Some of the limiting cases are treated in this chapter (secs. 10.3, 10.5, and 10.6), others in the following chapters (11 and 12).

The restriction to non-absorbing particles is dropped only where the equations are sufficiently simple to formulate the result at once for the more general assumption that σ has a non-zero value.

10.1. A Survey of the m-x Domain

All formulae derived in the preceding chapter have two parameters, m and x. The size parameter x can have values from 0 to ∞. The refractive index m can have values between 1 and ∞ for a sphere in vacuum and smaller than 1 if the surrounding medium is not vacuum (e.g., air bubbles in water). For convenience we neglect the latter possibility and show in Fig. 20 a schematic diagram in which any combination of x and m is represented by a point within a square. To each point belongs a particular scattering diagram, a definite value of the efficiency factors for extinction and for radiation pressure, etc., but the points for which such complete computations have been made are scanty in spite of the many computations that have been made lately. Mainly for that reason it is important to investigate what simpler ways to treat the problem exist and in which parts of the m-x domain these approximate methods are legitimate.

The theoretical structure of the m-x plane is made clear if we pay attention to three main divisions. The first one is the division according to the size of the scattering sphere. There is a gradual change from the typical scattering properties of small spheres to those of big spheres. The importance of this division is well known and is evident from the rigorous formulae. We have already applied it in chaps. 6 and 8 for particles of arbitrary forms. The second division is a division according to the value of m. The simplifications that arise for small m have been known for a long time.   The opposite is $m = \infty$, i.e., total reflectors.   Again there is a gradual transition.

The third division is a division according to *phase shifts*.   The change of the phase of a light ray passing through the sphere along a diameter is

$$2a \cdot (m - 1) \cdot 2\pi/\lambda = 2x(m - 1).$$

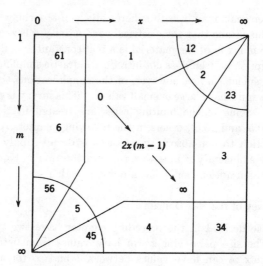

Fig. 20. Survey of limiting cases in the m-x domain.

The scattering process appears to be fundamentally different according as these phase shifts are small or large compared to 1.   The dividing line is approximately the diagonal of the square in Fig. 20.   This division is not a sharp one either.   It is curious that it has escaped attention in most treatises on the subject, although in the related quantum mechanical problems it is of prime importance.[1]   This division was used in chap. 7.

The simplest mathematical reductions and the clearest physical interpretation of the scattering formulae are obtained in points of the m-x domain which are far from two dividing lines. i.e., points which satisfy at least two of the following inequalities:

$$x \ll 1 \quad \text{or} \quad x \gg 1,$$
$$m - 1 \ll 1 \quad \text{or} \quad m - 1 \gg 1,$$
$$x(m - 1) \ll 1 \quad \text{or} \quad x(m - 1) \gg 1.$$

[1] Approximations based on the assumption that the phase shifts are small, are there called "Born's approximation."

The possibilities for this to occur are shown in Table 8, where s means small, l means large, and arb means arbitrary.

Table 8. Boundary Regions of the m-x Domain

Region	x	$m-1$	$x(m-1)$	Chapter or Section	Extinction Formula
61	s	s	s		$Q = (32/27)(m-1)^2 x^4$
1	arb	s	s	7.2 (Rayleigh-Gans)	
12	l	s	s		$Q = 2(m-1)^2 x^2$
2	l	s	arb	11 (anomalous diffraction)	
23	l	s	l		$Q = 2$
3	l	arb	l	12 (large spheres)	
34	l	l	l		$Q = 2$
4	arb	l	l	10.6 (total reflector)	
45	s	l	l		$Q = (10/3)x^4$
5	s	l	arb	10.5 (optical resonance)	
56	s	l	s		$Q = (8/3)x^4$
6	s	arb	s	6.3 (Rayleigh scattering)	

The table shows and Fig. 20 illustrates that there are six *boundary regions* (1 to 6) in which one of the parameters may have an arbitrary value ranging from near 0 to infinity. The simple formulae, or simplified methods of computation, that hold in these regions are discussed in the sections shown in the table.   Two by two they overlap in the regions that are numbered as 61, 12, 23, etc.   These *corner regions* give rise to even simpler formulae than most boundary regions.   The formula of Q, the efficiency factor for extinction, is given for these corner regions in the table.   In each corner region identical results are obtained from the somewhat different approaches that hold for the adjoining boundary regions. For instance, the extinction formula for region 12 is

$$Q = 2x^2(m-1)^2,$$

as follows independently from the Rayleigh-Gans theory (region 1) and from the quite different approach of anomalous diffraction (region 2).

We present now a brief description of the relevant features of the scattering phenomenon in all boundary regions, in anticipation of the full discussion given in later sections.   We make the round of the m-x domain starting from and returning to the upper left corner.

The left side (region 6) provides the simplest pattern.   Here both x and the phase shift are small.   One term, with coefficient a_1, is predominant in the Mie formulae and corresponds to electric dipole scattering.   This is the Rayleigh scattering, discussed in more general context in chap. 6.   Passing into the lower left corner both b_1 and terms with $n > 1$ become important. There are sharply defined values of x for which one of these terms gives a large contribution; this indicates a resonance phenomenon.   Passing from this optical resonance region (region 5) along the lower side of the square we enter the region of the totally reflecting spheres (4).   The fields inside the sphere may be neglected, and the Mie coefficients are particularly simple; sufficient numerical results are available to follow the gradual change from small spheres to large ones.   We then enter the region of geometrical optics (region 3) that occupies the right side of the square. The scattering diagram is the combination of a diffraction pattern (dependent on x but not on m), and a reflection + refraction pattern (dependent on m but not on x).   There $Q = 2$, as discussed in sec. 8.22. At the lower end of this region reflection prevails, and, if we rise along a vertical line in the diagram, more and more radiation is refracted by the sphere.   At the top end most of the light is transmitted after two refractions without an inner reflection. If now $m - 1$ decreases still further, the transmitted light becomes less divergent and its intensity increases. Finally it can compete with the intensity of the diffracted light.   Here optical interference appears and we enter the upper right corner (region 2). In this region the light is scattered in diffraction rings with anomalous sizes and distribution of brightness.

The limiting case of small phase shifts brings us to the upper side of the square.   All phase shifts are small.   The Rayleigh-Gans scattering that occurs in this region has been discussed in a more general context in chap. 7.   It can be understood as simultaneous Rayleigh scattering by all volume elements of the sphere.   Purely geometrical interference effects cause a scattering diagram with successive bright and dark rings.   Most of the light is strongly forward-directed for large x.   With decreasing x the angles between the rings widen, and one by one the rings vanish at $\theta = 180°$.   Finally, when x becomes $\ll 1$, all interference effects are gone and we have returned to ordinary Rayleigh scattering.   Our round is thus completed.

The central part of the m-x plane is left.   Here the royal road along the Mie formulae should be taken.[2]   Even so, correct methods of interpolation between numerical results can be devised only if we first have a good understanding of the effects found in the boundary regions.

---

[2] Rigorous transformations of the Mie formulae into other expansions may be made (sec. 12.35), but without approximations they are not useful.

10.2. An Important Simplification

10.21. Introduction of Phase Angles

A real value of m means *no* absorption, so that we know beforehand that $Q_{sca} = Q_{ext}$. This is not at once evident from the rigorous formulae in sec. 9.32. It becomes very clear after the following transformation.

Let m be real, then x and $y = mx$ are real. Also the functions $\psi_n(x)$, $\psi_n(y)$, and $\chi_n(x)$ are real, but $\zeta_n(x)$ is complex, so that the coefficients a_n and b_n are also complex. We define the *real angles* α_n and β_n by

$$\tan \alpha_n = - \frac{\psi_n'(y)\psi_n(x) - m\psi_n(y)\psi_n'(x)}{\psi_n'(y)\chi_n(x) - m\psi_n(y)\chi_n'(x)},$$

$$\tan \beta_n = - \frac{m\psi_n'(y)\psi_n(x) - \psi_n(y)\psi_n'(x)}{m\psi_n'(y)\chi_n(x) - \psi_n(y)\chi_n'(x)}.$$

The coefficients as derived before (sec. 9.22) then are

$$a_n = \frac{\tan \alpha_n}{\tan \alpha_n - i},$$

$$b_n = \frac{\tan \beta_n}{\tan \beta_n - i},$$

and by a simple geometrical transformation

$$a_n = \tfrac{1}{2}(1 - e^{-2i\alpha_n}),$$

$$b_n = \tfrac{1}{2}(1 - e^{-2i\beta_n}).$$

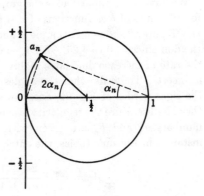

The locus of a_n and b_n in the complex domain is a circle with center $\tfrac{1}{2}$ and radius $\tfrac{1}{2}$ (Fig. 21). We can see geometrically and analytically that

Fig. 21. The locus of the coefficients a_n and b_n in the complex domain is a circle.

$$\mathrm{Re}\,(a_n) = |a_n|^2 = \sin^2 \alpha_n = \tfrac{1}{2}(1 - \cos 2\alpha_n),$$

$$\mathrm{Re}\,(b_n) = |b_n|^2 = \sin^2 \beta_n = \tfrac{1}{2}(1 - \cos 2\beta_n),$$

which proves (sec. 9.32) that Q_{ext} and Q_{sca} have the common value:

$$Q = \frac{2}{x^2} \sum_{n=1}^{\infty} (2n + 1)(\sin^2 \alpha_n + \sin^2 \beta_n).$$

The same transformation might be made for complex values of m (absorbing spheres). The main advantage is then lost, for α_n and β_n are complex, and the real part of a_n is no longer equal to the square of its modulus, so that also

$$Q_{sca} \neq Q_{ext}.$$

The definition of the angles α_n and β_n is completed by the requirement that they are 0 for $x = 0$ and are continuous functions of x. They also are 0 for $m = 1$ and continuous functions of m. Discontinuities occur, however, when we pass from very large values of m to $m = \infty$ (see sec. 10.6).

The change from the complex coefficients a_n and b_n to the real angles α_n and β_n means an important simplification. The angles α_n and β_n may be plotted as ordinates in a graph against m, or x as abscissae. Such graphs permit us to check numerical accuracy and to trace errors of computation at the stage of the computation where it is most needed, for after the summation over n has been performed errors are hard to trace. These graphs also lend us better insight into the way in which the solutions for various limiting cases originate from the general solution.

We have used in particular the graphs showing the values of α_n and β_n for all values of n as functions of x at a fixed value of the refractive index m. Figure 22 shows an example of such a graph constructed for $m = 2$ in the domain $x = 0$ to 4.25. All α_n and β_n increase with increasing x, but the rate of increase fluctuates highly, and at regular intervals the curves intersect. The order that the angles have at one value of x, e.g., along the right border of Fig. 22, does not show any apparent regularity. In the older literature the regularity is even more hidden because the complex numbers a_n and b_n are given with additional complex factors. For instance, in Lowan's tables, the entries are

$$A_n = i^{2n+1} \frac{2n+1}{n(n+1)} \cdot \tfrac{1}{2}(1 - e^{-2i\alpha_n}),$$

$$P_n = i^{2n-1} \frac{2n+1}{n(n+1)} \cdot \tfrac{1}{2}(1 - e^{-2i\beta_n}).$$

(The angles α_n and β_n have been computed backwards from these tables for the construction of Fig. 22.) The same coefficients are given in Blumer's tables. It is not surprising that a clear regularity in the real and imaginary parts of these complex numbers does not show up, except for the alternating sign due to the factor $i^{2n} = (-1)^n$.

10.22. Properties of Nodes

The most interesting feature of the curves shown in Fig. 22 is the existence of points at which

$$\alpha_n = \beta_n.$$

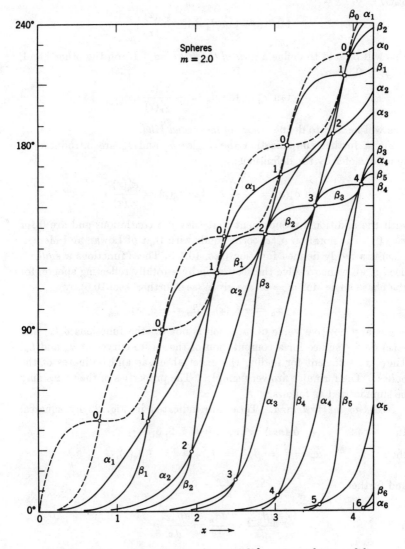

Fig. 22. Graphs of the phase angles α_n and β_n versus x for $m = 2.0$.

Any such point we shall call a node. A look at the formulae (at the beginning of sec. 10.21) shows that there are nodes of two kinds. If $\psi_n(y) = 0$, then

$$\tan \alpha_n = \tan \beta_n = -\frac{\psi_n(x)}{\chi_n(x)} ;$$

this will be said to define a *node of the first kind*. If, on the other hand, $\psi'_n(y) = 0$, then

$$\tan \alpha_n = \tan \beta_n = -\frac{\psi'_n(x)}{\chi'_n(x)} ;$$

this will be said to define a *node of the second kind*.

For a further simplification the angles δ_n and ε_n are introduced as functions of x by the definitions:

$$\tan \delta_n(x) = \frac{\psi_n(x)}{\chi_n(x)} , \quad \tan \varepsilon_n(x) = \frac{\psi'_n(x)}{\chi'_n(x)} ,$$

with the additional requirement that they are continuous and are 0 for $x = 0$. The notation $\delta_n(x)$ corresponds with that of Lowan and Morse; $\varepsilon_n(x)$ is a newly defined function (sec. 10.4). These functions are *all* we need if we want to solve the scattering by a totally reflecting sphere, for the phase angles for $m = \infty$ are given (see, further, sec. 10.6) by

$$\alpha_n = -\varepsilon_n(x), \quad \beta_n = -\delta_n(x).$$

For any positive value of m which is not ∞, the functions $\delta_n(x)$ and $\varepsilon_n(x)$ do *not* suffice for a computation of the entire curves of α_n and β_n. They are sufficient for finding the exact abscissae and ordinates of the nodes. The formulae are very simple. The properties of the nodes may be summarized as follows.[3]

Nodes of the first kind. These are indicated in Fig. 22 by squares:

If $$\delta_n(mx) = k\pi, \quad (k = 1, 2, 3, \cdots)$$

then $$\alpha_n = \beta_n = \beta_{n-1} = \beta_{n+1} = k\pi - \delta_n(x),$$

and further $$\frac{d\beta_n}{dx} = 0$$

and $$\frac{d\beta_{n-1}}{dx} = \frac{d\alpha_n}{dx} = \frac{d\beta_{n+1}}{dx} .$$

---

[3] These expressions for α_n and β_n follow from the expressions for $\tan \alpha_n$ and $\tan \beta_n$ given earlier. The term $k\pi$ follows from the consideration that all angles must be 0 for $m \to 1$. The additional properties for the nodes of the first kind were first noted in the numerical work and then found to be rigorously correct by use of recursion formulae for the Bessel functions.

Nodes of second kind.   These are indicated in Fig. 22 by circles:

If $\varepsilon_n(mx) = k\pi,$ $(k = 0,1,2,3, \cdots)$

then $\alpha_n = \beta_n = k\pi - \varepsilon_n(x).$

It is conceivable that further simple relations exist; e.g., in the analogous cylindrical case also the partial derivatives with respect to m and x in the nodes can be read from available tables (sec. 15.31).   Those presented above are already very helpful in detecting any computation errors. Especially, the properties of the nodes of the first kind are remarkable. In such a node 4 curves come together, of which 3 have a common, sloping tangent and 1 has a horizontal tangent.   For example, the node in the figure at $x = 2.88$ is traversed by the three sloping curves of β_1, α_2, and β_3 which are tangent to each other but do not cross over.   Further the curve of β_2 has a horizontal tangent.   All these angles are $137°$ at the node.

The figure shows by dotted curves the values of $\alpha_0 = \beta_{-1}$ and $\beta_0 = \alpha_{-1}$. They follow the same rules as the other phase angles but are not needed in the computation because the series expansions in the Mie formulae start with $n = 1$.

10.23.   The Functions $\delta_n(x)$ and $\varepsilon_n(x)$

This section has the character of an appendix, in which some useful formulae are collected for ready reference.   Derivations are omitted.

The defining equations are (sec. 10.22)

$$\tan \delta_n = \psi_n(x)/\chi_n(x), \quad \tan \varepsilon_n = \psi_n'(x)/\chi_n'(x).$$

The functions $\psi_n(x)$ and $\chi_n(x)$ have been defined in sec. 9.22.   For $n = 0$:

$$\psi_0(x) = \sin x, \qquad \psi_0'(x) = \cos x,$$

$$\chi_0(x) = \cos x, \qquad \chi_0'(x) = -\sin x,$$

from which

$$\delta_0(x) = x, \qquad \varepsilon_0(x) = x - \pi/2.$$

For $n = 1$:

$$\psi_1(x) = \frac{\sin x}{x} - \cos x, \quad \psi_1'(x) = \left(1 - \frac{1}{x^2}\right) \sin x + \frac{\cos x}{x},$$

$$\chi_1(x) = \frac{\cos x}{x} + \sin x, \quad \chi_1'(x) = \left(1 - \frac{1}{x^2}\right) \cos x - \frac{\sin x}{x},$$

from which

$$\delta_1(x) = x - \text{arc tan } x, \quad \varepsilon_1(x) = x - \text{arc tan } \frac{x}{1 - x^2}$$

For general n:

$$\psi_n'(x)\chi_n(x) - \chi_n'(x)\psi_n(x) = 1.$$

$$\left.\begin{aligned}\psi_{n+1}(x) &= \frac{2n+1}{x}\,\psi_n(x) - \psi_{n-1}(x)\\[2mm]\psi_n'(x) &= -\frac{n}{x}\,\psi_n(x) + \psi_{n-1}(x)\end{aligned}\right\} \quad \text{and similarly for } \chi_n(x)$$

The angles $\delta_n(x)$ and $\varepsilon_n(x)$ show a different behavior according as $(n + \tfrac{1}{2})/x$ is small, approximately equal to 1, or large. We have

$$\varepsilon_n(n) = \varepsilon_n(n + 1).$$

Between these values $\varepsilon_n(x)$ reaches a minimum at $x = \sqrt{[n(n+1)]}$. For $n \to \infty$ we have

$$\varepsilon_n(n + \tfrac{1}{2}) \to -\frac{\pi}{6} = -30°, \qquad \delta_n(n + \tfrac{1}{2}) \to \frac{\pi}{6} = 30°.$$

For $(n + \tfrac{1}{2})/x = \cos\tau < 1$ the asymptotic expressions are

$$\delta_n(x) = \frac{\pi}{4} + xf, \qquad \varepsilon_n(x) = -\frac{\pi}{4} + xf,$$

where f is defined by $f = \sin\tau - \tau\cos\tau$ (compare sec. 12.33). For small $(n + \tfrac{1}{2})/x$ the symptotical expansions based on Hankel's expansions of the Bessel functions are useful:

$$\left.\begin{aligned}\delta_n(x) &= -\frac{n\pi}{2} + x + \frac{u}{x} + \frac{u^2 - 3u}{6x^3} + \cdots\\[2mm]\varepsilon_n(x) &= -\frac{(n+1)\pi}{2} + x + \frac{u}{x} + \frac{u^2 + 3u}{6x^3} + \cdots\end{aligned}\right\} \quad u = \frac{n(n+1)}{2}.$$

Approximate expressions for the zeros of $\tan\delta_n(x)$ and $\tan\varepsilon_n(x)$ are

$$\delta_n(x) = k\pi \qquad \text{at } x = \left(k + \frac{n}{2}\right)\pi - \frac{n(n+1)}{2x} + \cdots$$

$$\varepsilon_n(x) = k\pi \qquad \text{at } x = \left(k + \frac{n}{2} + \tfrac{1}{2}\right)\pi - \frac{n(n+1)}{2x} + \cdots$$

Numerical values of $\delta_n(x)$ and $\varepsilon_n(x)$ are given in Table 9. Those of $\delta_n(x)$ may be found more extensively in the tables of Morse and others. The function $\delta_n'(x)$ given in those tables is *not* the same as $\varepsilon_n(x)$. The values of $\varepsilon_n(x)$ have been computed from Morse's tables by a formula given in sec. 10.4. The values of the zeros of $\tan\delta_n(x)$ and $\tan\varepsilon_n(x)$ are given in Table 10.

Table 9a. Values of $\delta_n(x)$ in Degrees

x	$n=0$	1	2	3	4	5	6	7	8	9	10	11	12	13
0.0	0.0	0.0												
0.5	28.6	2.1	0.0											
1.0	57.3	12.3	1.0	0.0										
1.5	85.9	29.6	5.4	0.4	0.0									
2.0	114.6	51.2	15.1	2.3	0.2	0.0								
2.5	143.3	75.0	29.8	7.4	1.0	0.1								
3.0	171.9	100.3	48.2	16.7	3.5	0.4	0.0	0.0						
4	229.2	153.2	91.9	46.4	17.7	4.5	0.7	0.1	0.0					
5	286.5	207.8	140.8	86.2	45.1	18.4	5.3	1.0	0.1	0.0				
6	343.7	263.2	192.4	131.7	81.9	44.1	19.0	6.0	1.3	0.2	0.0			
7	401.1	319.2	245.6	180.6	124.7	78.6	43.3	19.5	6.6	1.6	0.3	0.0	0.0	
8	458.4	375.5	299.8	221.6	171.2	119.1	76.0	42.7	19.9	7.2	1.9	0.4	0.1	0.0
9	515.7	432.0	354.8	284.1	220.2	163.5	114.5	73.8	42.1	20.2	7.7	2.2	0.5	0.1
10	573.0	488.7	410.1	337.5	270.9	210.6	157.0	110.6	71.9	41.7	20.5	8.1	2.4	0.6

Table 9b. Values of $\varepsilon_n(x)$ in Degrees

x	$n=0$	1	2	3	4	5	6	7	8	9	10	11	12	13
0.0	−90.0	0.0	0.0											
0.5	−61.4	−5.0	−0.1											
1.0	−32.7	−32.7	−1.7	0.0										
1.5	−4.1	−43.9	−11.6	−0.7	0.0									
2.0	24.6	−31.7	−31.7	−4.1	−0.2	0.0								
2.5	53.3	−11.3	−39.6	−15.1	−1.5	−0.1								
3.0	81.9	12.4	−31.3	−31.3	−6.2	−0.6	0.0							
4	139.2	64.1	5.6	−31.1	−31.1	−7.8	−1.0	−0.1	0.0					
5	196.5	118.2	52.4	1.0	−30.9	−30.8	−9.3	−1.4	−0.2	0.0				
6	253.7	173.5	103.3	44.1	−2.2	−30.8	−30.8	−10.5	−1.9	−0.3	0.0			
7	311.1	229.3	156.2	92.0	37.6	−4.7	−30.8	−30.8	−11.5	−2.3	−0.4	0.0		
8	368.4	285.6	210.3	142.4	82.9	32.5	−6.7	−30.7	−30.7	−12.4	−2.8	−0.5	0.0	
9	425.7	342.1	265.0	194.6	131.2	75.6	28.4	−8.4	−30.6	−30.7	−13.3	−3.2	−0.6	0.0
10	483.0	398.7	320.2	247.9	181.7	122.0	69.4	24.9	−9.7	−30.6	−30.6	−14.0	−3.7	−0.7

Table 10. Positions of the Nodes for Spheres

$n =$	0	1	2	3	4	5	6	7
$\delta_n = 180°$	3.142	4.493	5.764	6.988	8.183	9.36	10.52	11.68
$\delta_n = 360°$	6.283	7.725	9.095	10.417	11.705	12.97		
$\delta_n = 540°$	9.425	10.904	12.323	13.698	15.040	16.36		
$\delta_n = 720°$	12.566	14.067	15.515	16.924	18.301			
$\delta_n = 900°$	15.708	17.221	18.689	20.122	21.526			
$\delta_n = 1080°$	18.849	20.371				Roots of $\psi_n(y) = 0$		
$\varepsilon_n = 0°$	1.571	2.744	3.870	4.973	6.062	7.18	8.27	9.30
$\varepsilon_n = 180°$	4.712	6.117	7.443	8.722	9.968	11.20	12.42	
$\varepsilon_n = 360°$	7.854	9.317	10.713	12.064	13.380	14.69		
$\varepsilon_n = 540°$	10.996	12.486	13.921	15.314	16.674	18.02		
$\varepsilon_n = 720°$	14.137	15.644	17.103	18.524	19.916			
$\varepsilon_n = 900°$	17.279	18.796				Roots of $\psi_n'(y) = 0$		

10.3. Computation by Means of Series Expansion

Numerical results from the Mie formulae are required in view of many applications. Usually, numerical data for a fixed refractive index m and a range of values of x are requested because of variations in size and/or wavelength. The accuracy required, the question whether extinction coefficients or scattered intensities are sought, the availability of computing machines, and similar considerations may influence the choice of a particular method of computation. However, the following recommendations may be found useful:

A. If both x and mx are below 0.8, use series expansion (this section).

B. If x is larger and the computation is feasible at all, use phase angles (sec. 10.4).

C. Frequently approximations may be useful, as sketched in sec. 10.1 and worked out in further chapters.

In the range of small x and mx Rayleigh scattering, with coefficient a_1 and phase angle α_1, is the dominant term. It is convenient to compute more accurate results from a series expansion. The expansions as derived in the literature are the following:

Write

$$s = \frac{2}{3}\frac{m^2 - 1}{m^2 + 2}, \qquad t = \frac{3}{5}\frac{m^2 - 2}{m^2 + 2},$$

$$u = \frac{1}{30}(m^2 + 2), \qquad w = \frac{1}{10}\frac{m^2 + 2}{2m^2 + 3};$$

then

$$\alpha_1 = sx^3(1 + tx^2 + \cdots), \qquad \beta_1 = sux^5 + \cdots,$$
$$\alpha_2 = swx^5 + \cdots, \qquad \beta_2 = \cdots$$

so that

$$Q_{ext} = Q_{sca} = 6s^2x^4[1 + 2tx^2 + \cdots]$$
$$= \frac{8}{3} x^4 \left(\frac{m^2 - 1}{m^2 + 2}\right)^2 \left[1 + \frac{6}{5} \frac{m^2 - 1}{m^2 + 2} x^2 + \cdots\right] ;$$

to the same accuracy we have

$$\overline{\cos \theta} = x^2(u + w) + \cdots = \frac{x^2}{15} \frac{(m^2 + 3)(m^2 + 2)}{2m^2 + 3} + \cdots$$

and

$$Q_{pr} = 6s^2x^4[1 + (2t - u - w)x^2 + \cdots]$$
$$= \frac{8}{3} x^4 \left(\frac{m^2 - 1}{m^2 + 2}\right)^2 \left[1 - \frac{m^6 - 29m^4 + 34m^2 + 120}{15(m^2 + 2)(2m^2 + 3)} x^2 + \cdots\right].$$

These formulae are useful if the second term between brackets does not surpass 0.10 or 0.20. If it does, more terms ought to be taken into account, and it is advisable to switch to the method of phase angles.

It may be noted that to the same approximation the Mie coefficients are given by

$$a_1 = isx^3(1 + tx^2 - isx^3), \qquad b_1 = isux^5,$$
$$a_2 = iswx^5, \qquad b_2 = \cdots.$$

These formulae hold also for complex values of m, so that they give directly the correct extinction cross section for absorbing spheres (sec. 14.21).

Also the scattered intensities may be found by means of series expansion in this range of x. The dominant term of a_1 gives Rayleigh scattering with

$$i_1 = \left(\frac{m^2 - 1}{m^2 + 2}\right)^2 x^6 \qquad \text{and} \qquad i_2 = \left(\frac{m^2 - 1}{m^2 + 2}\right)^2 x^6 \cos^2 \theta,$$

in agreement with sec. 6.31. The scattering pattern for Rayleigh scattering is perfectly symmetric, and the light scattered at right angles is fully polarized ($i_2 = 0$ for $\theta = 90°$). These properties change if the further terms with t, u, and w are taken into the formulae.

The first few functions of θ are

$$\pi_1(\cos \theta) = 1, \qquad \pi_2(\cos \theta) = 3 \cos \theta,$$
$$\tau_1(\cos \theta) = \cos \theta, \qquad \tau_2(\cos \theta) = 3 \cos 2\theta.$$

When these expressions and the previous expressions for a_1, b_1, and b_2 are entered into the general formula for S_1 (sec. 9.31) we obtain

$$S_1 = \tfrac{3}{2}(a_1\pi_1 + b_1\tau_1) + \tfrac{5}{6}(a_2\pi_2 + b_2\tau_2) + \cdots,$$

which after some reductions is transformed into

$$S_1 = \frac{isx^3}{2}\{3 + (3t + 3u\cos\theta + 5w\cos\theta)x^2 - 3isx^3 + \cdots\}.$$

Likewise the equation for S_2 becomes

$$S_2 = \tfrac{3}{2}(a_1\tau_1 + b_1\pi_1) + \tfrac{5}{6}(a_2\tau_2 + b_2\pi_2) + \cdots$$

$$= \frac{isx^3}{2}\{3\cos\theta + (3t\cos\theta + 3u + 5w\cos 2\theta)x^2 - 3isx^3\cos\theta + \cdots\}.$$

The neglected terms within braces are of the order of x^4 and higher. By calculating $i_1 = |S_1|^2$ and $i_2 = |S_2|^2$ and making the necessary integrations over θ the expressions for Q_{ext}, $\overline{\cos\theta}$, and Q_{pr} given above can be found again.

These formulae show that the scattering pattern begins to deviate from the familiar Rayleigh pattern as soon as the second-order terms containing t, u, and w become of importance. These deviations are striking in two respects.

1. The intensity of forward scattering becomes stronger than the intensity of backward scattering, for direct substitution of $\theta = 0°$ gives

$$S_1(0) = S_2(0) = \frac{isx^3}{2}\{3 + (3t + 3u + 5w)x^2 - 3isx^3 + \cdots\},$$

whereas substitution of $\theta = 180°$ gives

$$S_1(180°) = -S_2(180°) = \frac{isx^3}{2}\{3 + (3t - 3u - 5w)x^2 - 3isx^3 + \cdots\}.$$

The scattered intensities are proportional to the squares of the absolute values of these amplitudes. The ratio of these squares is

$$\frac{\text{Intensity of forward scattering}}{\text{Intensity of backscattering}} = 1 + 4x^2(u + \tfrac{5}{3}w) + \cdots$$

$$= 1 + \frac{4x^2}{15}\frac{(m^2+4)(m^2+2)}{2m^2+3} + \cdots$$

This expression is correct except for neglected terms of the order of x^4 and higher. It is always >1, so that the forward scattering is always stronger than the backscattering as long as the assumptions of this section hold. The case of $m = \infty$ with its stronger backscattering is not included, because it violates the assumption that mx is small (see sec. 10.61).

2. The polarization of the light scattered through a right angle ($\theta = 90°$) is still strong but no longer complete. This follows from the fact that for $\theta = 90°$ (again with the neglect of higher powers):

$$\frac{S_2}{S_1} = (u - \tfrac{5}{3}w)x^2 = \frac{x^2(m^2 - 1)(m^2 + 2)}{15(2m^2 + 3)}.$$

It is well known that the Rayleigh scattering, represented at 90° only by i_1, gives a scattered intensity proportional to x^4, i.e., to λ^{-4}. The light in the other direction of polarization is now seen to have an intensity, which is of the order of x^4 times as weak. This intensity, which can be observed by watching spherical particles scatter through 90° in the "wrong" polarization, is proportional to x^8, i.e., to λ^{-8}. The color is a much deeper blue that the blue corresponding to Rayleigh's λ^{-4} law, which causes the blue of the sky. Tyndall called this the "residual blue."

Almost complete polarization occurs for an angle $>90°$. There is an angle slightly over 90°, where $|S_2|$ goes through a minimum that is almost zero. The formula shows that $S_2 = 0$ to the second order of x, when

$$\cos \theta = -(u - \tfrac{5}{3}w)x^2 = -\frac{x^2(m^2 - 1)(m^2 + 2)}{15(2m^2 + 3)},$$

which is the same expression as given above but with the negative sign showing that $\theta > 90°$. For this angle S_2/S_1 is of the order of x^4, i.e., virtually zero as long as the approximation is permitted. This is a characteristic feature of the scattering pattern of a small sphere. The zero of S_2 (the point of maximum polarization) shifts with growing size from $\theta = 90°$ in the direction of $\theta = 180°$, and the position of this angle may be used to measure the size.

With still further growing size more coefficients than a_1, b_1, and a_2 enter the picture, and further powers of x become important. The scattering patterns become more complicated, and generally one has to resort to the full solution. This warning holds for all the formulae derived in this section and especially for the simple expressions for the ratio of forward to backward scattering and for the angle of maximum polarization. Some authors have attached great weight to these formulae. Aside from their simplicity they have little advantage. They describe the very first deviations from Rayleigh scattering, but further deviations appear very soon after the first have become prominent, so that the full Mie formulae have to be used.

A generalization of the method of series expansion for particles of arbitrary form is mentioned in sec. 16.13.

A numerical example may serve as an illustration. In Table 11 we compare the values computed from the above formulae with those contained in Lowan's tables, which are based directly on the Mie formulae. The agreement is entirely satisfactory.

An example, in which the deviations are more pronounced (but at the same time the approximation already unsatisfactory) is the scattering diagram for $m = 2$, $x = 1$, shown in Fig. 23. See also the diagrams in Fig. 25 (sec. 10.4).

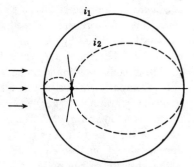

Fig. 23. Scattering diagram for spheres with $m = 2$, $x = 1$, showing characteristic preponderance of forward scattering. I_1 and I_2 have been plotted separately.

Table 11. Computation to the Next Higher Order than Rayleigh Scattering
(Example: $m = 1.33$, $x = 0.5$)

Quantity		From Above Formulae	Lowan
	s	0.1360	—
	t	−0.0368	—
	u	0.1256	—
	w	0.0960	—
Angles in radians $\left\{\vphantom{\begin{matrix}a\\b\\c\end{matrix}}\right.$	α_1	0.01684	0.01679
	β_1	0.000534	0.000525
	α_2	0.000244	0.000238
Q_{ext}		0.00681	0.00677
$\overline{\text{Forward}}$ $\left\{\vphantom{\begin{matrix}a\\b\end{matrix}}\right.$ amplitude ratio		1.119	1.116
$\overline{\text{Backward}}$ intensity ratio		1.252	1.245
$\lvert S_2/S_1 \rvert$ for $\theta = 90°$		0.00740	0.00755
Angle for which $i_2 \approx 0$		90° 25′	—

10.4. Computation by Means of the Phase Angles

For any x between 0.5 and 30, approximately, the method of the phase angles may be the fastest method to accurate and complete numerical results. The formulae, which are based on the rigorous solution, are given in the preceding pages and do not need any comment, except for two factors: the necessities to avoid errors and to save computing time. Both these factors are reduced or absent in computations with fully automatic machines but are important in deciding on a method with desk machines. The following method, which deviates strongly from traditional methods, was found to be handy and reliable.

1. Compute the position of all nodes in the required range, and plot them on a graph like Fig. 22. Needed for this step are tables of the functions $\delta_n(x)$ and $\varepsilon_n(x)$. These tables are provided in Tables 9 and 10 (sec. 10.23). The values of $y = mx$ for which $\delta_n(mx)$ or $\varepsilon_n(mx)$ are 0 can first be taken from Table 10. Dividing by m we find the value of x, which is the abscissa of the node. The value $\delta_n(x)$ or $\varepsilon_n(x)$ is then found by interpolation from Table 9, and the difference with $k\pi$ gives the phase angle $\alpha_n = \beta_n$, i.e., the ordinate of the node.

2. A rough outline of the complete curves can be drawn through these nodes, but usually fairly precise values of α_n and β_n for intermediate values of x are required. So the second step is the computation of α_n and β_n for a number of selected values of x. Recourse must be had to the full formulae (sec. 10.21) and to the tables of the functions $\psi_n(x)$ and $\chi_n(x)$ or of the spherical Bessel functions. The exact procedure depends on what tables are available.

Computations of this type may be speeded up by using the auxiliary tables constructed by Lowan, Morse, Feshbach, and Lax (1946), abbreviated as LMFL. These tables give among other functions the angles α_n^*, β_n^* (an asterisk is added in order to distinguish this notation from ours), γ_n, δ_n, and δ_n', which are functions of x defined by

$$\tan \alpha_n^*(x) = -\frac{xj_n'(x)}{j_n(x)},$$

$$\tan \beta_n^*(x) = -\frac{xn_n'(x)}{n_n(x)},$$

$$\tan \delta_n(x) = -\frac{j_n(x)}{n_n(x)},$$

$$\tan \delta_n'(x) = -\frac{j_n'(x)}{n_n'(x)} = \tan \delta_n(x)\frac{\tan \alpha_n^*(x)}{\tan \beta_n^*(x)},$$

$$\tan \gamma_n(x) = \tan \delta_n(x) \cos \beta_n^*(x) \sec \alpha_n^*(x).$$

Somewhat unfortunately for the present purpose, the angles are expressed in the spherical Bessel functions $j_n(x)$ and $n_n(x)$ and their derivatives, and not in terms of the Riccati-Bessel functions (cf. sec. 9.22)

$$\psi_n(x) = xj_n(x),$$
$$\chi_n(x) = -xn_n(x),$$

which have the derivatives

$$\psi_n'(x) = j_n(x) + xj_n'(x),$$
$$\chi_n'(x) = -n_n(x) - xn_n'(x).$$

Straightforward translation of the formulae for *our* α_n and β_n (which are the phase angles) in terms of these functions gives, after some reduction,

$$\tan \alpha_n = -\tan \delta_n(x) \frac{\tan \alpha_n^*(y) - 1 - m^2\{\tan \alpha_n^*(x) - 1\}}{\tan \alpha_n^*(y) - 1 - m^2\{\tan \beta_n^*(x) - 1\}},$$

and

$$\tan \beta_n = -\tan \delta_n(x) \frac{\tan \alpha_n^*(y) - \tan \alpha_n^*(x)}{\tan \alpha_n^*(y) - \tan \beta_n^*(x)}.$$

The fractions in both formulae can be made suitable for logarithmic computation by transforming them into the form:

$$-\tan \delta \frac{\tan \varphi - \tan \alpha^*}{\tan \varphi - \tan \beta^*} = -\tan \gamma \frac{\sin (\alpha^* - \varphi)}{\sin (\beta^* - \varphi)}.$$

Evidently we have to use

$$\tan \varphi = \frac{1}{m^2} \{\tan \alpha_n^*(y) + m^2 - 1\}$$

in the first formula and

$$\tan \varphi = \tan \alpha_n^*(y)$$

in the second formula. Analogous transformations are described in sec. 15.31 for circular cylinders.

It may still be remarked that the $\varepsilon_n(x)$ introduced here (sec. 10.23) does not appear in the LMFL tables. In the notation of these tables this function can be defined by

$$\tan \varepsilon_n(x) = \tan \delta_n(x) \frac{\tan \alpha_n^*(x) - 1}{\tan \beta_n^*(x) - 1}.$$

3. After this digression we proceed to the third step, which is: plotting of all computed values of α_n and β_n on the master graph and drawing the complete curves through these points and the known nodes. Effective use may be made in this step of the properties of the nodes summarized in

sec. 10.2. This entire step may be skipped in automatic machine computations. In ordinary work, however, it serves two purposes, namely, the detection of computational errors and a time saver if results for many values of x are required. For it is possible by reading the values α_n and β_n from these graphs to obtain the final results (step 4) for many values of x, although the greatest labor (step 2) is limited to relatively few values of x. Usually two points between any two nodes will define the curves with sufficient precision.

4. The final step may be the computation of the cross section Q (sec. 10.21) or of the scattered intensities i_1 and i_2 (secs. 9.31 and 10.21), whichever is required. In both cases the values of α_n and β_n are listed for a given combination (m, x) and for $n = 1, 2, \cdots$ until the angles become $<1°$. Then the cosines of $2\alpha_n$ and $2\beta_n$ are found, and Q is obtained by simple multiplications and additions. A numerical example (Table 12) may again illustrate the method.

Table 12. Computations by Means of Phase Angles

(Example : $m = 2$, $x = 3$)

n	α_n	β_n	$1 - \cos 2\alpha_n$	$1 - \cos 2\beta_n$	Same, with Factor $2n + 1$	
1	164°	154°	0.152	0.384	0.46	1.15
2	149°	136°	0.530	0.965	2.65	4.82
3	78°	145°	1.914	0.658	13.40	4.61
4	6°	5°	0.022	0.015	0.20	0.14
				Sum:	16.71	10.72

27.43

Therefore $Q = 27.43/x^2 = 27.43/9 = 3.05$.

In this way one point of the extinction curve is obtained. Figure 24 shows the extinction curves computed for a variety of values of m.

The number of decimals in Table 12 is about what is suitable in practice. Rounding α_n and β_n off to whole degrees usually gives sufficient accuracy. The cosines of the double angles are then off by 0.017 at the most, and as a consequence we find by simple probability calculus that the accidental error in each of the separate sums is about $0.02x^{3/2}$ for $x \gg 1$. This gives $0.03x^{3/2}$ for their sum and $0.03/\sqrt{x}$ for the mean error in Q due to the rounding off of α_n and β_n to whole degrees. For $x = 3$ this error is already smaller than 1 per cent of Q.

For the computation of $i_1(\theta)$ and $i_2(\theta)$ a table of the functions $\pi_n(\cos \theta)$ and $\tau_n(\cos \theta)$ is needed in addition to the values of the phase angles. A selection of numerical results obtained in this manner is represented in

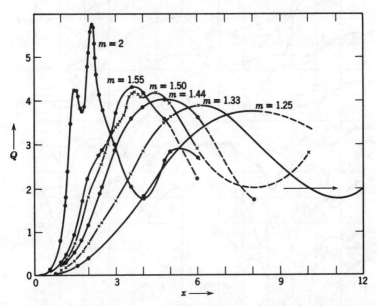

Fig. 24. Extinction curves of spheres for six values of the refractive index. Not all minor wiggles are shown (cf. Fig. 32).

graphical form in Fig. 25, which also permits a comparison with the scattering diagrams of very long cylinders, with radiation incident perpendicular to the axis (sec. 15.34).

10.5. The Region of Optical Resonance

The lower left corner of the m-x square (Fig. 20) is the region of optical resonance. It is the region where x is small and m is large. Under these conditions the spheres have certain modes of electric and magnetic vibration which are nearly self-sustained. If the value of $y = mx$ is favorable for the excitation of such a mode, a vibration of large amplitude can be sustained by an incident wave of relatively small amplitude, but for slightly smaller or larger y this is not true. The extinction curve thus has characteristic resonance peaks. Also the scattering diagram changes drastically if we let mx grow gradually and pass the resonance value.

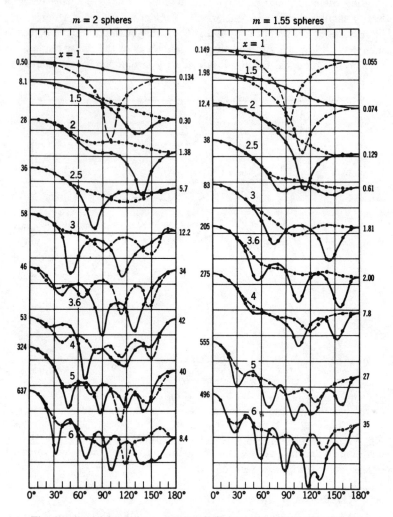

Fig. 25. A sample of the scattering diagrams computed by means of the rigorous formulae; m = refractive index, $x = 2\pi a/\lambda$. In all graphs the logarithms of the intensities (1 division = a factor 10) are plotted against the scattering angle (1 division = 30°).

Spheres (formulae in sec. 9.31): Solid curves i_1, dotted curves i_2. The values for $\theta = 0°$ and $180°$ are indicated in the margin. For $m = 2$, 1.55, and 1.33 they have been copied from Lowan's tables

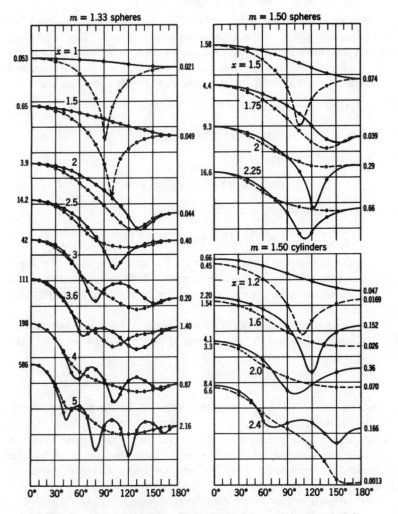

(1948), for $m = 1.50$ they have been obtained by squaring the moduli tabulated by Rayleigh (*Sci. Papers* 344, 1910).

Cylinders (formulae in sec. 15.13): Solid curves, case I ($E//$ axis); dotted curves, case II ($H//$ axis). The numbers in the margin are the squares of the moduli tabulated by Rayleigh (*Sci. Papers* 434, 1918). They equal $\{(2/\pi)|T_1(\theta)|\}^2$ and $\{(2/\pi)|T_2(\theta)|\}^2$.

10.51.   Free Vibration of a Sphere

Following the work in Debye's thesis, we first investigate under what conditions completely self-sustained vibrations occur. The conditions are that the denominator of the expressions for a_n or b_n is 0. This means that a solution of Maxwell's equations is found in which vibration and "scattering" occur without any incident wave. The equations are

For electric 2^n pole vibration:

$$\psi_n'(y)\,\zeta_n(x) - m\psi_n(y)\,\zeta_n'(x) = 0.$$

For magnetic 2^n pole vibration:

$$m\psi_n'(y)\,\zeta_n(x) - \psi_n(y)\,\zeta_n'(x) = 0.$$

From the relation (end of sec. 9.22)

$$\psi_n'(x)\,\zeta_n(x) - \psi_n(x)\,\zeta_n'(x) = i$$

we see that there are no roots for $m = 1$. For any other value of m these equations have complex values of x as roots. If these are close to a real value of x, resonance occurs for spheres of the exact size expressed by this real value.

We now limit our discussion to very large m, We see at once that one kind of roots exists for all large values of m and *also for* $m = \infty$. At $m = \infty$ they are the roots of

$$\zeta_n'(x) = 0 \text{ (electric)} \qquad \text{and} \qquad \zeta_n(x) = 0 \text{ (magnetic)}.$$

For $n = 1$ this gives, e.g., $x = 0.86 + 0.50i$, $x = -0.86 + 0.50i$ (both electric), and $x = +1.00i$ (magnetic). For $m \neq \infty$ they are slightly shifted, but the large imaginary part remains. This means that the scattering and extinction by spheres is hardly influenced at all by the existence of these modes. They do not give rise to a typical resonance phenomenon.

A second set of modes of vibration exists, however. These do not exist for $m = \infty$ because they occur at a fixed value of $y = mx$, and thus for $m \to \infty$ they occur at ever smaller values of x. For $x \ll 1$ we have $\zeta_n'(x) \approx -(n/x)\,\zeta_n(x)$. So the equations for the second set of roots become

Electric:
$$\psi_n'(y) + \frac{nm^2}{y}\,\psi_n(y) = 0;$$

Magnetic:
$$\psi_n'(y) + \frac{n}{y}\,\psi_n(y) = 0.$$

The second equation is identical to $\psi_{n-1}(y) = 0$. (Debye did not notice this point in his discussion.) Writing r_n for any root of $\psi_n(y) = 0$, we have the magnetic multipole vibrations exactly at $y = r_{n-1}$ and the electric multipole vibrations in first approximation at $y = r_n\left(1 - \dfrac{1}{m^2 n}\right)$.

The values of r_n are found in Table 10 of sec. 10.23. The roots thus found are real. The vibrations signified by them are virtually undamped, provided that m is so large that $x = y/m$ is very small. They can be pictured as ordinary standing waves that are kept inside the sphere because of the high internal reflection coefficient and suffer no damping when m is real. In this case a very high and very sharp resonance peak may be expected in the extinction curve. Computation shows that the peaks remain strong for any value of $m > 5$. The strongest one, corresponding to the magnetic dipole vibration, has $y = r_0 = \pi$. Its effect is visible as a bump on the extinction curve down to $m = 1.33$, where it is located near $x = 2.4$ (see Fig 24). There is no resonance in the electric dipole term with coefficient a_1.

We may remark, incidentally, that the modes of vibration of a spherical cavity may be found in a similar manner. This is an important technical problem. Here $y = 2\pi a/\lambda$, where a is the radius and λ the wavelength inside the cavity, and x refers to the wavelength in the wall of the cavity. Thus x is large and complex, and $m = y/x$ is very small and complex. The equations become in first approximation:

$$\psi_n'(y) = 0 \text{ for electric modes,}$$

$$\psi_n(y) = 0 \text{ for magnetic modes.}$$

A small table of roots is given in Stratton's book; further values are found in Table 10 (sec. 10.23). There is not a general rule by which any result for a sphere may be translated into a result for a cavity, or conversely. The equations are similar but cannot be reduced to each other.

10.52   Resonance Effects in the Mie Theory

We now pass to the scattering problem. It is illuminating to see in what manner the resonance phenomenon appears in the graphs of the phase angles α_n and β_n. As an example, in Fig. 26 the graph of the phase angles for $m = 9$ is sketched. This is the refractive index of water for very long wavelengths (dielectric constant = 81) and about the largest value of m encountered in practice (compare sec. 14.3).

The graph was constructed on a skeleton figure showing all nodes. The curves for α_0 and β_0 (not needed in the formulae) have a symmetric, recurrent pattern and were plotted from the exact formula. The other

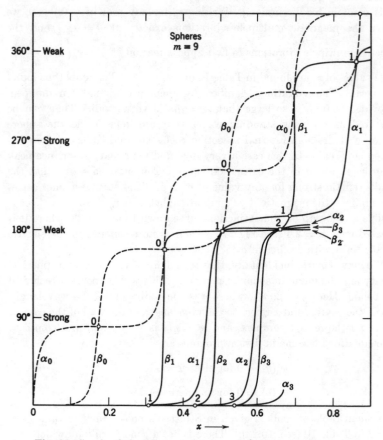

Fig. 26. Graph of the phase angles α_n and β_n versus x for $m = 9$. The words "weak" and "strong" refer to the contribution to the extinction (or scattering); strong contributions define the resonance points.

curves were drawn as free-hand curves through the nodes using all rules about the tangents in the nodes (sec. 10.3). The angles linger first near $0°$, then near $180°$, then $360°$, i.e., just near the angles for which the Mie coefficients

$$a_n = \tfrac{1}{2}(1 - e^{-2i\alpha_n}), \qquad b_n = \tfrac{1}{2}(1 - e^{-2i\beta_n})$$

are 0. However, in passing from $0°$ to near $180°$, etc., the angles swiftly pass $90°$, $270°$, where the coefficients are 1.

The a_n and b_n then swing rapidly around the entire circle in the complex domain (see Fig. 21, sec. 10.21). Their real parts reach the maximum values $a_n = 1$, or $b_n = 1$, when α_n or β_n are 90°, 270°, etc. The values of x for which this maximum is reached may be said to define the resonance points. We see from the graph that the resonance point of α_n precedes that of β_{n+1} and that both precede by a small amount the node of the first kind that is reached at $x = r_n$. This is in good agreement with the positions of the resonance points derived above from the natural modes of vibration of a sphere.

The treatment of resonance by means of the phase angles goes beyond the treatment by means of natural modes in several respects. First it is useful also if the conditions $m \gg 1$ and $x \ll 1$ are not strictly fulfilled. Even the values $m = 9$ and $x = 0.346$ of our example are not in the range in which we would ordinarily expect the asymptotic formulae to hold. In fact, we can see the whole deterioration of the resonance phenomenon if we let m gradually decrease from large to small values. It was empirically noted that most extinction curves have "bumps"; the values of x for the position of the first bumps are

$m =$	9	2	1.55	1.5	1.33
$x =$	0.346	1.6	2.0	2.0	2.4
$y = mx =$	3.11	3.2	3.1	3.0	3.2.

This first bump is clearly due always to the magnetic dipole term, with coefficient b_1. Asymptotically at $m \to \infty$ the resonance is at $y = 3.14$. At $m = 9$ the resonance character is still very well preserved. For the smaller values $m = 2$ or 1.5 the bump is just due to a sudden increase of β_1 without a full swing of b_1 around the circle in the complex domain.

Another advantage of using the phase angles is that we can easily determine how large the extinction is at the resonance peak. We now restrict ourselves to the asymptotic conditions $m \gg 1$ and $x \ll 1$. Retaining only one term in the formula for Q, with either a_n or b_n equal to 1 we find

$$Q = \frac{2(2n + 1)}{x^2}$$

and consequently the cross section

$$C = Q \cdot \pi a^2 = \frac{(2n + 1)\lambda^2}{2\pi}.$$

We thus find in the resonance point an extinction cross section which is independent of size and an efficiency factor Q which is far greater than 1. The Q in the magnetic dipole peak is $6/x^2$ (neglecting all further terms). For $m = 9$, $x = 0.346$ this gives $Q = 50$, which is a great deal higher than the highest value $Q = 5.7$, shown in Fig. 24 for moderate m. The width of the resonance peak may be determined from the slope at which the phase-angle graph passes the resonance point.

The scattering diagram in the resonance region assumes many weird forms, except in the resonance peaks where a pure electric or magnetic dipole or multiple radiation prevails. It is advisable to refer to the general formulae of $S_1(\theta)$ and $S_2(\theta)$ for full details.

Resonance is found also for complex values of m (sec. 14.31).

10.6. Totally Reflecting Spheres ($m = \infty$)

All formulae become particularly simple in the limiting case of $m = \infty$. This may mean the limiting case of large complex values of m or large real values of m; the result is the same in both situations.

The practical reason for large m usually is a large conductivity σ, and then

$$m = \sqrt{(\eta/2)} - i\sqrt{(\eta/2)}.$$

This formula applies to metals in the infrared and microwave region and $\eta \to \infty$ as the frequency goes to zero (sec. 14.41).

Because of its simplicity the case $m = \infty$ has often served as a practicing example in scattering problems. For instance, the scattering by a perfectly conducting sphere of arbitrary size had been solved several years before Mie gave the solution for arbitrary values of the refractive index. In looking over the results for $m = \infty$ it should be borne in mind that this practicing example is in some respects not representative. For instance, even the smallest spheres with $m = \infty$ do not give Rayleigh scattering. Blumer (1926) seemed to notice this with surprise in calculating his extensive collection of scattering diagrams. The reason for the deviation was noted in sec. 6.4; the conditions for Rayleigh scattering are

$$x \to 0 \qquad \text{and} \qquad mx \to 0$$

and cannot possibly be fulfilled for $m = \infty$.

10.61 Small Totally Reflecting Spheres

Rayleigh scattering and the scattering by spheres with $m = \infty$ form two quite different physical problems. In Rayleigh scattering we assume full and instantaneous penetration of the fields into the particle. By putting $m = \infty$ we assume that the fields do not penetrate into the particle at all. The transition between these extremes may be described as follows.

(a) If the transition is made via the values of m appropriate for absorbing (metallic) particles, the "skin depth" is $\gg a$ for Rayleigh scattering and $\ll a$ for $m \to \infty$. In the transition region the skin depth is comparable to the radius a of the sphere. This leads to complicated formulae, however, without typical resonance phenomena.

(b) If the transition is made via increasing real values of m, we encounter the resonance phenomena described in the preceding section (10.5). The large value of m means virtually total reflection of the incident wave so that the situation outside the resonance peaks is the same as for $m = \infty$. However, the large value of m also means that the waves inside the sphere are almost totally reflected against the boundary. This is the condition for the building up of standing waves, which are responsible for the resonance effect.

It thus is clear that very small spheres with $m = \infty$ have to be treated as a problem separate from that treated in chap. 6. The formulae for the phase angles are simply (sec. 10.22)

$$\alpha_n = -\varepsilon_n(x), \qquad \beta_n = -\delta_n(x).$$

The further results follow by substitution into the general formulae (secs. 9.31 and 9.32, with the substitution of 10.21); they can be given with few comments. The first few terms of the expansions for small x are

$$\alpha_1 = \frac{2}{3} x^3 + \frac{1}{5} x^5 - \cdots, \qquad \beta_1 = -\frac{1}{3} x^3 + \frac{1}{5} x^5 - \cdots,$$

$$\alpha_2 = \frac{1}{30} x^5 - \cdots, \qquad \beta_2 = -\frac{1}{45} x^5 + \cdots$$

These lead in turn to the following expansions:

$$Q_{sca} = Q_{ext} = Q = \frac{10}{3} x^4 + \frac{4}{5} x^6 - \cdots$$

$$Q \overline{\cos \theta} = -\frac{4}{3} x^4 + \frac{26}{45} x^6 - \cdots$$

$$Q_{pr} = (1 - \overline{\cos \theta})Q = \frac{14}{3} x^4 + \frac{2}{9} x^6 - \cdots$$

The scattering diagram does not have the symmetrical form of Rayleigh scattering. This is one of the few examples in which the scattering is predominantly backward, as is shown by the negative value of $\overline{\cos \theta} = -0.4$. As a result the efficiency factor for radiation pressure exceeds that for extinction. The scattering diagram for very small particles with $m = \infty$ is shown in Fig. 27. The graph gives a polar diagram of

$$i_1(\theta) = |S_1(\theta)|^2 \qquad \text{and} \qquad i_2(\theta) = |S_2(\theta)|^2,$$

where

$$S_1(\theta) = ix^3(1 - \tfrac{1}{2} \cos \theta),$$

$$S_2(\theta) = ix^3(\cos \theta - \tfrac{1}{2}).$$

The forward scattered intensity is 1/9 times the backward scattered intensity.

The reason for this strange pattern is that, besides electric dipole radiation, also magnetic dipole radiation is emitted. A derivation of the combined pattern of these crossed dipoles was given for an ellipsoid with arbitrary ratios of the axes in sec. 6.4.

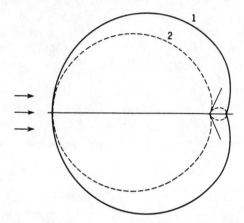

Fig. 27. Scattering diagram for very small totally reflecting spheres. The scattering is predominantly back to the source.

It is instructive to add a derivation for spheres based on the simplest equations. As in the derivation for Rayleigh scattering (sec. 6.1) we may consider the incident fields $\mathbf{E}_0$, $\mathbf{H}_0$ as homogeneous. The field of the surface charges induced by the field $\mathbf{E}_0$ is exactly that of an electric dipole $\mathbf{p}$ at the center; $\mathbf{p}$ is directed parallel to $\mathbf{E}_0$. Let γ be the angle of an arbitrary direction with this direction; then the dipole field has the components

$$E_r = \frac{2p \cos \gamma}{r^3}, \qquad E_\gamma = \frac{p \sin \gamma}{r^3};$$

further, the incident field has the components

$$E_{0,r} = E_0 \cos \gamma, \qquad E_{0,\gamma} = -E_0 \sin \gamma.$$

The boundary condition requires that the tangential components cancel at $r = a$, so

$$\mathbf{p} = a^3 \mathbf{E}_0.$$

So far the derivation is identical with that for Rayleigh scattering (sec. 6.31) if the transition $m \to \infty$ is made.

In addition to surface charges there are surface currents. These can exist only for $m = \infty$ (sec. 9.14). The magnetic field of the surface currents is exactly that of a magnetic dipole **m** at the center, parallel to $\mathbf{H}_0$. Let β the angle with the direction $\mathbf{H}_0$; then

$$H_r = \frac{2m \cos \beta}{r^3}, \qquad H_\beta = \frac{m \sin \beta}{r^3} ;$$

further

$$H_{0,r} = H_0 \cos \beta, \qquad H_{0,\beta} = -H_0 \sin \beta.$$

The boundary condition requires that the radial components cancel at $r = a$, so that

$$\mathbf{m} = -\tfrac{1}{2}a^3\mathbf{H}_0.$$

Therefore we have the superposed fields of a radiating electric dipole and a radiating magnetic dipole across it. Their amplitudes are in the ratio $-\tfrac{1}{2}$. Adding both fields we obtain the intensity pattern shown above. The minus sign causes a favorable interference in the backward direction.

10.62. Intermediate and Large Sizes

For any value of $x > 0.4$ the efficiency factors and scattering diagram of a totally reflecting sphere are best found by means of the rigorous formulae. The angles $\delta_n(x)$ and $\varepsilon_n(x)$ can be taken directly from the tables (sec. 10.23) or computed from the definitions. The further computation proceeds as with finite m, explained in sec. 10.4.

Some numerical results are shown in Table 13. Comparison of the

Table 13. Efficiency Factors for Totally Reflecting Spheres

x	Q	$Q \overline{\cos \theta}$	Q_{pr}	x	Q	$Q \overline{\cos \theta}$	Q_{pr}	x	Q
0.1	0.00034	−0.00013	0.00047	1.3	2.267	−0.113	2.380	15	2.043
0.2	0.0054	−0.0021	0.0075	1.4	2.204	0.018	2.186	20	2.033
0.3	0.028	−0.011	0.039	1.5	2.155	0.156	1.999	25	2.027
0.4	0.086	−0.031	0.117	1.6	2.115	0.286	1.829	30	2.023
0.5	0.218	−0.072	0.290	1.8	2.136	0.495	1.641	35	2.020
0.6	0.466	−0.143	0.609	2.0	2.209	0.623	1.586	55	2.013
0.7	0.795	−0.224	1.019	2.5	2.171	0.728	1.443	60	2.012
0.8	1.257	−0.320	1.577	3.0	2.172	0.851	1.321	70	2.011
0.9	1.696	−0.382	2.078	3.5	2.136	0.876	1.260	75	2.010
1.0	2.036	−0.385	2.421	4.0	2.140	0.929	1.211	80	2.010
1.1	2.230	−0.342	2.572	5.0	2.116	0.965	1.151	85	2.009
1.2	2.280	−0.242	2.522	10.0	2.061	1.006	1.055	90	2.009

older computations with each other and with the newly computed values in the table (up to $x = 10$) shows that errors of 1 or 2 per cent are quite common and that occasionally larger errors occur. The third decimal in Table 13 is not reliable. The extinction values for $x > 10$ are cited from

Boll, Gumprecht, and Sliepcevich (1954). Even these machine computations contain errors, for the values $Q = 2.006$, 2.014, 2.009, and 2.053 given by these authors for $x = 40$, 45, 50, and 65, respectively, are unacceptable. The other values, given in Table 13 define a smooth curve that may very well be represented from $x = 6$ to 90 by the empirical formula

$$Q = 2 + 0.50x^{-8/9},$$

or equally well by

$$Q = 2 + 0.07x^{-2/3} + 0.49x^{-1}.$$

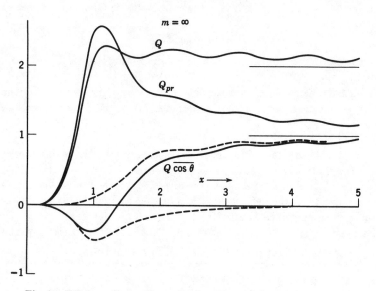

Fig. 28. Efficiency factors for extinction (Q) and for radiation pressure (Q_{pr}) for totally reflecting spheres. Their difference ($Q \overline{\cos \theta}$) and its two separate terms are also given.

The latter representation is found by plotting $x^{2/3}(Q - 2)$ against $x^{-1/3}$. Reasons for trying an empirical formula of this kind will become apparent in chap. 17.

The same results are plotted in a graphical form in Fig. 28. The extinction reaches its maximum value $Q_{ext} = 2.28$ at $x = 1.2$; the radiation pressure reaches its maximum $Q_{pr} = 2.57$ at $x = 1.12$. The sign of $\overline{\cos \theta}$ changes at $x = 1.38$. Beyond that value of x the scattering diagram has the usual preponderance of forward radiation over back radiation.

A more striking illustration of the change in the scattering pattern is shown in Fig. 29. Here we have plotted the ratios $|S_1(\theta)|/x$ and $|S_2(\theta)|/x$

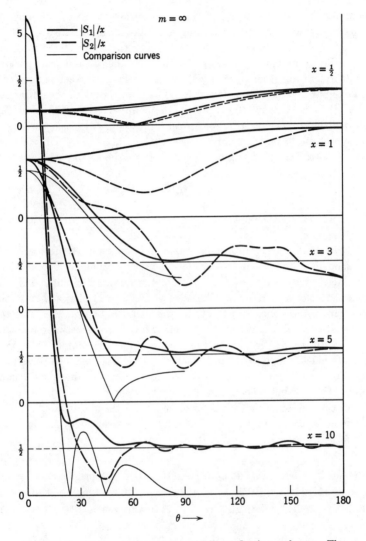

Fig. 29. Scattering diagrams for totally reflecting spheres. The ordinates are proportional to the amplitudes of the scattered light in polarizations 1 and 2. The graph for $x = \frac{1}{2}$ shows for comparison the law for very small, totally reflecting particles. The graphs $x = 3$, 5, and 10 show for comparison the Fraunhofer diffraction and the pattern due to simple reflection.

as a function of θ for the values $x = \frac{1}{2}$, 1, 2, 3, 5, and 10. Note that, except for the linear (instead of logarithmic) scale, these diagrams are comparable with the scattering diagrams shown in Fig. 25 (sec. 10.4). From $x = 3$ on the scattering diagrams may clearly be divided into two parts: the diffracted light (near the forward angles) and the reflected light (through all other angles). Quantitatively the agreement is quite good even at a value of x as low as 3. The central peak of the diffraction pattern has a height $S(0) = \frac{1}{2}x^2$, so that $|S(0)|/x = \frac{1}{2}x$. A thin line indicates the further distribution in the diffraction pattern according to the asymptotic formula for large x (secs. 8.31 and 12.32). The reflected light is isotropically distributed according to the asymptotic formula (secs. 8.42 and 12.44):

$$\left| S_1(\theta) \right| = \left| S_2(\theta) \right| = \tfrac{1}{2}x,$$

also in perfect agreement with the average run of the curves shown in Fig. 29.

The fact that the asymptotic formulae give a fair representation of the scattering diagram for values of x as low as 3 is exceptional. The situation for refractive indices like $m = 1.2$, 1.33, or 1.5 is quite different. This can be theoretically understood. The theory of chap. 11 shows that in the latter case interference effects of the diffracted light and the refracted light persist to large values of x. A similar interference between reflected light and diffracted light occurs for $m = \infty$, but it is confined to a small range of angles (e.g., $\theta = 20°$ to $60°$ for $x = 10$ in Fig. 29). In particular, at $\theta = 0$ no such interference in the ordinary sense of the word occurs. Instead, the grazing reflection that goes with the diffracted light in the forward direction modifies the diffraction in a more subtle manner. This effect will be described as the effect of the "optical edge" in chap. 17.

References

The survey of the m-x domain (sec. 10.1) and the properties of phase angles and nodes (sec. 10.2) were taken, with improvements, from

H. C. van de Hulst, "Thesis Utrecht," *Recherches astron. Obs. d'Utrecht*, **11**, part 1 (1946).

The series expansions (sec. 10.3) were given by

E. Schoenberg and B. Jung, *Astron. Nachr.*, **253**, 261 (1934)

and in improved form by

C. Schalén, *Meddelande Uppsala Astron. Obs.*, **64**, (1936).

The expansion for Q_{pr} was also given by

P. Debye, *Ann. Physik*, **30**, 59 (1909).

The resonance domain (sec. 10.5) was first explored by Debye (loc. cit.). The first statement that for small, totally reflecting spheres Rayleigh scattering had to be completed by magnetic dipole radiation (sec. 10.61) was by

J. J. Thomson, *Recent Researches in Electricity and Magnetism*, p. 437, Oxford, Oxford Univ. Press, 1893.

Further results for $m = \infty$ (sec. 10.62) were computed by

K. Schwarzschild, *Sitzber. bayer. Akad. Wiss. Math.-naturw. Kl.*, **31**, 293 (1901),
P. Debye, *Ann. Physik*, **30**, 59 (1909).
J. W. Nicholson, *Monthly Notices Roy. Astron. Soc.*, **70**, 544 (1910),
J. Proudman, *Monthly Notices Roy. Astron. Soc.*, **73**, 535 (1913).
J. Proudman, A. T. Doodson, and G. Kennedy, *Phil. Trans. Roy. Soc.*, **A217**, 279 (1918).

and by the authors cited in the following bibliography.

A short bibliography of numerical computations by the Mie theory

Auxiliary Tables

In sec. 10.4 reference is made to

A. N. Lowan, P. M. Morse, H. Feshbach, and M. Lax, *Scattering and Radiation from Circular Cylinders and Spheres, Tables of Amplitudes and Phase Angles* Washington, U.S. Navy, 1946.

They cover the arguments $x = 0$ (0.1) 10.0.[4] Application to cylinders is described in sec. 15.31. An abstract of these tables is contained in

P. M. Morse and H. Feshbach, *Methods of Theoretical Physics*, Vol. 2, pp. 1565, 1576, 1928–1933, New York, McGraw-Hill Book Co., (1953).

Some related functions for cylinders, used in sec. 15.52, are tabulated by

M. Lax and H. Feshbach, *J. Acoust. Soc. Amer.*, **20**, 108 (1948).

The most important direct table of Riccati-Bessel functions for real arguments, obtained by electronic machine computation (ENIAC), is

R. O. Gumprecht and C. M. Sliepcevich, *Tables of Riccati-Bessel Functions for Large Arguments and Orders*, Willow Run Research Center, Univ. of Michigan Press 1951.

It covers the arguments $x = 1$ (1) 10 (5) 100 (10) 200 (50) 400, and about 60 intermediate and larger values up to 640. The orders go up to $n = 420$.

Tables of the angular function π_n and τ_n are

R. O. Gumprecht and C. M. Sliepcevich, *Tables of Functions of First and Second Partial Derivatives of Legendre Polynomials*, Willow Run Research Center, Univ. of Michigan Press, 1951,

---

[4] This notation indicates a table covering arguments with constant interval 0.1 between 0 and 10.

with the range $n = 1$ (1) 420, $\theta = 0°$ (1°) 10° (10°) 180°, and

F. T. Gucker and S. H. Cohn, *J. Colloid Sci.*, **8**, 555 (1953),

with the range $n = 1$ (1) 32, $\theta = 0°$ (2.5°) 180°.

Tables for Real Values of m

A list of the computations that have come to the author's attention is given below. Computations of minor interest before 1925 have been omitted. It includes results for $m = \infty$ (totally reflecting spheres, sec. 10.62). For *absorbing spheres* see the list in sec. 14.22. For *coated spheres*, i.e., concentric spheres of different material, see sec. 16.11. The papers may give any or all of the following data.

C: Mie coefficients in one form or other.
A: Complex amplitude of scattered light for two polarizations.
I: Intensity of scattered light for two polarizations.
Q: Total scattering or extinction efficiency.
P: Efficiency factor for radiation pressure.

Minor differences in notation that necessitate simple multiplication in order to make the numbers comparable are ignored in this list. If A is given for $\theta = 0$, Q is implicitly given.

Reference	m	x	Tabulated Quantities
P. Debye, *Ann. Physik*, **30**, 59 (1909)	1.33 1.5 2 ∞	1 2 3 4 1 2 3 4 13 values $\leq$ 3.6 0.5 (0.5) 3	P only
M. A. Schirmann, *Sitzber. Akad. Wiss. Wien, Math. naturw. Kl. Abt. IIa*, **127**, 1559 (1918)	2.0	0.70 to 1.49 (8 values)	I (11 angles, 0° to 180°)
Lord Rayleigh, *Proc. Roy. Soc.* **A84**, 25 (1910) (*Sci. Papers* 344)	1.5	1, 1.5, 1.75, 2, 2.25	I (0°, 60°, 90°, 120°, 180°)
B. B. Ray, *Proc. Indian. Assoc. Cultivation Sci.*, **7**, 10 (1921); **8**, 23 (1923)	1.333 1.466	12 5	C, I (25 angles, 0° to 180°) C, I (13 angles, 0° to 180°)
W. Shoulejkin, *Phil. Mag.*, **48**, 307 (1924)	1.32	1, 3	C, I (0°, 20°, 40°, 60°, 90°, 120° 140°, 160°, 180°)
H. Blumer, *Z. Physik*, **32**, 119 (1925); **38**, 304 (1926)	1.25 1.33 1.5 ∞	0.4, 0.8, 1.6, 4, 8 1.5, 3, 12 4 0.1, 0.5, 1, 3, 5, 10	C, I, (0°, 10°, 20°, 30°, 45°, 60°, 70°, 80°, 90°, 100°, 110°, 120°, 135°, 150°, 160°, 170°, 180°)
J. A. Stratton and H. G. Houghton, *Phys. Rev.*, **38**, 159 (1931)	1.33	1 to 15	Q, graph only
T. Casperson, *Kolloid-Z.* **60**, 151 (1932); **65**, 162 (1933)	1.50 1.56 1.63	1 to 3.16 1 to 3.16 1 to 3.89	C, graphs of $I(\theta)$

Reference	m	x	Tabulated Quantities
F. W. P. Götz, *Astron. Nachr.*, **255**, 63 (1935)	∞	0.1 to 2.0 (14 values), 3, 4, 5, 6, 8, 10, 12, 20	Q
H. Engelhard and H. Friess, *Kolloid-Z.*, **81**, 129 (1937)	1.44	0.4, 1, 1.5, 2, 2.5, 3, 4, 6, 8	I (25 angles from 0° to 180°)
J. L. Greenstein, *Harvard Obs Circ.*, **422** (1938)	2	0.1, 0.3, 0.5, 0.8, 1, 2, 3, 4, 5, 10	Q
	∞	0.1, 0.4, 0.6, 0.8, 1, 2, 3, 4, 5, 10	
G. R. Paranjpe, J. G. Naik, and P. S. Vaidya, *Proc. Indian Acad. Sci.*, **A9**, 333 (1939)	1.33	4 (1) 10, 12, 15, 20, 30	I (every 10°)
R. Ruedy, *Can. J. Research*, **A21**, 79 and 99 (1943); **A22**, 53 (1944)*	1.33	$\frac{1}{8}, \frac{1}{4}, \frac{3}{8}, \frac{1}{2}, \frac{3}{4}, 1$	Q
H. Holl, *Optik*, **1**, 213 (1946)	$\frac{4}{3}$	0.3 to 4.5	C, I, Q
M. D. Barnes, A. S. Kenyon, E. M. Zaiser, and V. K. La Mer, *J. Colloid Sci.*, **2**, 349 (1947)	1.5	0.5 to 12.0 (28 values)	Q
H. Holl, *Optik*, **4**, 173 (1948)	$\frac{4}{3}$	4.8, 5.4, 6.0	C, I (every 10°), Q
		6.6 to 18.0 (24 values)	$C, I,$ (0°, 90°, 180°), Q
A. N. Lowan, "Tables of Scattering Functions for Spherical Particles," *Natl. Bur. Standards,* (*U.S.*), *Appl. Math. Series 4*, Washington, Govt. Printing Office (1948)	1.33 1.44 1.55 2.00	0.5 to 6.0 (15 values)	C, A and I (every 10°), Q
	1.50	0.5 to 12.0 (32 values)	C, Q

* Quoted by M. L. Kerker, *J. Opt. Soc. Amer.*, **45**, 1081 (1955).

Reference	m	x	Computed
H. G. Houghton and W. R. Chalker, J. Opt. Soc. Amer., 39, 955 (1949)	$\frac{4}{3}$	7 to 24 (33 values)	Q
J. D. Riley, "Calculations of Light Intensity Functions," U.S. Naval Research Laboratory, Radio Division III, ORB Information Bull. 9, 8 June 1949†	1.486	0.5 (0.1) 3.0	I (every 10°), Q
E. J. Durbin, NACA Techn. Note 2441, August, 1951	1.20	0.1 (0.1) 0.6 (0.2) 1.2	I (20°, 30°, 40°, and 140°, 150°, 160°), Q
R. O. Gumprecht and C. M. Sliepcevich, Light-Scattering Functions for Spherical Particles, Willow Run Research Center, Univ. of Michigan Press, 1951	1.2 1.33 1.4 1.44 1.5 1.6	1 to 4 (many values), 5, 6, 8, 10 (5) 100 (10) 200 (50) 400	C, I (90° only), Q
E. de Bary, Optik, 9, 319 (1952)	$\frac{4}{3}$	4.8 to 15	I (30°, 60°, 120°, 150°)
R. O. Gumprecht, Neng-lun Sung, Jin. H. Chin, and C. M. Sliepcevich, J. Opt. Soc. Amer., 42, 226 (1952)	1.33	6, 8, 10 (5) 40	A and I [0°(1°) 10° (10°) 180°]
R. O. Gumprecht and C. M. Sliepcevich, J. Phys. Chem., 57, 90 (1953)	1.20 1.33 1.44	20, 80 20, 30, 40, 60, 80, 100, 200, 400 20, 80, 150	A and I [0° (0.2°) 1° for large x, 0° to 7° for smaller x]

† Quoted by M. L. Kerker, loc. cit.

Reference	m	x	Tabulated Quantities
B. Goldberg, *J. Opt. Soc. Amer.*, **43**, 1221 (1953)	1.33	0 to 30 with very small intervals	Graph of Q only
M. L. Kerker and H. E. Perlee, *J. Opt. Soc. Amer.*, **43**, 49 (1953)	2.00	1.30 to 2.80 (12 values not in Lowan's tables)	C, A and I (90° only)
R. H. Boll, R. O. Gumprecht, and C. M. Sliepcevich, *J. Opt. Soc. Amer.*, **44**, 18 (1954)	0.80 0.90 0.93	1 to 110 (18 values) 1 to 200 (20 values) 1 to 200 (24 values)	Q only
	∞	1 (½) 4, 5, 6, 8, 10 (5) 90	
J. C. Johnson, R. G. Eldridge, J. R. Terrell, *Sci. Rept.* 4, 1954, M.I.T. Dept. of Meteorology	1.29	1.0 to 19.3 (22 values)	Q
	∞	0.5, 1, 2, 4, 8.2, 12, 25	
W. Heller and W. J. Pangonis, *J. Chem. Phys.*, **22**, 948 (1954)	1.05 1.10 1.15 1.20 1.25 1.30	0.2 (0.2) 7.0	Q‡
W. Heller, *J. Chem. Phys.* **23**, 342 (1955)	1.05 1.10 1.15 1.20 1.25 1.30	8 (1) 15	Q‡

‡ The values actually tabulated are 1153 Q/x. Complete data appear in *J. Chem. Phys.*, 1957.

Reference		Range	Tabulated
M. L. Kerker and A. L. Cox, *J. Opt. Soc. Amer.*, **45**, 1080 (1955); *Document 4677*, American Documentation Institute	2.00	3.3 to 12.5 (11 values) 1.3 to 12.5 (23 values)	*C* *I* (40° only)
F. T. Gucker, Office of the Publication Board, U.S. Dept. of Commerce, Report No. PB107016. Available from Library of Congress, Photoduplication Service	1.33	3.3 to 18.5 (53 values)	*C, Q*
R. Penndorf, *J. Meteorol.*, **13**, 219 (1956)	1.33	0.1 (0.1) 8.0 (0.2) 30.0, 30 (1) 45	*Q*, smoothed
R. Penndorf, "Tables of Mie Scattering Functions for Spherical Particles," *Geophysical Research Paper* No. 45, in preparation, Cambridge Air Force Research Center	1.33 1.40 1.44 1.486 1.50	0.1 (0.1) 30.0 0.1 (0.1) 30.0	*C, I* (every 5°), *Q* *C, I* (every 10°), *Q*
W. J. Pangonis, W. Heller, and A. W. Jacobson, "Tables of Light Scattering Functions for Spherical Particles," in preparation, Detroit, Wayne University Press	1.05 1.10 1.15 1.20 1.25 1.30	0.2 (0.2) 7.0 (1.0) 15.0, and some selected values	*C, I* (every 45°), *Q*

11. SPHERES WITH REFRACTIVE INDEX NEAR 1

11.1. Distinction of Limiting Cases

Scattering by a particle is dependent on the fact that its refractive index m differs from the refractive index of the outer medium, which is put equal to 1 in all our derivations. So for $m = 1$ there is *no* scattering, and for m near 1 scattering is small. If $m - 1$ is very small, it makes no difference in the scattering pattern whether $m - 1$ is <0 or >0. For convenience we assume $m > 1$.

Earlier references to asymptotic formulae for $m \to 1$ have been made in sec. 6.22 and in all of chap. 7 (Rayleigh-Gans scattering). Yet the discussion in that chapter is by no means complete. The survey of the m-x domain given in sec. 10.1 has shown that there are *two important limiting cases*, both involving $m \to 1$. These two are the Rayleigh-Gans scattering (treated for spheres in sec. 7.2) and the anomalous diffraction, discussed in this chapter. The second possibility had escaped attention in all earlier work on the problem, and the term "anomalous diffraction" is proposed here as a distinctive term for any theory based on the assumptions

$$m - 1 \ll 1, \quad x \gg 1.$$

The reason for the existence of separate limiting cases is the following. Let us write $\rho = 2x(m - 1)$. Rayleigh gave x an arbitrary (fixed) value and made the transition $m \to 1$. This implies that $\rho \to 0$, which was the condition made in chap. 7. It also implies that $Q_{ext} \to 0$, so that the theory of chap. 7 is restricted to relatively small efficiency factors.

However, for a fixed value of m there always is a value of x near which Q_{ext} rises from near 0 to a value of the order of 2, its limiting value for $x \to \infty$ (sec. 8.22). This value of x shifts to higher values, the smaller we make m, and it is found (Fig. 24) that the place of rise and the place of the characteristic maxima and minima in the curve occur at a fixed value of ρ. The theory of anomalous diffraction is based on keeping ρ fixed and making the transition $m \to 1$. This implies $x \to \infty$, gives a definite value of Q_{ext} (sec. 11.2), and the scattering pattern runs through a set of "homologous diagrams" (sec. 11.3).

In many applications only particles whose extinction efficiency is relatively large are important. In such applications the anomalous-diffraction theory is the relevant theory for $m \to 1$.

Table 14 summarizes the possibilities. It may be regarded as a part of the survey in sec. 10.1 (Fig. 20), and the five columns are denoted accordingly. The proper physical description in the Rayleigh-Gans domain is: interference of light scattered independently by all volume elements. In the anomalous-diffraction domain it is: straight transmission and subsequent diffraction according to Huygens' principle. Throughout this domain the scattered intensity is concentrated near the original direction of propagation.

Table 14. Survey of Limiting Cases in which $m \to 1$

(61)	(1)	(12)	(2)	(23)
x small ρ small (Rayleigh scattering)		x large ρ small (intermediate case)		x large ρ large (geometrical optics plus diffraction)

Rayleigh-Gans scattering (sec. 7.2) — Anomalous diffraction (secs. 11.2 and 11.3)

Although anomalous diffraction is itself a limiting case ($m \to 1$ and $x \to \infty$), it has again two limiting cases, namely, $\rho \to 0$ and $\rho \to \infty$. The formulae for these two, which correspond to columns (12) and (23) in the table, may be derived from other chapters and will serve as a check on the formulae for arbitrary ρ.

Intermediate case[1] (ρ very small): For small θ, which angles alone are important, sec. 7.21 gives

$$S_1(\theta) = S_2(\theta) = ix^3(m - 1) \left(\frac{2\pi}{u^3}\right)^{1/2} J_{3/2}(u)$$

$$= \frac{2}{3} ix^3(m - 1)\, G(u),$$

where $u = x\theta$ and $G(u)$ is tabulated in sec. 7.4. The corresponding total scattering, or extinction, is found from sec. 7.22 as

$$Q_{sca} = Q_{ext} = 2(m - 1)^2 x^2 = \tfrac{1}{2}\rho^2.$$

---

[1] The transition from Rayleigh-Gans scattering to anomalous diffraction for cylinders and disks was briefly sketched in secs. 7.32 and 7.33. In particular it was shown that, starting from these two disciplines, the same formulae for the intermediate case were obtained.

Geometrical optics plus diffraction (ρ very large): The complete formulae for $S_1(\theta) = S_2(\theta)$, that follow from the theory in chap. 12, are given in sec. 11.31. In the forward direction they become

$$S_1(0) = S_2(0) = \tfrac{1}{2}x^2 - \frac{ix^2 e^{-i\rho}}{\rho}.$$

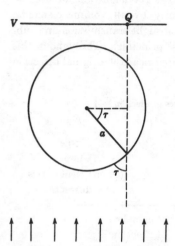

Fig. 30. Ray passing through sphere.

The first term arises from diffraction, the second term from the centrally transmitted light, i.e., from the beam that goes through the sphere along a diameter and is slightly refracted. Since $\rho \gg 1$, the first term is the dominant term. The corresponding extinction value is $Q_{ext} = 2$. The oscillating second term makes the actual value of Q_{ext} fluctuate around this limit. At the same time the formulae for $\theta \neq 0$ (sec. 11.3) show that the pattern corresponding to the second term is wider than that corresponding to the first term. The energy scattered in both patterns is equal; the diffracted and refracted light are thus separable in the sense described in sec. 8.1.

If ρ is gradually diminished, the refracted light becomes more and more intense and at the same time more concentrated near $\theta = 0$ so that the two patterns tend to become more nearly equal in strength *and* in width. The interference effects become stronger, and finally the entire separation breaks down. This will be demonstrated in sec. 11.32.

11.2.   The Extinction Curve

11.21.   General Formula

The assumption that $x = 2\pi a/\lambda$ is very large means that we can trace a ray all through the sphere. The additional assumption that m is very close to 1 means that the ray suffers hardly any deviation at the two boundaries it crosses, so that it is virtually straight. Moreover, the energy reflected at the boundaries is negligible because the Fresnel reflection coefficients go to 0 when $m \to 1$. Thus the field at a point Q beyond the sphere (Fig. 30) is not changed in amplitude by the presence of the sphere, but only in phase. The path traveled in the medium with refractive index m (instead of 1) is $2a \sin \tau$. So the phase lag at Q is

$$2a \sin \tau \cdot (m-1) \cdot 2\pi/\lambda = \rho \sin \tau,$$

where $\rho = 2x(m - 1)$. The physical meaning of ρ is the phase lag suffered by the central ray that passes through the sphere along a full diameter.

The field in the entire plane V, that must be taken not too far beyond the sphere, is now known. If we put it equal to 1 in all points outside the geometrical shadow circle, it is $e^{-i\rho \sin \tau}$ at a point that is at a distance $a \cos \tau$ from the center of this circle, for a phase *lag* corresponds to a *negative* imaginary exponent (compare secs. 3.12 and 4.1). A direct application of Huygens' principle now gives both the extinction (this section) and the scattering diagram (sec. 11.3).

The field of the original plane wave is 1 in the entire plane V. Inside the geometrical shadow circle it is replaced by $e^{-i\rho \sin \tau}$. So the field that is *added* to the field of the original wave is $e^{-i\rho \sin \tau} - 1$ (only inside the shadow circle). This added field determines the scattered wave. For an opaque body only the term -1 is present and gives according to sec. 8.22

$$S(0) = \frac{k^2}{2\pi} \cdot G$$

where

$$G = \iint dx\, dy = \pi a^2;$$

G is the area of the geometrical shadow. The generalization in the presence of the additional term due to rays penetrating the sphere is

$$S(0) = \frac{k^2}{2\pi} \iint (1 - e^{-i\rho \sin \tau}) dx\, dy.$$

This result may be derived along exactly the same lines as in sec. 8.21.

The integral can be expressed in terms of simple functions. We use polar coordinates inside the shadow circle. The element of area then becomes the area of a ring:

$$-2\pi a \cos \tau\, d(a \cos \tau) = 2\pi a^2 \cos \tau \sin \tau\, d\tau.$$

So:

$$S(0) = k^2 a^2 \int_0^{\pi/2} (1 - e^{-i\rho \sin \tau}) \cos \tau \sin \tau\, d\tau.$$

By partial integration we find the definite integral

$$K(w) = \int_0^{\pi/2} (1 - e^{-w \sin \tau}) \cos \tau \sin \tau\, d\tau$$

$$= \tfrac{1}{2} + \frac{e^{-w}}{w} + \frac{e^{-w} - 1}{w^2}.$$

In terms of this function the result is

$$S(0) = x^2 K(i\rho)$$

and (compare sec. 9.32)

$$Q_{ext} = \frac{4}{x^2} \, \mathrm{Re}\,\{S(0)\} = 4\,\mathrm{Re}\,\{K(i\rho)\}.$$

The derivation holds for complex as well as real values of m, provided that

$$|m - 1| \ll 1.$$

We shall discuss both possibilities below.

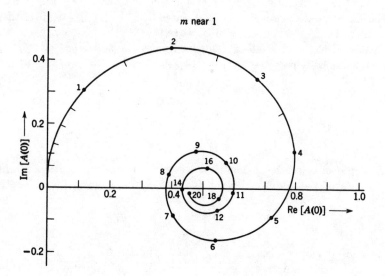

Fig. 31. Plot of $A(0) = x^{-2}S(0)$ in the complex domain for refractive indices very close to 1; running number $= \rho$.

11.22.   Non-Absorbing Spheres (real m)

The curve described by

$$x^{-2}S(0) = K(i\rho)$$

in the complex domain is shown in Fig. 31.   Four times the real part of this function is

$$Q_{ext} = 2 - \frac{4}{\rho}\sin\rho + \frac{4}{\rho^2}(1 - \cos\rho).$$

This is one of the most useful formulae in the whole domain of the Mie theory, because it describes the salient features of the extinction curve not only for m close to 1 but even for values of m as large as 2. This is brought out in Fig. 32, where the extinction curves for $m = 1.5$, $m = 1.33$, $m = 0.93$, $m = 0.80$, and $m \to 1$ are plotted with a common scale of ρ.

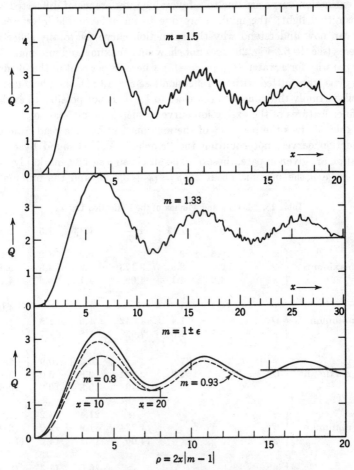

Fig. 32. Extinction curves computed from Mie's formulae for $m = 1.5$, 1.33, 0.93, and 0.8. The scales of x have been chosen in such a manner that the scale of $\rho = 2z|m - 1|$ is common to these four curves and to the extinction curve for $m = 1 \pm \varepsilon$.

The scale of ρ at the bottom holds for all five graphs. The corresponding scales of x depend on m. For $m = 1.05$ or 0.95 the range shown in the figure corresponds to the range of x from 0 to 200.

The most striking feature is the existence of a sequence of maxima and minima. The maxima occur when the term $-e^{-i\rho \sin \tau}$ enhances the term

1 in the integral from which Q_{ext} was computed. Since one term describes the transmitted (= twice refracted) light and the other the diffracted light (in the sense defined in sec. 8.1) we may also say: The maxima in the extinction curve are due to a favorable interference of diffracted and transmitted light; the minima are due to an unfavorable interference. We can now understand why the extinction curve for totally reflecting spheres (sec. 10.62, Fig. 28) does not show such maxima and minima. We also see why for a water drop ($m = 4/3$), where 88 per cent of the incident energy is transmitted without reflection (see sec. 13.11), the phenomenon is still virtually the same as for m close to 1. It is not possible to explain all finer features of the extinction curve in such a simple manner.

Table 15 gives the positions of the maxima and minima read from the extinction curves, not counting the "ripple." Values found by extrapolation are given in parentheses. The abscissae ρ agree remarkably well. For large ρ the maxima occur at $\rho = (k + 3/4)2\pi$ and the minima at

Table 15. Maxima and Minima of the Extinction Curves

	m	0.8	0.93	$1 \pm \varepsilon$	1.33	1.5	2
1st maximum	x	10.5	29		6.2	4.2	2.2
	Q	2.4	2.9	3.173	3.9	4.3	5.8
	ρ	4.2	4.1	4.09	4.1	4.2	4.4
1st minimum	x	18.3	54		11.5	7.7	4.0
	Q	1.4	1.4	1.542	1.6	1.8	1.7
	ρ	7.3	7.6	7.63	7.6	7.7	8.0
2nd maximum	x		77		16.4	10.9	5.4
	Q		2.3	2.404	2.8	3.0	3.1
	ρ		10.8	10.79	10.8	10.9	10.8
2nd minimum	x				21.4	14.2	(7.0)
	Q			1.734	1.9	2.0	(2.0)
	ρ			14.00	14.1	14.2	(14.0)
3rd maximum	x				26.0	17.3	(8.6)
	Q			2.246	2.5	2.6	(2.7)
	ρ			17.16	17.2	17.3	(17.2)
3rd minimum	x				30.5	20.3	
	Q			1.814	2.0	2.0	
	ρ			20.33	20.3	20.3	
4th maximum	x				(35.6)	23.5	
	Q			2.178	(2.3)	2.5	
	ρ			23.52	(23.6)	23.5	

$\rho = (k + 1/4)2\pi$, where k is an integer.   The ordinates Q are systematically higher for larger m.   This effect, due to grazing reflection, is discussed in secs. 13.42 and 17.26.

Expansion of $K(w)$ for small w gives

$$K(w) = \frac{w}{3} - \frac{w^2}{8} + \frac{w^3}{30} - \cdots,$$

so

$$Q_{ext} = 4 \operatorname{Re} \{K(i\rho)\} = \tfrac{1}{2}\rho^2 + \cdots,$$

for the odd powers of w give a purely imaginary value.   This expression was also found as the asymptotic expression of Rayleigh-Gans scattering, when $x \to \infty$ (sec. 7.22).   It constitutes the intermediate case referred to in sec. 11.1.

11.23.   Complex Values of m; Black Body

Let the refractive index be denoted by $m = n - in'$.   The requirement that $|m - 1| \ll 1$ means that both $n - 1$ and n' have to be $\ll 1$.   In order to obtain positive absorption, n' has to be positive; $n - 1$ may have either sign, but we shall assume it to be positive.   We denote by $\tan \beta$ the ratio

$$n'/(n - 1) = \tan \beta;$$

this ratio may have any value from 0 to ∞.   As before, we introduce the real parameter

$$\rho = 2x(n - 1);$$

the phase shift of a ray passing through the center of the sphere then is

$$\rho^* = 2x(m - 1) = \rho(1 - i \tan \beta).$$

The real part denotes an actual phase shift, and the imaginary part a decay of the amplitude.

The efficiency factor for extinction is

$$Q_{ext} = 4\operatorname{Re} \{K(i\rho + \rho \tan \beta)\},$$

which after some reduction turns out to be

$$Q_{ext} = 2 - 4e^{-\rho \tan \beta} \frac{\cos \beta}{\rho} \sin (\rho - \beta)$$

$$-4e^{-\rho \tan \beta} \left(\frac{\cos \beta}{\rho}\right)^2 \cos (\rho - 2\beta) + 4 \left(\frac{\cos \beta}{\rho}\right)^2 \cos 2\beta.$$

Numerical values computed by means of this equation are given in Table 16 for various values of β.   For $\beta = 0$ the expression reduces to the one

Table 16. Extinction and Absorption by Partially Absorbing Spheres with m close to 1

ρ	Q_{ext}				Q_{abs}
	$\beta =$ $0°$ $\tan \beta =$ 0	$15°$ 0.27	$30°$ 0.58	$45°$ 1.00	$45°$ 1.00
0	0.00	0.00	0.00	0.00	0.00
0.2	0.02	0.09	0.10	0.26	0.23
0.4	0.08	0.21	0.35	0.52	0.40
0.6	0.18	0.36	0.54	0.76	0.53
0.8	0.31	0.53	0.74	0.97	0.63
1.0	0.47	0.71	0.93	1.16	0.70
1.2	0.66	0.90	1.11	1.32	0.76
1.4	0.88	1.09	1.28	1.46	0.80
1.6	1.11	1.28	1.43	1.58	0.84
1.8	1.35	1.47	1.57	1.68	0.86
2.0	1.60	1.64	1.69	1.76	0.89
2.2	1.84	1.80	1.80	1.82	0.90
2.4	2.08	1.95	1.89	1.87	0.92
2.6	2.31	2.08	1.96	1.91	0.93
2.8	2.51	2.19	2.02	1.94	0.94
3.0	2.70	2.28	2.06	1.96	0.95
3.5	3.03	2.42	2.13	1.98	0.96
4.0	3.17	2.46	2.14	2.00	0.97
4.5	3.11	2.43	2.13	2.00	0.97
5.0	2.88	2.34	2.10	2.00	0.98
5.5	2.55	2.23	2.08	2.00	0.98
6.0	2.19	2.14	2.05	2.00	0.99
6.5	1.87	2.07	2.04	2.00	0.99
7.0	1.64	2.02	2.03	2.00	0.99
7.5	1.55	1.99	2.02	2.00	0.99
8.0	1.58	1.99	2.02	2.00	0.99
8.5	1.71	2.00	2.02	2.00	0.99
9.0	1.91	2.02	2.02	2.00	0.99
9.5	2.12	2.03	2.02	2.00	0.99
10.0	2.29	2.04	2.02	2.00	1.00

for nonabsorbing spheres (sec. 11.22).   Figure 33 shows the same data in a graph.   The curve marked $\beta = 0$ is the same as the lower curve of Fig. 32.

The energy absorbed inside the sphere is also computed easily.   The phase-shift factor $\exp(-i\rho^* \sin \tau)$ contains a factor $\exp(-2xn' \sin \tau)$ for

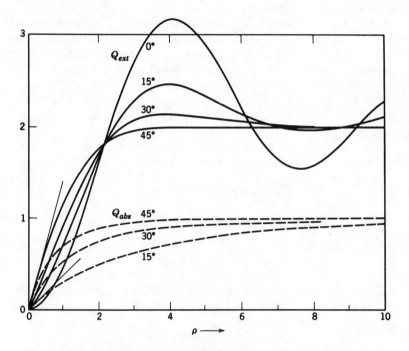

Fig. 33. Influence of an imaginary term in the refractive index upon the extinction curve for m close to 1. The refractive index is
$$1 + \varepsilon - i\varepsilon \tan \beta \quad (\varepsilon \text{ small}).$$

the decrease in amplitude.   Consequently the decrease in intensity is $\exp(-4xn' \sin \tau)$, and the fraction

$$1 - e^{-4xn' \sin \tau}$$

of the original ray is absorbed within the sphere.   The integration over all incident rays leads again to an integral of the type $K(w)$.   Upon integration the absorbed fraction of the total energy incident on the sphere, which is by definition denoted as Q_{abs}, is found to be

$$Q_{abs} = 2K(4xn').$$

The Q_{abs} computed from this formula are represented by dotted curves in

Fig. 33; the values for one value of β are given in Table 16.   The argument may also be written as

$$4xn' = 2\rho \tan \beta = 2a\gamma,$$

where a is the radius of the sphere and γ the absorption coefficient of its material per unit length (sec. 14.1).

The form of the curves in Fig. 33 is easily understood.   The absorption inside the sphere reduces the amplitude of the transmitted light.   Thus both the favorable and unfavorable interference with the diffracted light becomes less striking: the maxima are reduced and the minima are filled in.   When hardly any light gets through the sphere at all, as is indicated by the fact that Q_{abs} approaches 1, nothing interferes with the ordinary diffraction, and we have simply $Q_{ext} = 2$, which holds for any large opaque body (sec. 8.22).

It is also instructive to investigate the behavior of the curves for small values of ρ.   The series expansion of $K(w)$ gives

$$Q_{ext} = \tfrac{4}{3}\rho \tan \beta + \tfrac{1}{2}\rho^2(1 - \tan^2 \beta),$$

$$Q_{abs} = \tfrac{4}{3}\rho \tan \beta - \rho^2 \tan^2 \beta,$$

and by subtraction

$$Q_{sca} = \tfrac{1}{2}\rho^2(1 + \tan^2 \beta).$$

This agrees in all respects with what we know from earlier chapters.   The dominant term of the extinction is due to absorption.   The expression as given in sec. 7.12 reduces to

$$Q_{abs} = \tfrac{8}{3}n'x = \tfrac{4}{3}\rho \tan \beta = \tfrac{4}{3}a\gamma,$$

from which $C_{abs} = \gamma V$, in agreement with sec. 6.22.

As was explained in sec. 7.12, this formula holds throughout the Rayleigh-Gans region and therefore also in the intermediate case (in the sense of sec. 11.1) for which we now have confirmed it.   The main scattering term in the intermediate case may by a generalization (for complex m) of sec. 7.22 be found to be

$$Q_{sca} = \tfrac{1}{2} \left| \rho^* \right|^2 = \tfrac{1}{2}\rho^2(1 + \tan^2 \beta),$$

which also agrees with the result derived above.

The value reached by Q_{abs} for an arbitrary non-zero value of $\tan \beta$ if $\rho \to \infty$ is 1.   This means, physically, that here we have a body that is large compared to the wavelength and that absorbs all the incident radiation, i.e., a classical *black body*.

Conversely, we may inquire what properties a homogeneous body must have in order to be "black." In order to talk of the "geometrically incident" radiation at all, its transverse dimensions must be much larger than λ, i.e., $x \gg 1$. Further, in order to avoid reflection at the surface we must have $m \to 1$, including both $n \to 1$ and $n' \to 0$. Finally, in order to make the body opaque, its dimension in the direction of the beam should be sufficient to absorb all radiation, i.e., $xn' \gg 1$. This set of conditions is exactly the one discussed above. So this set is *the* representation of a classical black body in Maxwell's theory. The argument is sufficiently general to serve also for bodies of different shapes. An important consequence is that the *screen* in diffraction theory can never be black and very thin at the same time (cf. note on p. 329).

A black body as defined above has an "albedo" $\frac{1}{2}$, as defined by the ratio Q_{sca}/Q_{ext}. This is a consequence of our including the diffraction pattern in the scattering pattern, as we are forced to do if we want to preserve continuity from small to large particles. If the diffracted light is neglected, the albedo is 0.

11.3. Anomalous Diffraction Patterns

The interference between diffracted and transmitted light affects not only the amplitude function $S(0)$ (sec. 11.2) but also the function $S(\theta)$ for all other values of θ for which the diffracted light is appreciable. The result is a set of queer scattering diagrams. They bear some resemblance to the Fraunhofer diffraction coronae (sec. 8.31) in the successive bright and dark rings for small values of θ, but differ from those in the intensity distribution. For water drops ($m = 1.33$) these have been called the anomalous-diffraction patterns. The theory for water drops is more complicated (sec. 13.41). The present theory is confined to $m \to 1$.

11.31. Homologous Scattering Diagrams

Again the parameter $\rho = 2x(m - 1)$ is of prime importance. Let us consider the example of a water droplet ($m = 4/3$) of diameter 3μ scattering orange light ($\lambda/2\pi = 0.1\mu$). Then $x = 15$, and the phase shift of the central ray is $\rho = \frac{2}{3}x = 10$. This is not large enough to neglect interference effects, as is clearly visible from the extinction curve (Fig. 32). Next, we imagine a sphere with $m = 7/6$ and diameter 6μ exposed to light of the same wavelength. Here $m - 1$ has half the value and x has twice the value of the preceding example, so that the phase shift ρ is the same. This implies that the interference effects have a similar character. The two examples are said to define *homologous scattering diagrams*. The diagram in the second example is two times as narrow and four times as intense as in the first one, but the anomalies in the pattern are (very nearly) the same.

The mathematical formulation of this idea is as follows: When $m - 1$ and θ are varied proportionally to $1/x$, so that[2]

$$\rho = 2x(m - 1) \qquad \text{and} \qquad z = x\theta$$

are constants, then the functions

$$x^{-2}S_1(x, m, \theta) \qquad \text{and} \qquad x^{-2}S_2(x, m, \theta)$$

approach for $x \to \infty$, $m \to 1$, and $\theta \to 0$, a common limit that is a function of ρ and z only. This function will be denoted by $A(\rho,z)$. For $z = 0$, i.e., also $\theta = 0$, it must reduce to

$$A(\rho,0) = S(0)/x^2 = K(i\rho),$$

where $K(i\rho)$ is defined in sec. 11.21. For simplicity we confine the computation to real values of m and ρ. Three methods of computing $A(\rho,z)$ may be suggested.

First method: We apply Huygens' principle to a plane V just beyond the sphere (Fig. 30). Let the ζ-axis be in the direction of propagation, and let the direction in which the scattered light is sought be in the $\zeta O\xi$ plane, making a small angle θ with the ζ-axis. Then, with a simple additional factor to the formula in sec. 11.21 we have

$$S(\theta) = \frac{k^2}{2\pi} \int \int (1 - e^{-i\rho \sin \tau}) e^{-ik\xi\theta} \, d\xi \, d\eta.$$

Here the integral has to be taken over the area of a circle with radius a; further, τ is defined by $a^2 \cos^2 \tau = \xi^2 + \eta^2$. We change to polar coordinates by $\xi = a \cos \tau \cos \varphi$, $\eta = a \cos \tau \sin \varphi$, $d\xi \, d\eta = a^2 \cos \tau \, d\varphi \, d(\cos \tau)$. The integration over φ can be performed directly by means of the integral:

$$\frac{1}{2\pi} \int_0^{2\pi} e^{-iz \cos \tau \cos \varphi} \, d\varphi = J_0(z \cos \tau),$$

where $x = ka$, $z = x\theta$. Thus we obtain

$$A(\rho,z) = \frac{S(\theta)}{x^2} = \int_0^{\pi/2} (1 - e^{-i\rho \sin \tau}) J_0(z \cos \tau) \cos \tau \sin \tau \, d\tau.$$

Second method: We may also start from Mie's rigorous solution. In the next chapter we shall derive the asymptotic formulae for α_n and β_n for

[2] It is strictly indifferent whether we write $x\theta$, $x \sin \theta$, or $2x \sin \frac{1}{2}\theta$, or any such function. The theory of this section borders that of chap. 7, where $2x \sin \frac{1}{2}\theta$ is the proper argument, and of chap. 8, where $x \sin \theta$ is the proper argument.

$x \to \infty$. When $m \to 1$, only the term with $p = 1$ (the transmitted light) is retained in these formulae, and we find after some reduction

$$\alpha_n = \beta_n = \tfrac{1}{2}\rho \sin \tau.$$

The further reduction is fully analogous to the usual derivation of the diffraction pattern of a sphere (sec. 12.32). The spherical harmonics are approximated by Bessel functions near $\theta = 0$ and the sum replaced by an integral. The result is the same integral just derived.

Third method (approximate): It has been suggested that the simplest method of arriving at $A(\rho,z)$ is to add the amplitude functions for the diffracted light and for the transmitted light:

$$A(\rho,z) = A_{diff}(\rho,z) + A_{trans}(\rho,z).$$

The radiation diffracted by an opaque disk or sphere is under the conditions of this section (large particles, small angles) given by the formula derived in secs. 8.31 and 12.32:

$$S_1(\theta) = S_2(\theta) = x^2 J_1(z)/z$$

which gives

$$A_{diff}(\rho,z) = \frac{J_1(z)}{z}.$$

The transmitted radiation has

$$A_{trans} = \frac{-i\rho}{y^2}\, e^{-iy},$$

where $y = (\rho^2 + z^2)^{1/2}$. This expression follows from the one derived in sec. 12.43:

$$S_1(\theta) = S_2(\theta) = -\frac{x \sin^2 \tau}{2\mu}\, i e^{-2x\mu i/\sin \tau},$$

by applying the relation $\theta = 2\mu \cot \tau$ and by making the substitutions

$$A = x^{-2}S, \qquad \rho = 2x\mu, \qquad z = x\theta,$$

from which also follows that $y = \rho/\sin \tau$.

This completes the computation of both terms A_{diff} and A_{trans}. However, the suggestion that their sum is the correct expression of $A(\rho,z)$ is not confirmed by the computation from the exact formula, which we shall now sketch.

11.32. Expansion of the Integral

The integral expression for $A(\rho,z)$ obtained by the first (or second) method is exact for any complex value of m in the limit $m \to 1$; therefore ρ may be complex. The following discussion is restricted to real values of

m and ρ. Let Re A and Im A denote the real and imaginary parts of $A(\rho,z)$. By the substitution $\gamma = (\pi/2) - \tau$ we find

$$\text{Im } A = \int_0^{\pi/2} \sin(\rho \cos \gamma) J_0(z \sin \gamma) \sin \gamma \cos \gamma \, d\gamma,$$

which is Sonine's second integral with $n = \frac{1}{2}$, $m = 0$. It is equal to

$$\text{Im } A = \frac{\rho}{y^2} \left(\frac{\pi y}{2}\right)^{1/2} J_{3/2}(y) = \frac{\rho}{y^2} \psi_1(y),$$

where $y^2 = \rho^2 + z^2$ as before and $\psi_1(y)$ is the Riccati-Bessel function defined in sec. 9.22.

The real part can be expressed in known functions by means of series expansions. We shall give two different expansions, which are useful for small and large values of ρ, respectively. The first series is found by expanding $1 - \cos(\rho \cos \gamma)$ in powers of $\rho \cos \gamma$. The separate terms are integrated according to Sonine's first integral, and the resulting series is

$$\text{Re } A = \rho^2 \frac{1}{z^2} J_2(z) - \frac{\rho^4}{1 \cdot 3} \frac{1}{z^3} J_3(z) + \frac{\rho^6}{1 \cdot 3 \cdot 5} \frac{1}{z^4} J_4(z) + \cdots,$$

which converges for any combination of ρ and z.

This expansion permits us to discuss (and compute) the deviations from Rayleigh-Gans scattering. If z has a fixed non-zero value and ρ is made very small, y is approximately equal to z. The imaginary part is proportional to ρ and the real part to ρ^2 so that the imaginary part prevails and the combined result is in a first approximation

$$A(\rho,z) = i\rho \left(\frac{\pi}{2z^3}\right)^{1/2} J_{3/2}(z),$$

which is the result of sec. 7.21, if we use the fact that θ is small, so $u \approx z$. The deviations from the Rayleigh-Gans formula if ρ is not very small are seen to be the following: the argument u (or z) must be replaced by y and a real part must be added.

The expansion of Re A for large values of ρ is less obvious. It is suggested by the separation into diffracted and refracted light rather than found by a straightforward asymptotic expansion. The diffraction term results from the term 1 in the integrand, as is clearly seen by comparing the integral in sec. 11.31 with that in 8.21. It will show up again in the discussion of the asymptotic form of Mie's formulae (sec. 12.32).

For the refracted part we expect a term in $A(\rho,z)$ which has the asymptotic form $i\rho y^{-2} \exp(-iy)$ for large values of ρ and has the imaginary part derived above for all values of ρ. A simple expression satisfying these conditions is

$$A'_{trans}(\rho,z) = \frac{i\rho}{y^2} \left(\frac{\pi y}{2}\right)^{1/2} H^{(2)}_{3/2}(y) = \frac{i\rho}{y^2}\, \zeta_1(y).$$

Its real part

$$\frac{\rho}{y^2} \left(\frac{\pi y}{2}\right)^{1/2} N_{3/2}(y) = -\frac{\rho}{y^2}\, \chi_1(y)$$

is to be expected as a term in Re A. The notation $\zeta_1(y)$ and $\chi_1(y)$ is taken from sec. 9.22. The method of finding the further terms will not be described in detail. The author simply made a doubly infinite expansion in powers of ρ and z and then grouped all terms with the same power of ρ. The result is, including the two postulated terms:

$$\mathrm{Re}\, A = \frac{1}{z} J_1(z) + \frac{\rho}{y^2}\left(\frac{\pi y}{2}\right)^{1/2} N_{3/2}(y)$$

$$+ \frac{1}{\rho^2} J_0(z) + \frac{1.3}{\rho^4} z J_1(z) + \frac{1\cdot 3\cdot 5}{\rho^6} z^2 J_2(z) + \cdots$$

This series converges like a geometrical series with the ratio $-z^2/\rho^2$, if $\rho > z$. Beyond its domain of convergence it is still useful as a semi-convergent series. The formula for Im A was checked by a similar expansion, which turns out to be much simpler as only even powers of $y = (\rho^2 + z^2)^{1/2}$ occur.

The results derived above show conclusively that we are not permitted to add simply the diffraction and refraction terms. By replacing $A_{trans}(\rho,z)$ (end of sec. 11.31) with the new expression $A'_{trans}(\rho,z)$ as defined above, we can make the imaginary part of the sum exactly correct for all ρ, but the real part still requires the addition of a "remainder" starting with $(1/\rho^2)J_0(z)$. This conclusion is not surprising. A precise examination of the conditions on which the geometrical-optics approximation is based (sec. 12.3) shows that there is no reason at all to expect for moderate values of ρ a simple addition of diffracted light and light scattered according to the laws of geometrical optics.

Going to $z = 0$ we obtain the result for forward scattered radiation:

$$A(\rho,0) = \tfrac{1}{2} + \frac{i}{\rho}\, \zeta_1(\rho) + \frac{1}{\rho^2}\, ;$$

the three terms are respectively due to diffraction, the generalized refraction term, and the "remainder." On expressing $\zeta_1(\rho)$ in trigonometrical functions this expression becomes

$$A(\rho,0) = \frac{1}{2} - \frac{i}{\rho} e^{-i\rho} + \frac{1}{\rho^2}(1 - e^{-i\rho}),$$

which is identical to the function $K(i\rho)$, as it should be (sec. 11.21).

11.33. Numerical Results

The results of this section may now be summarized. Throughout this section we have dealt with very large particles $(x \gg 1)$, of which the refractive index is very close to 1 $(m - 1 \ll 1)$. All these particles have scattering diagrams with high, non-polarized intensities concentrated around the forward direction, and their scattering properties depend apart from scale factors in angles and intensities, only on the parameter $\rho = 2x(m - 1)$. This is expressed by the function $A(\rho,z)$, where $z = x\theta$. For very small ρ the Rayleigh-Gans theory is correct, with a series of minima at $z = 4.49$, 7.73, etc. (secs. 7.21 and 7.4). For very large ρ the Fraunhofer diffraction theory is correct, with a series of minima at $z = 3.83$, 7.02, etc. (secs. 7.4 and 8.31). The present section provides the formulae for computing the transitional forms, the anomalous-diffraction patterns that occur for moderate values of ρ. The numerical results are presented in the form of an altitude chart in Fig. 34. The coordinates are z (proportional to the scattering angle) and ρ (proportional to size). The distance of a point from the origin (upper left corner) is y. The function whose absolute value is given by the contours is $1000A(\rho,z)$. The intensities are proportional to $|A|^2$. Remembering that

$$S_1(\theta) = S_2(\theta) = x^2 A(\rho,z),$$

we find (compare secs. 4.22 or 9.4) that the intensity scattered by a single particle (independently of the polarization of the incident light) is

$$I(\theta) = \frac{k^2 a^4 |A|^2}{r^2} I_0.$$

The figure was constructed by selecting a number of values of z and ρ and by computing Im A and Re A separately from the formulae given above. For instance, $z = 2$, $\rho = \sqrt{20}$, $y = \sqrt{24}$ gave Im $A = -0.0694$; Re $A = 0.4690$ using 8 terms of the first expansion; Re $A = 0.4691$ using the two main terms plus 6 terms of the "remainder" in the second expansion. These values give

$$|A|^2 = (\text{Im } A)^2 + (\text{Re } A)^2 = 0.225; \qquad |A| = 0.474.$$

In constructing this chart special attention was given to the positions of maxima, minima, saddle points, and zeros.

The chart may be read as follows: On a horizontal line the scattering pattern of a single particle is found; the particle is characterized by the phase shift ρ of the centrally transmitted ray. Vertical lines correspond to fixed points in the classical diffraction pattern. A quarter circle around the center O joins all points with constant y. The rays arriving on

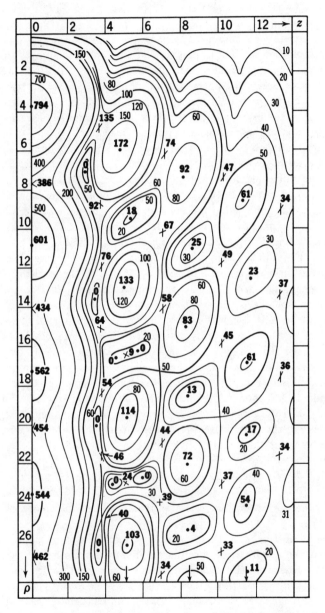

Fig. 34. Altitude chart showing the values of the amplitude function
$|A| = |x^{-2}S|$ in the region of anomalous diffraction; $z = x\theta$,
$\rho = 2x(m - 1)$.

the points of such a circle after two refractions according to Snell's law have suffered equal phase shifts.

The maxima and minima alternate in a regular way. They are separated on the one hand by the vertical lines corresponding to the dark rings in the asymptotic diffraction pattern ($z = 3.8$, 7.0, 10.2, 13.3) and on the other hand by the circles on which the phase of the transmitted light is perpendicular to the phase of the diffracted light ($y = 2\pi, 3\pi, 4\pi, 5\pi, \cdots$). The saddle points are situated near the points of intersection of these circles and vertical lines.

In interpreting the relative heights of maxima and minima we must distinguish three cases:

a. The diffracted light prevails. This applies to the central peak ($z = 0$) and, moreover, to all bright rings if ρ approaches very large values.

b. The transmitted light prevails. In the domain of Fig. 34 this applies to the maxima and minima at the position of the second bright ring ($z = 8.4$) and of the third one ($z = 11.6$).

c. At the position of the first diffraction ring ($z = 5.1$) we observe the gradual transition from case *b* to case *a*. At $\rho = 10$ the transmitted light is still somewhat stronger than the diffracted light, and their superposition causes a deep minimum, but at $\rho = 16$ the situation is reversed. Here the transmitted light does not outweigh the diffracted light at the top of its first maximum but does so a little distance beside the top. Therefore the altitude graph shows a saddle point flanked by two zeros. A similar situation recurs at all further values of ρ where the amplitudes of diffracted and transmitted light have opposite signs. The altitudes of the saddle points approach the final value of 0.066 contributed by diffraction only, and the zeros diverge to the positions $z = 3.83$ and 7.02 of the dark rings in the pure diffraction figure. A similar series of zeros runs beside the central maximum. The first one is situated

$$\rho = 7.2, \qquad z = 2.9, \qquad y = 7.73.$$

The further zeros in this series approach more and more the line $z = 3.83$.

The practical value of the diagram just explained is restricted by the fact that the condition of a refractive index very close to 1 is seldom fulfilled. However, the diagram gives a very good indication of the peculiarities in the scattering pattern that are to be expected for values of m even as high as 1.33 (water). This is illustrated by some detailed examples in sec. 13.41. See in particular Figs. 51 and 52, where the full-drawn lines have been constructed on the basis of Fig. 34 and further symbols show the values for water drops.

It thus is seen that the simple theory of this chapter is quite adequate for an approximate treatment of the anomalous-diffraction phenomena encountered in nature.

11.4. Extinction in the Region of a Spectral Line

Suppose that a spherical particle consists of a material that has a nearly constant index of refraction through a wide wavelength region with the exception of an absorption line or an absorption band.

The index of refraction m in this absorption line has an imaginary part, and also the real part differs from the common value. It is not easy to judge how these changes will affect the scattering properties of a particle of any given size. Generally, the answer cannot be found without complete computations according to the rigorous formulae for complex values of m (chap. 14). However, the simple form of the equations for m close to 1 enables us to judge at least in two situations how the absorption line affects the extinction by the particle. These are: 1. *very small particles*; 2. *weak absorption lines in "soft" particles of a fairly large size*.

Let the material follow the classical dispersion theory. Its refractivity is given by

$$F = \frac{3(m^2 - 1)}{m^2 + 2} = \frac{4\pi e^2}{m_e} \sum_j \frac{N_j f_j}{\omega_j^2 - \omega^2 + i\gamma_j \omega}.$$

Here m is the complex refractive index; e and m_e are charge and mass of the electron, and ω is the circular frequency. The summation has to be extended over all absorption lines, each of them characterized by its proper frequency ω_j, its damping constant γ_j, and its oscillator strength f_j; N_j denotes the number of molecules per cm^3 that can give rise to the absorption line. This formulation applies also to materials in which molecules of various kinds are mixed together.

11.41. Small Particles

A sphere with radius $a \ll \lambda$ has the cross section for absorption

$$C_{abs} = \pi a^2 Q_{abs} = \pi a^2 \cdot \frac{4}{3} x \operatorname{Im}(-F) = \frac{2\pi V}{\lambda} \operatorname{Im}(-F),$$

where V is the volume of the absorbing sphere. This cross section has a close relation to the atomic cross section for absorption:

$$\sigma_j = \frac{\pi e^2}{m_e c} \frac{\gamma_j f_j}{(\omega - \omega_j)^2 + (\gamma_j/2)^2}.$$

The relation can be derived by writing $\lambda/2\pi = c/\omega$, where c is the velocity of light, and by further making the usual assumption that the line extends over a region which is small in comparison with the proper frequency itself. A simple reduction then gives

$$C_{abs} = V \sum_j N_j \, \sigma_j.$$

Here $N_j V$ is the total number of molecules in the sphere that can absorb the line j; thus a particle which is small compared to the wavelength absorbs exactly the same energy as its molecules would do together in the gaseous state. Evidently the lattice frequencies, which the solid particle possesses in excess of the frequencies of the separate molecules, are not included in this statement.

The particles for which this rule holds have a very small scattering cross section. Their extinction cross section inside the spectral line is due mainly to true absorption and exceeds the extinction outside the line by a very large factor. Measuring the extinction in the usual way, by spectral photometry of a continuous light source seen through a cloud of these particles, one finds an "absorption" line that is indeed due to actual absorption.

11.42. Large Particles with Refractive Index Near 1

Let the refractivity near the spectral line or band consist of two parts: a constant real part F_c due to remote proper frequencies and a changing part due to the proper frequency ω_0. We may then replace the dispersion formula by

$$F = F_c + F_a/(i - v),$$

where $v = 2(\omega - \omega_0)/\gamma$ and F_c and F_a are real constants. This equation is a quite useful approximation to the actual refractive index for weak and strong, wide or narrow absorption bands. Our further computation is restricted to F_c and F_a both small compared to 1. Then this is true also for $|m - 1|$, and we may write

$$2(m - 1) = F_c + F_a/(i - v).$$

The important parameter $2x(m - 1)$ has the value $\rho = xF_c$ outside the line and

$$\rho^* = \rho \left\{ 1 + \frac{F_a}{F_c(i - v)} \right\}$$

inside the line. Throughout the line the theory of sec. 11.23 is applicable.

The locus of $1/(i - v)$ in the complex domain is a circle having the center $-\frac{1}{2}i$ and the radius $\frac{1}{2}$. Thus, starting at low frequencies (negative v) we see first a gradual increase of the real part, corresponding to normal dispersion.   At $v = -1$ the imaginary part equals the real part, and the latter begins to decrease.   This lasts until $v = +1$ (anomalous dispersion), while the imaginary part passes its maximum at $v = 0$; finally, for large positive values of v the real part increases again until it reaches its normal value.   The effect on the extinction can be best illustrated by means of a numerical example.   Such an example is given in Fig. 35.   The value assumed for F_a/F_c is $\frac{1}{4}$.   If, for instance $\rho = 3$, then Fig. 33 (or Table 16) shows that $Q_{ext} = 2.70$ outside the line ($v = \pm \infty$).   Inside the line, at $v = 1$ one obtains $\rho^* = 2.62 - 0.38i = 2.62(1 - i \tan 8°)$.   The corresponding Q_{ext} found by interpolation from Table 16 is 2.23.   In this way the curves of Fig. 35 have been constructed.

The results are certainly surprising at first sight.   They can be readily understood, however, if one observes that the extinction is affected by changes both in n and n', the real and imaginary parts of the refractive index.   The partial derivatives of Q_{ext} with respect to n and n' may have either sign (depending on the value of ρ), so that various kinds of curves may result.   The familiar absorption-line contour is shown only by very small particles (sec. 11.41).   At $\rho = 2$, Q_{ext} is virtually independent of n', so that the curve resembles the dispersion curve.   Finally, at $\rho = 4$ the extinction inside the absorption line is smaller than it is outside the line.   If this case were realized in interstellar space, an absorption line of the interstellar particles might be mistaken for an emission line in the star.

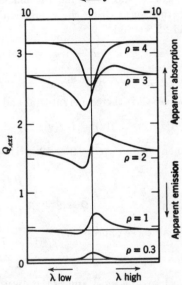

Fig. 35. Variation of the extinction by spheres of various sizes across an absorption line. The absorption sometimes determines a decrease in the extinction by the sphere.

11.5.   Mixtures of Particles with Different Sizes

The extinction coefficient γ in a medium consisting of particles with different sizes has been given in sec. 9.4.   There is no reason to come back to it in this chapter, more than in any other chapter, except for the fact that the resulting expressions become very simple for various distributions of radii.   This may be illustrated by the following examples.   Let

1 cm^3 contain $N_0 f(a/a_1)\, d(a/a_1)$ particles with radii between a and $a + da$. The total geometrical cross section per cm^3 then is

$$G = N_0 \pi a_1^2 \int_0^\infty f(u) u^2\, du$$

and the total actual cross section for extinction is

$$\sigma = N_0 \pi a_1^2 \int_0^\infty Q_{ext}(u) f(u) u^2\, du.$$

The ratio

$$\sigma/G = \bar{Q}$$

may be said to define the efficiency factor of the entire medium for extinction. Using the expression for Q_{ext} given at the beginning of sec. 11.22 we find

(a) If $\qquad f(u) = 1/u \quad (u < 1) \qquad$ and $\qquad 0 \quad (u > 1)$,

then $\qquad \bar{Q} = 2 + 8\rho_1^{-2}(-1 + \cos \rho_1 + C + \log \rho_1 - \text{Ci } \rho_1)$,

where Ci is the cosine integral and $C = 0.577$ is Euler's constant.

(b) If $\qquad f(u) = 1 \quad (u < 1) \qquad$ and $\qquad 0 \quad (u > 1)$,

then $\qquad \bar{Q} = 2 + 12\rho_1^{-2}(1 + \cos \rho_1) - 24\rho_1^{-3} \sin \rho_1$.

(c) If $\qquad\qquad\qquad f(u) = e^{-u}$,

then $\qquad \bar{Q} = 2 + 2(1 + \rho_1^2)^{-1} - 4(1 + \rho_1^2)^{-2}$.

In all these expressions

$$\rho_1 = 2k a_1 (m - 1).$$

Perhaps the most appropriate way of defining the effective radius of the mixture is

$$a_{eff} = \frac{3}{4} \cdot \frac{\text{total volume}}{\text{total geometrical cross section}}.$$

This effective radius is $(2/3)a_1$, $(3/4)a_1$ and $3a_1$, in examples a, b, and c, respectively. Figure 36 compares the extinction factors $\bar{Q}$ for the distributions a, b, and c with that for uniform size, all plotted against

$$\rho_{eff} = 2k a_{eff} (m - 1)$$

as the abscissa. As expected, the curves show how the typical features are washed out if particles of a wide range of sizes are present.

11.6. Extensions to Not Very Small Values of $m - 1$

The great attention given to the limiting case $m \to 1$ in the literature and in chaps. 7 and 11 does not arise from purely mathematical interest. The more practical reason is that many applications deal with particles that have real values of m between 1.1 and 1.5 (water drops in air, proteins in water, etc.). Many investigators, beginning with Lord Rayleigh, hoped that the rigorous solution of the limiting case $m \to 1$ might be a first step towards finding simple formulae valid for values of $m - 1$ that are not infinitesimal.

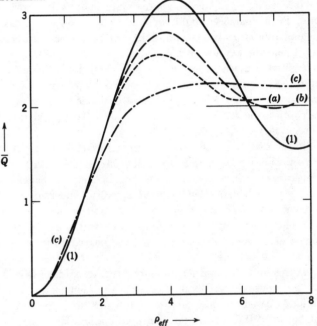

Fig. 36. Sample extinction curves for mixtures of particles of various sizes, all with m close to 1; (1) = uniform size.

The systematic exploration of this possibility has been made only in recent years, in the field of quantum mechanics. The approximation denoted by us as Rayleigh-Gans scattering in optical problems is called the Born approximation in quantum mechanics. This approximation is found as the first term of the *Born expansion* if the scattering problem is formulated by means of an integral equation and this equation solved by means of the iteration-perturbation method. This expansion and its various modifications are treated in detail by Morse and Feshbach (1953). It is difficult to go beyond the second approximation, either analytically or numerically.

The strict distinction between the limiting cases treated in chap. 7 (Rayleigh-Gans scattering) and in chap. 11 (anomalous diffraction) vanishes if higher orders of $m - 1$ are considered. However, the fact remains that the latter theory is more useful in most optical applications. Rayleigh-Gans scattering is confined to the region where $Q < 1$. It occurs at small values of x so that a few terms of the Mie formulae suffice for a rigorous solution. The region of anomalous diffraction, however, includes the entire region of x in which the extinction shows its large fluctuations. As the first maximum occurs near $x = 2.0/(m - 1)$, fairly high values of x are involved, and an approximate analytical answer would be welcome. Such an answer might be found from an expansion in which the anomalous-diffraction result would be the first term. So far, an expansion of this type has not been given.

Intuitively, it would seem that one obvious extension is needed. The derivation in the preceding sections has to be completed by the inclusion of reflection and refraction effects. If this extension is made by means of geometrical optics, the results are not very good. This problem is considered in secs. 13.41, 13.42, and 17.26.

A partially successful attempt based on the rigorous formulae has been made in two papers by Hart and Montroll (1951). These authors consider first the case of scalar waves. A good part of their first paper and all of the second paper (which deals with cylinders, prolate spheroids, and thin disks) are devoted to the scalar problem. Hart and Montroll start from the rigorous solution in terms of the infinite series with the coefficients a_n and b_n (our notation, sec. 9.22) for the scattered field and the coefficients c_n and d_n for the internal field. They replace the numerators by exactly equivalent expressions, which are simple for the internal field (sec. 9.22), but adopt an approximate expression for the denominators. This approximation is based on an approximation which holds for $n \ll x$ and thus might be called the "central-incidence approximation" (sec. 12.31). However, they hope that the approximation may be useful for other n, as in the limit of $m \to 1$ it gives correct values for any n.

It is then found that the sum can be evaluated rigorously, giving (in the scalar case and in our notation, sec. 4.1)

$$S^*(\theta) = - \frac{im(m - 1)x^3(2\pi)^{1/2}}{1 - r^2 e^{-4imx}} \left\{ e^{i(m-1)x} \frac{J_{3/2}(w)}{w^{3/2}} - r e^{i(3m-1)x} \frac{J_{3/2}(v)}{v^{3/2}} \right\},$$

with $r = (m - 1)/(m + 1)$ (Fresnel reflection coefficient, sec. 12.21),

$x = ka,$ $w = x(1 + m^2 + 2m \cos \theta)^{1/2},$

$$v = x(1 + m^2 - 2m \cos \theta)^{1/2}.$$

The asterisk means the complex conjugate; Hart and Montroll use a different sign convention for i.

Hart and Montroll then do not use $S(0)$ but find the total extinction by integration over all angles. This is not possible in a closed form unless further approximations are made. They neglect all terms with r, thus approximating $S^*(\theta)$ by

$$S^*(\theta) = -im(m-1)x^3(2\pi)^{1/2}e^{i(m-1)x}\frac{J_{3/2}(w)}{w^{3/2}}.$$

This expression, properly squared and integrated, gives without further approximations the result (our notation):

$$Q_{sca} = \pi x^2 m(m-1)^2\{\Phi(mx-x) - \Phi(mx+x)\},$$

where

$$\Phi(u) = \frac{1}{u}\{J_{1/2}^2(u) + J_{3/2}^2(u)\} = \frac{2}{\pi u^2}\left\{1 - \frac{2\sin 2u}{2u} + \frac{2(1-\cos 2u)}{2u}\right\}.$$

This is the final result for scalar waves. In the limit for $m \to 1$ if $\rho = 2x(m-1)$ is fixed, the second term becomes negligible and the form is reduced to the form derived in sec. 11.22. If, however, x is fixed and $m \to 1$, it does not give the correct total for the Rayleigh-Gans scattering (for scalar waves). Nor does it give the limit $Q = 2$ for m fixed, $x \to \infty$. The result for electromagnetic waves, which is similar, has the same defects. Also, it gives a scattered intensity independent of polarization. All this is easy to understand from the approximations made and remains far from a successful theory for non-infinitesimal values of $m-1$. Even the agreement in the limit with the result of sec. 11.22 is somewhat puzzling, for, if the central-incidence approximation should be made in our formulae, incorrect results would be obtained in this limit.

Perhaps the most valuable idea in the work of Montroll and Hart is in their second paper. They observe that, quite rigorously for any kind of particles, the field at a point outside the particle may be expressed by a volume integral over the fields inside the particle. The proof is most easily given for scalar waves and for a particle that does not have a discrete boundary, but in which m is a continuous function of the space coordinates and is 0 outside a certain volume V.

Let the total field be $\psi = \psi_0 + \psi_s$, where ψ_0 is the incident plane wave and ψ_s the additional field, which, outside V, is the same as the field of the scattered wave. Then we have

$$\nabla^2\psi + m^2k^2\psi = 0,$$
$$\nabla^2\psi_0 + k^2\psi_0 = 0,$$

and by subtraction

$$\nabla^2\psi_s + k^2\psi_s = (k^2 - m^2k^2)\psi,$$

which has the solution

$$\psi_s(\mathbf{r}) = -\frac{1}{4\pi} \int \frac{e^{ik|\mathbf{r}-\mathbf{r}_1|}}{|\mathbf{r}-\mathbf{r}_1|} \, k^2(1-m^2)\psi \, dV.$$

Evidently the integrand is zero outside V, and in the limit for a homogeneous particle the factor $k^2(1-m^2)$ can be put outside the integral. The ψ in the integrand is the total field at any point inside the particle.

This formula has a similar rigor as, e.g., the integral formula in Kirchhoff's diffraction theory, but its usefulness is similarly restricted, for the internal field never is given but first has to be found. It is quite possible, however, that approximate assumptions for the internal field may be more readily made than for the scattered field, and then the formula is useful. Montroll and Hart demonstrate this by applications to soft cylinders (infinite and finite, prolate spheroids, and thin disks), all in the scalar case. They also show that their earlier approximations for a sphere are consistent with this formula.

A generalization to electromagnetic waves can easily be made, and perhaps this formula may be a fruitful start for fresh attempts at solving the problem of the not very soft spheres. Hart's and Montroll's attempt itself cannot be called successful. The numerical examples show that the extinction for $m \to 1$ has to be increased by an amount that it is nearly constant and does not vanish for $x \to \infty$. The resulting Q values are reasonably accurate in the range

$$1 < m < 1.5, \qquad \tfrac{1}{2} < x < 6.$$

It is suggested in sec. 17.26 that the actual increase, illustrated by Fig. 32, has a clear physical explanation, namely, the influence of grazing reflection at the sphere's "edge."

References

Most of the results of this chapter were first derived in

H. C. Van de Hulst, "Thesis Utrecht," *Recherches astron Obs. d'Utrecht*, **11**, part 1 (1946); **11**, part 2 (1948).

The anomalous diffraction as a separate limiting case seems to have escaped notice in earlier work. The correct formulae for Q in the regions of overlap (61), (12), and (23) were first given by

G. Jöbst, *Ann. Physik*, **78**, 157 (1925).

References to the numerical results in Fig. 32 are found at the end of chap. 10.

Sonine's first and second integrals in sec. 11.32 may be found in

G. N. Watson, *Theory of Bessel Functions*, pp. 373 and 376, Cambridge, Cambridge Univ. Press, 1922.

The calculations of secs. 11.4 (spectral line) and 11.5 (size distribution) were made in the author's 1948 paper, cited above, with a view to application to interstellar particles.   See also the references at the end of chap. 21.

Section 11.6 discusses formulae for soft particles derived by

R. W. Hart and E. W. Montroll, *J. Appl. Phys.*, **22**, 376 (1951),
E. W. Montroll and R. W. Hart, *J. Appl. Phys.*, **22**, 1278 (1951),

and summarized by

E. W. Montroll and J. M. Greenberg, *Proc. Symposia Applied Math.*, Am. Math. Soc., **5**, 103 (1954).

A systematic account of the equivalent problems in quantum mechanics is found in

P. M. Morse and H. Feshbach, *Methods of Theoretical Physics*, part II, chap. 9, New York, McGraw-Hill Book Co., 1953, and in
L. I. Schiff, *Phys. Rev.*, **103**, 443 (1956).

12. VERY LARGE SPHERES

12.1. Survey of Problems

The laws of geometrical optics are asymptotic laws on electromagnetic waves, valid in the limit of very small wavelength. It is evident for this reason that the present chapter, which deals with the scattering by spheres very large compared to the wavelength, will have as a subject the approach to the laws of geometrical optics.

Yet it would be incorrect to expect that the scattering diagram of such spheres is fully rendered by the laws of geometrical optics. First, a necessary condition besides the large size $(x \gg 1)$ is that the phase shift $\rho = 2x(m-1)$ must be large, i.e., that the refractive index of the particle must differ sufficiently from its surroundings (sec. 11.1). Second, it is necessary to include the phases in order to compute the interference effects between various emerging rays. Third, the geometrical-optics approximation fails for arbitrarily large sizes at the angles of rainbows and glory. Fourth, only half the total scattering is due to reflection and refraction by the sphere. The other half arises from diffraction around the sphere and forms the Fraunhofer diffraction pattern.

The separation of these two parts on the basis of Huygens' principle was discussed in sec. 8.1. It is found from the Mie solutions for spheres in sec. 12.32. The main points of distinction may be summarized as follows. The geometrical-optics pattern is wide and of moderate intensity; it arises from reflection and refraction of the rays that hit the sphere. The diffraction pattern is narrow, very intense, and concentrated near the forward direction; it arises from the incompleteness of the wave front passing the sphere. The total radiation contained in both patterns (for non-absorbing spheres) equals the energy incident on the geometrical cross section, πa^2.

The inclusion of the diffraction pattern in the overall scattering pattern is not an arbitrary decision. Starting from Rayleigh scattering, where no such distinction is possible, and going to ever larger particles, we meet *no* distinct size at which a separation becomes possible. In principle, the separation remains always vague. The Mie theory gives the rigorous pattern of both effects combined.

We shall now specify more completely what happens to the rays that hit the sphere. From the part of the wave that hits the sphere we can isolate a narrow beam, the width of which is much larger than λ and yet small compared to the radius a of the sphere. Such a beam is called a "ray," as it is in geometrical optics. Figure 37 presents an example. When the ray first hits the surface, it gives a reflected and refracted ray. The direction of the refracted ray follows from Snell's law. The intensity and phase of both rays follow from the Fresnel coefficients, which will be specified in the next section. A similar separation occurs when the refracted ray reaches the surface again: part of it is refracted and leaves the sphere whereas the other part is reflected internally. This goes on ad infinitum. The final result is that the entire energy available in the incident ray is distributed in a definite manner among the outgoing rays (or partially absorbed on the track inside the sphere). This distribution may be computed. If the same computation is made for all rays that hit the sphere and the results added, we obtain the scattering pattern according to the laws of geometrical optics. The computation is given in sec. 12.2. As all emerging rays are coherent in phase, there is interference between all rays emerging in the same direction. This effect will be included in the geometrical-optics pattern.

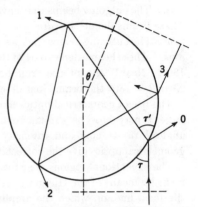

Fig. 37. Path of a light ray through the sphere according to geometrical optics.

This simple scheme of ray-tracing has one exception: *The intensity in or near a focal line* (or a focus) *is not rendered correctly by the laws of geometrical optics.* In fact, geometrical optics gives infinite intensity in a focal line, so that we can simply say that these laws are *not correct whenever the result would be infinity (or very large).* The reason why geometrical optics breaks down in a focal line was explained in sec. 3.22. The high intensity close to the focal line means that light has to be collected from so wide an area of the initial wave front that the wave front in this area is not sufficiently characterized by its two radii of curvature. Moreover, differences of amplitude over this area may become serious.

Spheres exposed to a plane wave of incident light present two sets of focal lines:

a. Any point of intersection of two adjacent rays in a meridional cross section is a point of a focal curve. The full focal curve is a circle around the axis in a plane perpendicular to the axis.

b. Any point where a ray intersects the axis is a point of a focal line because corresponding rays in other meridional sections have the same point of intersection. The full focal line is the full axis, both before and beyond the sphere.

All points close to these curves and line, both inside and outside the sphere, are points at which the intensity is high and cannot be computed from geometrical optics. As the main object of this book is to determine the scattered waves at a very large distance from the particle, we have to find *the focal lines at infinity*, i.e., the outgoing beams with a wave front that has one infinite radius of curvature at some point. Two of the finest natural scattering phenomena can be explained from this consideration. Corresponding to sets *a* and *b* for finite distances we have

a. The outgoing beams that have a wave front with a point of inflexion. These beams define the *rainbows*.

b. The outgoing beams that are parallel to, but not coincident with, the axis. These beams define part of the scattering near the angles $\theta = 0°$ and $180°$. Near $0°$ the intense Fraunhofer diffraction drowns all other effects. Near $\theta = 180°$ the beams just described give rise to the *glory*.

The use of geometrical optics in order to find the scattering pattern for large x may now be summarized as follows. Barring the exceptional angles,[1] ray-tracing and intensity computations based on the laws of geometrical optics give the correct scattering pattern (secs. 12.2 and 12.3).

The exceptional regions near focal lines or foci admit of a simple treatment by means of the Huygens-Fresnel principle. A wave front far from the focal line, on which the amplitude and phase may be computed by means of the ray-optics method, serves as the given front. This method is entirely adequate for very big spheres and gives correct expressions for the scattering pattern, including the exceptional angles (secs. 13.2 and 13.3).

On the other hand we have the Mie formulae. They give the rigorous scattering pattern for spheres of any size, including those for which the parameter $x = 2\pi a/\lambda$ is very large. These formulae may be called upon as soon as any doubt of the validity of the ray-optics approximation arises. Our task is thus fourfold:

1. Demonstrate that the Mie formulae, asymptotically for $x \to \infty$, admit of a principle for localizing rays and of a fundamental separation of diffracted, refracted and reflected light.

2. Prove that the rays incident on the sphere give a pattern identical in intensity and phase with that found from geometrical optics.

---

[1] Mathematically, rainbows occur within an arbitrarily small angle $d\theta$ from any angle θ. However, most of them are due to a very large number of internal reflections and have negligible intensity. Only those with $p = 2$ (1 internal reflection) and $p = 3$ (2 internal reflections) are observed in practice.

3. Estimate the size, i.e., the value of x, beyond which these asymptotic formulae give a fair approximation. Some features of the asymptotic pattern will show up for much smaller values of x than others; values of $x = 3$ (for the isotropic scattering by totally reflecting spheres, Fig. 29) to $x = 5000$ (for the rainbow, Fig. 47) are found. It is foolish to look for a rainbow in the results of rigorous computations for $x = 30$!

4. Investigate whether methods exist, other than a complete numerical computation by Mie's formulae, by which fairly accurate results may be derived for values of x below these limits. This task is particularly important and particularly difficult for $\theta = 0$, and includes the problem of finding an adequate extinction formula.

12.2. Intensity and Phase from Ray-Optics

Our purpose in this section is to derive the scattering pattern for very large spheres from ray-optics with proper account of the phases. In sec. 12.3 we shall derive the same asymptotic formulae from the Mie solution. Either way we find the complex amplitudes $S_1(\theta)$ and $S_2(\theta)$ as defined in sec. 4.42. The index 1 refers to incident light and scattered light with electric vibration perpendicular to the scattering plane. The index 2 refers to electric vibration in this plane. These cases are treated separately.

12.21. Intensity

Let the sphere have the real refractive index m. In Fig. 37 the illumination comes from below. The place of incidence of a certain ray is characterized by an azimuth angle φ and by its distance from the axis. The latter is written as $a \cos \tau$, where a is the radius of the sphere and τ is the angle between the incident ray and the surface; $\tau = 90°$ means central incidence and $\tau = 0°$ means grazing incidence. The choice of this angle instead of its complement was first made by Debye and gives somewhat simpler formulae.

A finite pencil of light is characterized by $d\varphi$ and $d\tau$. Let the light in this pencil be plane-polarized in one of the two main directions, and let I_0 denote its intensity. The flux of energy in this pencil is

$$I_0 a^2 \cos \tau \sin \tau \, d\tau \, d\varphi.$$

This energy is divided, by successive reflection or refraction, among the single rays denoted by $p = 0, 1, 2$, etc., as shown by Fig. 37. The angle τ' between the surface and the internal parts of the path is given by Snell's law:

$$\cos \tau' = \frac{1}{m} \cos \tau;$$

all emerging rays make the angle τ with the surface.

The Fresnel reflection coefficients are

$$r_1 = \frac{\sin (\tau - \tau')}{\sin (\tau + \tau')}, \qquad r_2 = \frac{\tan (\tau - \tau')}{\tan (\tau + \tau')}.$$

They are more useful in the forms:

$$r_1 = \frac{\sin \tau - m \sin \tau'}{\sin \tau + m \sin \tau'}, \qquad r_2 = \frac{m \sin \tau - \sin \tau'}{m \sin \tau + \sin \tau'},$$

to which they can be reduced by means of Snell's law.

These coefficients express the ratio of the amplitude of the reflected wave to the amplitude of the incident wave if a plane wave, plane-polarized in one of the two principal directions, is incident from vacuum on the plane surface of a medium with refractive index m.

The reflected fractions of the energy are r_1^2 and r_2^2. The same fractions are reflected at an internal reflection, but the sign of r_1 and r_2 is then reversed. The refracted parts of the energy are $1 - r_1^2$ and $1 - r_2^2$, both at a refraction from inside and from outside. The total energy of an incident pencil, with polarization 1 thus is divided in the parts r_1^2 for the ray $p = 0$, $(1 - r_1^2)^2$ for the ray $p = 1$, $r_1^2(1 - r_1^2)^2$ for $p = 2$, etc. We shall denote these parts by ε_1^2 and define

$$\varepsilon_1 = r_1 \qquad\qquad \text{for } p = 0$$

$$\varepsilon_1 = (1 - r_1^2)(-r_1)^{p-1} \qquad \text{for } p = 1, 2, 3, \cdots$$

The same definition with index 2 holds for the other polarization.

The pencil of light emerging in one of these directions is characterized by a small range $d\theta$ around the scattering angle θ. We may derive from Fig. 37 that the total deviation from the original direction is

$$\theta' = 2\tau - 2p\tau',$$

which defines the scattering angle in the interval $(0, \pi)$ by

$$\theta' = k \cdot 2\pi + q\theta,$$

where k is an integer and $q = +1$ or -1. Differentiation with use of Snell's law gives

$$\frac{d\theta'}{d\tau} = 2 - 2p \frac{\tan \tau}{\tan \tau'},$$

which defines

$$d\theta = \left| \frac{d\theta'}{d\tau} \right| d\tau.$$

The emergent pencil spreads into a solid angle $\sin \theta \, d\theta \, d\varphi$, i.e., over an area r^2 times this solid angle at a large distance r from the sphere. Dividing the emergent flux by this area we obtain the intensity

$$I_1(p,\tau) = \frac{\varepsilon_1^2 I_0 a^2 \cos \tau \sin \tau \, d\tau \, d\varphi}{r^2 \sin \theta \, d\theta \, d\varphi} = \frac{a^2}{r^2} I_0 \varepsilon_1^2 D,$$

where

$$D = \frac{\sin \tau \cos \tau}{\sin \theta \, |d\theta'/d\tau|}$$

and similarly for $I_2(p,\tau)$. D is sometimes called the divergence.

It is possible to add a factor to take account of a small absorption coefficient γ of the material of the sphere. Since the track in the sphere has a total length $2pa \sin \tau'$, the extra factor is $e^{-2p\gamma a \sin \tau'}$.

Amplitude and phase of the scattered light may be expressed by means of the "amplitude functions" $S_1(\theta)$ and $S_2(\theta)$, which are defined in sec. 4.42 and used throughout this book. Only the squares of the moduli

$$i_1(\theta) = |S_1(\theta)|^2 \quad \text{and} \quad i_2(\theta) = |S_2(\theta)|^2$$

follow from the intensities just derived. Comparison with the formulae of secs. 4.42 or 9.4 shows that

$$i_1(\theta) = x^2 \varepsilon_1^2 D, \quad i_2(\theta) = x^2 \varepsilon_2^2 D.$$

For the very large drops discussed in this chapter another concept will also be found convenient. It is the "gain" relative to an isotropic scatterer. This "gain" is defined as the ratio of the scattered intensity to the intensity that would be found in any direction if the drop scattered the entire incident energy isotropically.

The average gain over the entire solid angle 4π is 1:

$$\frac{1}{4\pi} \int G_1(\theta) \, d\Omega = \frac{1}{4\pi} \int G_2(\theta) \, d\Omega = 1.$$

Isotropic scattering would mean the distribution of the radiation $I_0 \cdot \pi a^2$ over the surface of a sphere $4\pi r^2$. So

$$I_1(\theta,r) = \frac{I_0 a^2}{4r^2} G_1(\theta), \quad I_2(\theta,r) = \frac{I_0 a^2}{4r^2} G_2(\theta),$$

and by comparison with the formulae above

$$G_1 = \frac{4i_1}{x^2} = 4\varepsilon_1^2 D, \quad G_2 = \frac{4i_2}{x^2} = 4\varepsilon_2^2 D.$$

The gain for incident natural light is the average (compare sec. 4.42 for details):

$$G = \tfrac{1}{2}(G_1 + G_2).$$

We thus have completed the computation for a single ray, characterized by (τ, p). Many such rays emerge under the same angle. The common procedure is to add their intensities

$$I_1 = \Sigma I_1(p, \tau), \qquad I_2 = \Sigma I_2(p, \tau),$$

or similarly to make sums of i_1 and i_2 or G_1 and G_2. It may be noted that one value of p may give 0, 1, or more terms in these sums corresponding to different angles of incidence τ that give the same scattering angle.

12.22. Phase

Adding the intensities of the separate outgoing beams is not strictly correct. All beams originate from the same incident wave and are coherent in phase, so that their complex amplitudes have to be added and the squared modulus of the combined amplitude is the correct intensity. If the phase effects are averaged out, which may happen by a small spread in sizes, the average result is indeed simple addition of intensities. If they are not averaged out, the optical interference between different outgoing beams gives rise to numerous maxima and minima in the scattering pattern with angles of the order of $180°/x \approx 30°\lambda/a$ between maxima. The rigorous Mie formulae render these maxima and minima correctly and fully. We shall now compute the phases from ray optics.

Let us write

$$S_1(\theta) = \sqrt{i_1}\, e^{i\sigma_1} \qquad \text{and} \qquad S_2(\theta) = \sqrt{i_2}\, e^{i\sigma_2}.$$

We have to find the phases σ_1 and σ_2. As defined in sec. 4.1, σ_1 and σ_2 are $\pi/2$ plus the advance in phase of the actual ray with respect to a hypothetical ray scattered without phase lag at the center of the sphere. Three physical effects determine this phase.

1. *Change of phase at reflection.* By definition the phase of a wave which changes its direction by reflection or refraction is referred to the component perpendicular to both directions, i.e., to the **E** in polarization 1 and to the **H** in polarization 2. It turns out that refraction does not change the phase but reflection may change the sign of the amplitude and thereby introduce a phase shift π. This is expressed by the Fresnel coefficients (sec. 12.21). For instance, perpendicular incidence gives

$$r_1 = \frac{1 - m}{1 + m}, \qquad r_2 = \frac{m - 1}{m + 1},$$

both of which means that in external reflection ($m > 1$) **E** changes sign but **H** keeps its sign. Beyond the Brewster angle, at which $r_2 = 0$, both coefficients are negative; they approach -1 for grazing reflection. For internal reflection the opposite signs hold. All these possible changes of sign have already been taken into account in the definition of the factors ε_1 and ε_2.

2. *Phase due to length of optical path.* The reflected ray ($p = 0$) has a shorter path than the reference ray; this gives a positive phase shift. The refracted rays ($p = 1, 2, 3, \cdots$) have a long path and therefore a negative phase shift. Simple geometry, together with the fact that the internal path has to be multiplied by $2\pi m/\lambda$ instead of $2\pi/\lambda$, gives the expression

$$\delta = 2x(\sin \tau - pm \sin \tau')$$

as a contribution to σ. As a check we find by differentiation the formula $d\delta = x \cos \tau \, d\theta'$. This can be shown to express the condition that emerging rays have a perpendicular wave front.

3. *Phase shifts due to focal lines.* At the passage of any focal line the phase advances by $\pi/2$ (sec. 3.21); thus we must count the number of focal lines encountered along the entire path. It is found that the rays pass

$$p - \tfrac{1}{2}(1 - s)$$

focal lines of type a described in sec. 12.1 and

$$-2k + \tfrac{1}{2}(1 - q)$$

focal lines of type b. These expressions hold for any pencil characterized by (p, τ). The integers p, k, q, are defined above, and s denotes $+1$ or -1, equal to the sign of $d\theta'/d\tau$. The total phase shift arising from focal lines is

$$\frac{\pi}{2} (p - 2k + \tfrac{1}{2}s - \tfrac{1}{2}q).$$

Combined phase. Collecting the shifts due to these three effects we obtain

$$(\sigma_1 \text{ or } \sigma_2) = \frac{\pi}{2} + (\text{phase of } \varepsilon_1 \text{ or } \varepsilon_2) + \delta + \frac{\pi}{2} (p - 2k + \tfrac{1}{2}s - \tfrac{1}{2}q),$$

and with this phase the entire amplitude function becomes

$$S_1(\theta) = x \left| \varepsilon_1 \right| D^{1/2} e^{i\sigma_1}$$

and similarly for $S_2(\theta)$.

12.3. Intensity and Phase from Mie's Formulae

We now make a fresh start from the rigorous expressions for $S_1(\theta)$ and $S_2(\theta)$ derived in sec. 9.31.

The aim is to obtain the same final formula as in the preceding section. It is sufficient to give a formal solution based on asymptotic expansions of the Bessel functions and spherical harmonics for large x and n. It is more gratifying, however, to show also the physical meaning of each step in this solution. Our proof will be mathematically incomplete; we shall make plausible on physical grounds that various terms may be omitted, although this should be proved on the basis of the asymptotic expansions. The reader is invited to consult original papers for those proofs.

12.31. The Localization Principle

The Mie results consist of series with x fixed and n integer ranging from 1 to ∞. The experience with numerical computations is that the phase angles α_n and β_n are large for $n < x$, drop sharply for n near x, and have become virtually zero when n exceeds x by 2 or 3. This is confirmed by the asymptotic expressions. The terms of order n derive from Bessel functions with order $n + \frac{1}{2}$. These have very different asymptotic expressions according as $n + \frac{1}{2}$ is smaller or larger than x. For $n + \frac{1}{2} < x$ they have the character of oscillating functions, whereas for $n + \frac{1}{2} > x$ the main property is an exponential decrease with increasing $n - x$.

This behavior can be clarified by the *localization principle*, which states that *a term of the order n corresponds to a ray passing the origin at a distance* $(n + \frac{1}{2})\lambda/2\pi$. For $n + \frac{1}{2} = x$ this distance is exactly the radius of the sphere: the terms with $n + \frac{1}{2} < x$ correspond to rays hitting the sphere, whereas the other terms, which tend to zero, correspond to rays passing the sphere.

This localization principle is implicit in Debye's asymptotic formulae developed in 1908, for the terms with a certain value of n give asymptotic expressions containing the Fresnel reflection coefficients for a certain angle of incidence, as we shall see below. It is understandable that Debye himself did not comment on this correspondence between terms and more or less localized rays. However, after the advent of quantum mechanics it is very tempting to do so, as it shows the exact analogue of well-known effects in wave mechanics. The wave equation of an electron colliding with a center of disturbance is the Schroedinger equation. The solution has the form of a series with integer values of l, the quantum number of angular momentum. The de Broglie wavelength is $\lambda = h/mv$, where m is the mass, v the velocity, and h Planck's constant. If we think of the electron as localized and passing the center at a distance d, the angular momentum $lh/2\pi$ must be equal to mvd. This gives $d = l\lambda/2\pi$. Actually, no exact localization prevails, but the average value of d is $(l + \tfrac{1}{2})\lambda/2\pi$. The meaning of this formula is vague, as are all models in which the electron is thought to be localized. Yet it comes in handy as a guide to practical computations. To the same extent the localization of the Mie terms is not a strict law but a useful guiding principle.

12.32. The Diffraction Part

Each Mie coefficient

$$a_n = \tfrac{1}{2}(1 - e^{-2i\alpha_n}), \qquad b_n = \tfrac{1}{2}(1 - e^{-2i\beta_n})$$

consists of two terms: one independent of the nature of the particle and another dependent on it. This separation corresponds to the separation of the scattering pattern explained in secs. 8.1 and 12.1. The term 1 gives the Fraunhofer diffraction pattern, and the term $e^{-2i\alpha_n}$ or $e^{-2i\beta_n}$ the scattering by reflection and refraction. This separation is useful only if α_n and β_n are large and should, therefore, be made only for the terms with $n + \tfrac{1}{2} < x$, corresponding to rays falling on the sphere.[1]

The Fraunhofer diffraction pattern is found in the following way. The asymptotic formulae for the spherical harmonics if

$$u = (n + \tfrac{1}{2})\theta$$

is fixed and n goes to ∞ are

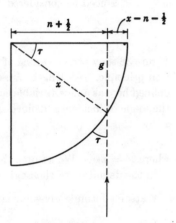

Fig. 38. Ray and angle of incidence τ attributed to the term of the order n according to the localization principle.

$$\pi_n(\cos\theta) = \tfrac{1}{2}n(n + 1)\{J_0(u) + J_2(u)\},$$

$$\tau_n(\cos\theta) = \tfrac{1}{2}n(n + 1)\{J_0(u) - J_2(u)\}.$$

Writing

$$a_n = b_n = \begin{cases} \tfrac{1}{2} \text{ for } n + \tfrac{1}{2} < x \\ 0 \text{ for } n + \tfrac{1}{2} > x, \end{cases}$$

we find

$$S_1(\theta) = S_2(\theta) = \Sigma(n + \tfrac{1}{2})J_0\{(n + \tfrac{1}{2})\theta\},$$

[1] If the angles α_n and β_n are small even for these terms, we are in the domain of Rayleigh-Gans scattering; the separation then has no meaning, and no trace of Fraunhofer diffraction is left.

where the sum has to be extended from $n = 1$ to an integer close to x. The sum may be replaced by an integral, and the result is

$$S_1(\theta) = S_2(\theta) = \int_0^x (n + \tfrac{1}{2}) J_0\{(n + \tfrac{1}{2})\theta\} \, dn = x^2 \frac{J_1(x\theta)}{x\theta} \, .$$

This is identical with the result derived in sec. 8.31, apart from the difference between the arguments $x \sin \theta$ and $x\theta$. This means a difference in the higher orders of θ, which were neglected.

12.33. The Reflected and Refracted Light

The second part requires a longer reduction. Again only terms with $n + \tfrac{1}{2} < x$ need be considered. We define τ and g by

$$n + \tfrac{1}{2} = x \cos \tau,$$

$$g = x \sin \tau.$$

Figure 38 shows the meaning of these symbols on the basis of the localization principle. All lengths are expressed in terms of $\lambda/2\pi$. The angle τ, defined here as a new variable substituted into Mie's formulae, agrees with the angle of incidence, defined in sec. 12.21. In the same way we define τ' by

$$n + \tfrac{1}{2} = x' \cos \tau',$$

where $x' = mx$. We assume that m is real. This definition of τ' agrees with the definition of the angle of refraction by Snell's law.

Watson's formula gives for large x

$$H_{n+\frac{1}{2}}^{(2)}(x) = \frac{g}{(n + \tfrac{1}{2})\sqrt{3}} e^{-i(\varphi + \pi/6)} H_{\frac{1}{3}}^{(2)}(z),$$

with

$$z = g^3/3(n + \tfrac{1}{2})^2 \quad \text{and} \quad \varphi = g - z - (n + \tfrac{1}{2})\tau.$$

At present we shall confine our attention to $z \gg 1$. By this assumption we exclude the terms for which g is of the order of $x^{2/3}$ and $x - n - \tfrac{1}{2}$ of the order of $x^{1/3}$ and smaller. This means that we exclude rays of nearly grazing incidence. For $x = 1000$ only 2 per cent of the incident radiation is thus left out of consideration, and for $x \to \infty$ this edge region may be omitted just as we have already omitted the terms corresponding to rays beyond the edge. A further discussion of these edge terms is found in sec. 17.2.

For $z \gg 1$ Watson's formula reduces to

$$H_{n+\frac{1}{2}}^{(2)}(x) = \left(\frac{2}{\pi x \sin \tau} \right)^{1/2} e^{-i(xf - \pi/4)},$$

where

$$f = \sin \tau - \tau \cos \tau, \qquad d(fx)/dx = \sin \tau.$$

The corresponding formula with argument x' contains

$$f' = \sin \tau' - \tau' \cos \tau'.$$

On taking the real and imaginary parts and omitting the factor $(2/\pi x)^{1/2}$ we have in the notation of sec. 9.22

$$(\sin \tau)^{1/2} \psi_n(x) = -(\sin \tau)^{-1/2} \chi'_n(x) = \cos (xf - \pi/4),$$

$$-(\sin \tau)^{1/2} \chi_n(x) = -(\sin \tau)^{-1/2} \psi'_n(x) = \sin (xf - \pi/4),$$

and corresponding formulae with x', f', τ'. Substitution for these functions in the formula for α_n in sec. 10.21 gives after some reduction

$$\tan \alpha_n = \frac{\sin (xf - x'f') - r_2 \cos (xf + x'f')}{-\cos (xf - x'f') - r_2 \sin (xf + x'f')},$$

where r_2 is the Fresnel reflection coefficient, defined in sec. 12.21; the same formula with r_1 gives $\tan \beta_n$. A further reduction gives

$$e^{-2i\alpha_n} = e^{2i(xf - x'f')} \frac{1 - ir_2 e^{2ix'f'}}{1 + ir_2 e^{-2ix'f'}}$$

and similarly for β_n with the coefficient r_1.

The nodes, where $\alpha_n = \beta_n$, have been investigated for non-absorbing spheres of arbitrary size in sec. 10.22. They still exist in the asymptotic case. If

$$x'f' = l\pi \pm \pi/4,$$

where l is an integer, the fraction in the formula just given is 1, and we find

$$\alpha_n = \beta_n = l\pi \pm \pi/4 - xf.$$

The upper sign corresponds to nodes of the second kind, the lower one to nodes of the first kind. Near central incidence ($n \ll x$, $\tau \approx \pi/2$) we have $f = f' = 1$, so that in the nodes

$$\alpha_n = \beta_n = (m - 1)x.$$

The fraction occurring in the asymptotic formula may be split into a geometrical series plus an extra term. Writing briefly

$$u = ie^{-2ix'f'},$$

we can expand as follows

$$e^{-2i\alpha_n} = -ie^{2ixf} [r_2 + (1 - r_2^2)u + (1 - r_2^2)(-r_2)u^2 + \cdots].$$

Again, replacing r_2 by r_1 the expression for β_n is found. In these expansions the coefficients ε_1 and ε_2 defined in the preceding section are recognized. Evidently, the separate terms correspond to the separate rays $p = 0, 1, 2$, etc., discussed in the preceding section.

Asymptotic formulae have to be inserted also for the spherical harmonics. For a fixed θ and $n \to \infty$ we have

$$\pi_n(\cos\theta) = \left(\frac{2n}{\pi \sin^3\theta}\right)^{1/2} \sin\left\{(n + \tfrac{1}{2})\theta - \frac{\pi}{4}\right\},$$

$$\tau_n(\cos\theta) = \left(\frac{2n^3}{\pi \sin\theta}\right)^{1/2} \cos\left\{(n + \tfrac{1}{2})\theta - \frac{\pi}{4}\right\}.$$

For any large n these formulae are insufficient in a region close to the forward direction ($\theta = 0$) and back direction ($\theta = \pi$). In the forward region they are complemented by the asymptotic formulae used in sec. 12.32. Similar formulae for the back region are used in sec. 13.32. More precisely, the present formulae hold for $\sin\theta \gg 1/n$, and those used previously hold for $\theta \ll 1$. The overlapping domain is

$$1/n \ll \sin\theta \ll 1.$$

Excepting the near-forward and the near-backward angles we find that τ_n has an order of magnitude $n \sin\theta$ times that of π_n; thus π_n may be neglected and we need retain only τ_n.

This solves a serious question. The formula of $S_1(\theta)$ obtained from geometrical optics (sec. 12.22) contains r_1, whereas $S_2(\theta)$ contains exclusively r_2. The Mie formulae for S_1 and S_2 contain both a_n and b_n and therefore r_1 and r_2 together. It is now seen that this paradox is resolved by the fact that the factor containing the "alien reflection coefficient" is multiplied by π_n, which is negligible with respect to τ_n under ordinary circumstances for very large spheres . In the exceptional range of angles near $\theta = 0°$ or $180°$ the presence of the alien coefficient can be understood on the basis of Huygens' principle, as will be shown in sec. 13.31.

The final step is the summation giving $S_1(\theta)$ and $S_2(\theta)$. We have

$$S_1(\theta) = \sum_p \sum_{n=1}^{\infty} \frac{2n + 1}{n(n + 1)} \cdot \tfrac{1}{2}ie^{2ixf}\varepsilon_1(ie^{-2ix'f'})^p$$
$$\cdot \left(\frac{2n^3}{\pi \sin\theta}\right)^{1/2} \cos\left\{(n + \tfrac{1}{2})\theta - \frac{\pi}{4}\right\}.$$

The cosine will be written as

$$\tfrac{1}{2}\sum_q e^{qi(n+\frac{1}{2})\theta - qi\pi/4},$$

where $q = -1$ and $+1$. Then with some obvious substitutions

$$S_1(\theta) = \sum_p \sum_q \sum_{n=1}^{\infty} \varepsilon_1 \left(\frac{x \cos \tau}{2\pi \sin \theta} \right)^{1/2} e^{iG_n},$$

where

$$G_n = (p + 1)\frac{\pi}{2} + 2(xf - px'f') + q(n + \tfrac{1}{2})\theta - q\frac{\pi}{4}.$$

We shall calculate this sum by the method of "stationary phase." The underlying idea is that the complex terms nearly cancel each other unless a number of terms have approximately equal phase. Therefore, regarding x, θ, p, and q as fixed quantities, we try to find a value of n most nearly satisfying the relation

$$G_n - G_{n-1} + k \cdot 2\pi = 0 \qquad\qquad (k = \text{integer}).$$

Replacing the difference by a derivative, and reducing it by

$$\frac{d}{dn} = \frac{1}{x}\frac{d}{d\cos\tau} = -\frac{1}{x\sin\tau}\frac{d}{d\tau}.$$

we find this condition to be

$$-2\tau + 2pr' + q\theta + k \cdot 2\pi = 0, \qquad\qquad (1)$$

which corresponds to the condition for a ray in geometrical optics. The intensity in this ray is found by summing the terms close to the value of n for which this condition holds. $S_1(\theta)$ is not altered is we replace G_n by $H_n = G_n + nk2\pi$. Stationary phase then occurs when $dH_n/dn = 0$; let n_0 be its root, which need not be an integer. We are now justified in replacing the sum by an integral. Since only the vicinity of n_0 contributes to this integral, all factors except the exponential are quasi-constants. The exponential integral that is left is in the second approximation a Fresnel integral. If it is extended from $-\infty$ to ∞, and s denotes ± 1, it has the value

$$\int e^{iH_n}\,dn = \int e^{iH_{n_0} + \frac{1}{2}is(n-n_0)^2\,|H''_{n_0}|}\,dn$$

$$= e^{iH_{n_0}}(2\pi/|H''_{n_0}|)^{1/2}\,e^{si\pi/4}.$$

Dropping the index 0 we find by differentiating the left-hand member of equation 1 with respect to n that

$$H''_{n_0} = \frac{1}{x\sin\tau}\frac{d\theta'}{d\tau}.$$

This shows that s has the sign of $d\theta'/d\tau$ and that a higher approximation going to H'''_{n_0} is needed in the rainbow theory.

The amplitudes may now be collected, but the phase needs one further reduction. On multiplying equation 1 by

$$n + \tfrac{1}{2} = x \cos \tau = x' \cos \tau'$$

and subtracting the product from H_n, we obtain

$$H_{n_0} = \delta + (p + 1)\pi/2 - q\pi/4 - k\pi.$$

The final result is that with omission of the summation sign

$$S_1(\theta) = x\varepsilon_1 \left\{ \frac{\sin \tau \cos \tau}{\sin \theta \, |d\theta'/d\tau|} \right\}^{1/2} \cdot e^{i\delta + i(\pi/2)(p + 1 - \frac{1}{2}q + \frac{1}{2}s - 2k)},$$

which is identical with the result derived in sec. 12.22. It should be noted that this result holds only for one particular set of the integers p, q, s and k. The results of all possible sets of these integers, i.e., of all possible rays emerging in the direction θ should be added to give the total amplitude function.

It is gratifying to see how all parameters which had to be introduced in the preceding section (sec. 12.22) also appear successively in the formal derivation of this section. Also the restrictions made in the preceding sections for obvious physical reasons were needed in this section for mathematical reasons.

12.34. Range of Validity of These Formulae; Possible Extensions

What is the minimum size for which the asymptotic formulae just derived from a good approximation to the actual amplitude function? Two serious requirements are: (1) the omission of further terms in the Taylor expansion of H_n should be permitted, and (2) the replacement of all further factors by constants should be permitted. The number of terms that effectively contributes to the sum is of the order of

$$\Delta n = \left(\frac{2\pi x \sin \tau}{|d\theta'/d\tau|} \right)^{1/2}.$$

The angles and reflection coefficients may be considered as quasi-constant in this interval if the corresponding range of τ is small, let us say <0.1 radian. This gives

$$x \sin \tau \left| \frac{d\theta'}{d\tau} \right| > 16.$$

So, in general, a reasonable approximation to the asymptotic pattern will be obtained near $x = 10$ or 20, i.e., for diameters 3 to 6 times the wavelength. Smaller values, as low as $x = 3$ or 5, are sufficient in the favorable case of isotropic scattering, occurring for $m = \infty$ (Fig. 29, sec. 10.62). However, rays incident close to the edge (with nearly grazing incidence) and those close to the rainbow angles will require much larger values of x. This will be shown in greater detail for the rainbow in sec. 13.24.

A similar result may be derived from a consideration of the Fresnel zones in a manner analogous to the considerations in sec. 3.21. A small elliptical area perpendicular to the incident beam may be said to be the effective area contributing to light scattered along any particular ray into a particular direction. The center of this area corresponds to the geometrical ray. Its axis in the radial direction has the length $\lambda/2\pi$ times Δn, and its axis in the azimuth direction is something like

$$\frac{\lambda}{2\pi} \left(\frac{x \cos \tau}{\sin \theta} \right)^{1/2}$$

This should be, say, smaller than one-fifth the radius of the sphere in order to avoid cross-polarization effects, i.e.,

$$x \sin \theta / \cos \tau > 25.$$

The cross-polarization effects are rigorously expressed by the terms with $\pi_n(\cos \theta)$, which we have neglected (compare sec. 13.32).

Summarizing, we may state that the ray-optics approximation is sufficient if the effective area is small both in the radial and in the azimuthal direction. It becomes relatively large in the radial direction if we approach the rainbow and relatively large in the azimuthal direction if we approach the glory. At those particular angles the approximation fails for any x, however large, and higher terms have to be used. They give in the first case the rainbow and in the second case the glory.

12.35. More Powerful Approaches

We have now seen that it is possible, by means of asymptotical expansions and the method of stationary phase, to derive from the Mie solution a set of formulae identical to those following from ray optics. Is it possible to do more and to find a set of formulae that is distinctly more accurate than the ray-optics approximation, yet equally suitable for very large sizes? This important challenge has been taken up by several authors, notably by van der Pol and Bremmer, by Ljunggrén, and by Franz.

The research of van der Pol and Bremmer (1937) has been mainly directed to the problem of radio wave propagation from a vertical dipole antenna around the spherical earth. In their paper they give expansions for the potential functions in this problem in four different ways.

1. A "harmonic series" with terms of integer order n, associated with an expansion in spherical harmonics and thus analogous with the Mie solution that is the basis of our discussions on the scattering problem.

2. A "residue series" with terms of integer order s, n_s being the sth (complex) value of n for which

$$N_n(x, y) = 0.$$

Here $N_n(x, y)$ is the denominator appearing in the nth term of the harmonic series, its definition being extended to non-integer, non-real values of n. The $x = ka$ and $y = mka$ correspond to our x and y. The residue series is obtained from the harmonic series by a transformation which can only briefly be indicated here. First, the sum is replaced by a contour integral of the same function with added factor $1/\cos \pi n$; the values $n + \frac{1}{2}$, where $n =$ integer, are the poles of $1/\cos \pi n$. Then the contour is changed into an equivalent one in which these poles are avoided, but the poles of the original function, i.e., the zeros of $N_n(x, y)$ are enclosed. *These zeros are also the zeros of the denominator of our coefficients a_n* (sec. 9.22), if this coefficient is considered as a function of the complex variable n[3].

3. A development of the coefficient $1/\cos \pi n$ in the residue series in the form

$$2e^{i\pi n} \sum_{m=0}^{\infty} e^{i\pi m(2n+1)}.$$

It can be shown that the integer m corresponds to the number of times the wave has traveled around the earth. In the radio case all terms except the one with $m = 0$ may be neglected.

4. A development of the coefficients of the original harmonic series with respect to an integer number K, which corresponds to our $p - 1$, the number of internal reflections against the sphere. The development in two initial terms and a geometric series is entirely analogous to that given in sec. 12.33. The important difference is that van der Pol and Bremmer transform the *rigorous* coefficients in this manner. The Fresnel reflection coefficients are replaced by the "spherical reflection coefficients."

The great importance of this work of van der Pol and Bremmer is evident, for in several respects it goes beyond the approach first suggested by Debye and followed in sec. 12.33. Also, the method has proved to be practical for obtaining numerical results in the radio case. Although their method is potentially more powerful, van der Pol's and Bremmer's direct contribution to the topic of this book does not go beyond a third-order approximation of the Mie formulae, which after due approximations gives Airy's theory (sec. 13.23).

---

[3] Our $\psi_n(x)$, a notation copied from Debye (sec. 9.22), is the same as that adopted by Ljunggrén, but van der Pol and Bremmer use the same symbol for $j_n(x) = \psi_n(x)/x$ (similarly for $\zeta_n^{(1)}(x)$ and $\zeta_n^{(2)}(x)$).

Ljunggrén's contribution (1948) is less important, as it does not solve the real mathematical problems. He starts the transformation of the exact solution by separating the Mie coefficients exactly in two initial terms plus a geometrical series. This expansion is analogous to the one under 4 on p. 216 and to the expansion made in sec. 12.33. It must become identical to the latter if Debye's asymptotic forms for the cylinder functions for $(n + \frac{1}{2})/x$ well below 1 are introduced. The only difference with sec. 12.33 then is that the asymptotic forms are introduced before the expansion and not after it. Upon summation, the first term gives the diffracted light, the second term the externally reflected light, the third term (the first term of the geometrical series) gives the twice refracted light, etc., as was explained in sec. 12.31 and subsequently. Further, a rainbow theory is presented, based upon a transformation of the sum into an integral, but the result does not go beyond Airy's approximaton (sec. 13.23). In all these results the edge terms (n near x) and the terms beyond the edge ($n > x$) have been neglected.

Ljunggrén's second paper (1949) is aimed expressly at computing the forward scattering, and an effort is made to include the edge terms correctly. This requires the use of the "exceptional series of Debye" in making the asymptotic expansions. Yet after many reductions and approximations the twice-refracted radiation appears to follow from a formula that (at any rate for θ near 0) is identical with the formula following from geometrical optics (secs. 12.21 and 12.22) or from Debye's approximation to the Mie formulae (sec. 12.33). Ljunggrén remarks that the saddle-point method cannot be applied for the reflected radiation. Instead, he proposes to add the numerical sum of terms for $(x - x^{1/3}) < n < x$ to the analytical sum of terms for smaller n, according to Debye's approximation. Unfortunately no numerical results are given. The objection may be made that the terms with $x < n < (x + x^{1/3})$ are also in the exceptional region and may be expected to give a contribution of the same order. The problem of the forward, or nearly forward, scattered light owing to grazing reflecting at the edge of the sphere is thus unsolved. This problem will be reconsidered in sec. 17.26.

The final step of Ljunggrén's work is to add the contributions of diffracted, reflected, and refracted light at and near $\theta = 0$ and thus to find theoretical intensity contours for the anomalous diffraction patterns. This problem and its application to the determination of drop sizes are relatively simple and are discussed in sec. 13.41.

The most important development is due to Franz (1954). His work is a direct continuation of the work of van der Pol and Bremmer with one major improvement. The integral over the order n, which is obtained by exact transformations from the series in integer n, is not at once changed into a residue series but is first separated into two parts. One part is left in the form of an integral over n and corresponds, asymptotically for large x, to the value obtained by means of geometrical optics. The second part is transformed in a residue series and is physically connected with the surface waves discussed in secs. 17.3 to 17.5. The advantage of this separation is not only that the agreement with the results derived from geometrical optics is more readily seen but also that the remaining

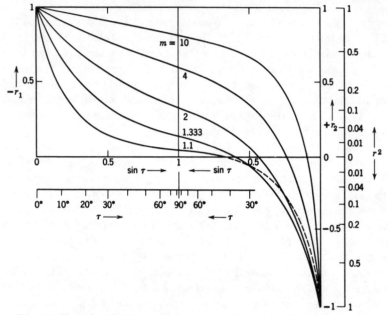

Fig. 39. Fresnel reflection coefficients r_1 and r_2 plotted against sin τ for external reflection against a body with refractive index 1.1, 1.333, 2, 4, or 10. The right-hand scale (r^2) gives the reflected fraction of the energy.

residue series are rapidly convergent. Franz's paper of 1954 contains full details of this method as applied to totally reflecting cylinders and spheres in the acoustical case. Beckmann and Franz have tackled optical diffraction by spheres and cylinders of arbitrary material. Full results are awaited with interest.

12.4. Some Special Results

The formulae for large spheres derived in the preceding sections are simple and straightforward. Yet the ray-tracing and the proper weighting of each ray with the Fresnel reflection coefficients are a laborious process. This is the reason why only for one refractive index, $m = 4/3$, computations of the geometrical-optical pattern have been made that are anywhere near complete.[4]   These results are reported in sec. 13.12, where a graphical illustration of the meaning of the parameters k, q, p, and s is also given (Fig. 41).   Miscellaneous results for other values of m will be reported here.

12.41.   Division of Energy among Rays

Numerical values of the Fresnel coefficients are given in Fig. 39 newly computed.   The presentation is unconventional but convenient for interpolation.   The coefficients for the two different directions of polarization are separately shown: the left half gives $-r_1$ and the right half gives $+r_2$, both plotted against a scale of $\sin \tau$.   Plotted this way the two curves form together one smooth curve.   The central part of the figure ($\sin \tau = 1$) corresponds with incidence at the center, i.e., perpendicular incidence; here $-r_1 = r_2 = (m-1)/(m+1)$.   The sides correspond to incidence at the edge, i.e., grazing incidence; here $r_1 = r_2 = -1$.   Scales of the angle of incidence, τ, and of the reflected fractions of the intensity r_1^2 and r_2^2 are added for convenience.   The Brewster angle, where $r_2 = 0$, can be found from the equation $\tan \tau = 1/m$.

It is of some interest to know just what fraction of the incident light is reflected against the surface of the sphere, what fraction is refracted by the sphere without an internal reflection, etc.   In the formulae these possibilities are denoted by the parameters $p = 0, 1, \cdots$ (Fig. 37).   The fraction f is found by means of

$$f_{1,2} = \int_0^1 \varepsilon_{1,2}^2 \, d(\cos^2 \tau).$$

Here ε^2 is the mentioned fraction for a narrow beam incident under the angle τ and $d(\cos^2 \tau)$ is proportional to the exposed fraction of the projected surface.   The integration has to be made separately for each direction of polarization and the results averaged:

$$f = \tfrac{1}{2}(f_1 + f_2).$$

Table 17 shows the values of f found by numerical integration for three values of m.   Errors in the numerical results are probably smaller than one-half per cent ($= 5$ units in the table).

Analytical integration is also possible, for substitution of the new variable $u = \sin \tau / \sin \tau'$ leads to rational integrals.   The result of the integration of the first two terms for polarization 1 is

$$\text{For } p = 0: f_1 = \frac{(3m+1)(m-1)}{3(m+1)^2}.$$

$$\text{For } p = 1: f_1 = \frac{8(5m^2 + 4m + 1)}{5(m+1)^4}.$$

[4] A corresponding computation for a big air bubble in water ($m = 0.750$) has since been made by G. E. Davis, J. Opt. Soc. Amer., **45**, 572 (1955).

Table 17. Distribution of the incident energy among light reflected externally ($p = 0$), twice refracted ($p = 1$), one internal reflection ($p = 2$), etc.
(The table gives $1000\, f$.)

	$m = 4/3$*			$m = 2$			$m = 4$		
	f_1	f_2	f	f_1	f_2	f	f_1	f_2	f
$p = 0$	102	30	66	260	62	161	520	212	366
$p = 1$	822	946	884	574	882	728	248	632	440
$p = 2$	62	20	41	120	52	86	118	118	118
$p = 3$	10	2	6	30	4	17	56	28	42
$p = 4$	2	1	2	8		4	26	8	17
$p > 4$	2	1	1	8		4	32	2	17

* More accurate values are given in Table 20 (sec. 13.11).

Numerical values based on these expressions agree satisfactorily with the results of numerical integration; they have been used in the construction of Table 17 and Table 20. Bucerius (1946) has given the same expression for $p = 0$ but a wrong formula for $p = 1$; his computation is also limited to polarization 1. Analytical integration for polarization 2 leads to complicated expressions.

Table 18. Distribution of the incident energy for refractive indices smaller and larger than 1
(The table gives $1000\, f$.)

$m =$	0	$\frac{1}{4}$	$\frac{1}{2}$	$\frac{3}{4}$	1	4/3	2	4	∞
Total Reflection	1000	937	750	438					
$p = 0$		23	40	37		66	161	366	1000
$p = 1$		27	182	497	1000	884	728	440	
$p = 2$		7	22	23		41	86	118	
$p = 3$		4	4	3		6	17	42	
$p = 4$		1	1	1		2	4	17	
$p > 4$		1	1	1		1	4	17	

12.42. Values of $m < 1$

The derivation given in sec. 12.21 did not include the possibility of a refractive index smaller than 1. This occurs, e.g., if the obstacle is an air bubble in water. The extension of the theory to this situation is relatively simple. The incident rays for which $\cos \tau > m$ give total reflection, for Snell's law gives an impossible value of $\cos \tau'$. The fraction f of the incident light that is totally reflected thus is $1 - m^2$. For the remaining fraction we use the rule that (apart from sign) the Fresnel coefficients remain the same if τ and τ' are interchanged and m replaced by $1/m$. So the fractions f for any p follow from the same integration as above; the only difference is that $\cos^2 \tau$ is replaced by $\cos^2 \tau'$, which by Snell's law is $m^{-2} \cos^2 \tau$. As a result, the *relative distribution* among various values of p

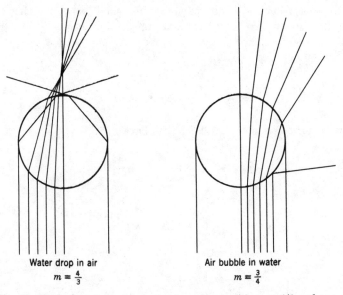

Water drop in air
$m = \frac{4}{3}$

Air bubble in water
$m = \frac{3}{4}$

Fig. 40. Equivalent geometrical rays in a sphere with $m = 4/3$ and one with $m = 3/4$.

is the same as it was for the inverse value of m, but the values of f are all scaled down by a factor m^2. Table 18 gives the resultant values.

The angular distribution of the reflected and refracted light cannot be found by adding a simple factor, for the angles θ', defined in sec. 12.21, become different if τ and τ' are interchanged. The only exception is in $p = 1$, where $\theta' = 2\tau - 2\tau'$. Here the angular intensity distribution is *exactly* the same for refractive indices m and $1/m$ except for a scale factor m^2 in the intensities. This is illustrated in Fig. 40.

12.43. Values of m Near 1

If m and $1/m$ closely approach 1, all light goes into $p = 1$. Then the *entire* scattering pattern becomes identical for these two values of the refractive index. This is a particular illustration (for large sizes) of the general rule that the scattering pattern for $m = 1 + \mu$ and $m = 1 - \mu$ are the same when $\mu \to 0$ (compare secs. 7.12 and 11.1).

The entire scattering pattern in this limit is computed as follows. We assume that $m = 1 + \mu$, where μ is positive and very small. The reflection coefficients tend to 0, and the terms with $p = 1$ remain and have $\varepsilon_1 = \varepsilon_2 = 1$. Neglecting higher orders of $\mu/\sin \tau$ we have

$$\tau' - \tau = \mu \cot \tau.$$

This approximation is not permitted for small $\sin \tau$, i.e., for grazing incidence, but the contribution of those rays is small. By further substitution we have

$$-\theta' = +\theta = 2\mu \cot \tau, \qquad k = 0, \qquad q = -1,$$

$$\frac{d\theta'}{d\tau} = \frac{2\mu}{\sin^2 \tau}, \qquad s = 1, \qquad \delta = -\frac{2x\mu}{\sin \tau}.$$

By the formulae in sec. 12.22 we then have

$$\sigma = -\frac{2x\mu}{\sin \tau} + \frac{3\pi}{2}$$

and

$$S_1(\theta) = S_2(\theta) = -\frac{x \sin^2 \tau}{2\mu} i e^{-2x\mu i/\sin \tau}.$$

Here τ is known as a function of θ; we can write

$$\sin^2 \tau = (1 + \theta^2/4\mu^2)^{-1}.$$

The final intensity distribution is found by squaring the modulus of S_1 and S_2:

$$i_1 = i_2 = \frac{4\mu^2 x^2}{(4\mu^2 + \theta^2)^2}.$$

This result shows that the refracted light forms a forward-directed lobe, which is narrower the smaller μ is. Further discussion of this case is found in sec. 11.31.

12.44. Values of m Near ∞; Metals

A quite different case occurs if m is so large that the reflection coefficients are

$$r_1 = -1, \qquad r_2 = 1$$

for all practical purposes and for all angles.[5] Then only the reflected light ($p = 0$) is important, and the further specifications in the formulae of sec. 12.2 are

$$\varepsilon_1^2 = \varepsilon_2^2 = 1,$$

$$k = 0, \qquad q = 1, \qquad s = 1,$$

$$\theta' = \theta = 2\tau, \qquad d\theta/d\tau = 2, \qquad \delta = 2x \sin \tau,$$

so that

$$-S_1(\theta) = S_2(\theta) = \tfrac{1}{2}ixe^{2ix \sin \tau}.$$

The squares of the moduli are $i_1 = i_2 = \tfrac{1}{4}x^2$, so the gain is constant $= 1$. This shows that the intensity of the reflected radiation is independent of direction: *a smooth totally reflecting sphere with radius large compared to the wavelength scatters light by reflection isotropically*. This rule holds for $m \to \infty$ as well as for $m \to 0$. It holds also for non-spherical particles with random orientation, provided that they are convex (sec. 8.42). In addition to the reflected light, however, we have the diffracted light with a narrow pattern in the forward angles.

In this connection the extension to metal spheres may be noted. All the derivations in this chapter were based on the assumption that m is real. The extension to complex m is not difficult. Two possibilities exist:

a. The absorption inside the sphere is so weak that a ray still has a measurable intensity after passing through the sphere. This means, in view of the assumed large size, that the complex part of m is $\ll 1$ (Type 3 in sec. 14.1). Ray-tracing and values of ε are then largely unaffected, but each ray is attenuated in accordance with the distance it has traveled through the sphere. The only change in the final formulae of secs. 12.2 or 12.3 is, therefore, the addition of a factor $e^{-2py a \sin \tau'}$ to the formulae for $S_1(\theta)$ and $S_2(\theta)$.

b. The other possibility is that the absorption is sufficiently large to absorb any refracted ray completely. This occurs if the sphere is metallic. Then only the reflected light ($p = 0$) remains, as it does for a totally reflecting sphere, but the amplitude and phase of the reflected light are functions of τ, i.e., of θ. They are found from Fresnel's formulae for r_1 and r_2 using the complex values of m. Perpendicular incidence gives partial reflection, but grazing incidence always gives reflection with $r_1 = -1$, $r_2 = 1$. The pattern of reflected light is therefore somewhat stronger in the forward direction but not nearly in such a pronounced way as the diffracted light.

---

[5] There still is a Brewster angle, for which $r_2 = 0$, but it occurs so close to grazing incidence that it is in a region where deviations from geometrical optics occur anyhow.

12.45. Forward Scattering

Mainly in view of the extinction formula, it is important to derive the formula for $S(0)$, complete with phase. This is made up of the following terms:

(a) Diffraction: $\frac{1}{2}x^2$ (secs. 8.31, 11.21, 12.32).

(b) Grazing reflection: $-\frac{1}{2}ix$ (sec. 12.44 with $r_1 = r_2 = -1$). However, this formula is incorrect, as it is not permitted to extend the ray-optics formulae to this case. See secs. 17.24 to 17.26 for better formulae.

(c) Nearly central rays: Here $r_1 = -r_2 = \dfrac{1 - m}{1 + m}$. Let $\tau = 90° - \gamma$, $\tau' = 90° - \gamma'$, where γ and γ' are very small. Then

$$\theta' = (1 - p)\pi + 2\gamma(-1 + p/m),$$

from which

$$k = \tfrac{1}{2}(1 - p), \quad q \text{ has the sign of } (-1 + p/m), \quad s = -q,$$

$$D = \frac{1}{4\left| -1 + p/m \right|^2}.$$

This gives for $p = 1$ (no internal reflection) and $m > 1$ a term of $S(0)$ equal to

$$-x \frac{2m^2}{(m + 1)(m^2 - 1)} ie^{-2ix(m - 1)},$$

which reduces to the form

$$-\frac{ix^2}{\rho} e^{-i\rho}$$

for $m \to 1$ (sec. 11.32). For $p = 3$ (two internal reflections) and $m < 3$ it gives

$$+xr_1^2(1 - r_1^2) \frac{1}{2\left(\dfrac{3}{m} - 1\right)} ie^{-2ix(3m - 1)}.$$

(d) Finally, a number of contributions to $\theta = 0$ may come from non-central and non-grazing rays, which requires at least two internal reflections, $p = 3$. These are the glory-like contributions (compare sec. 13.31), which are of small importance.

12.5. Radiation Pressure

Finally, *the radiation pressure* on large, dielectric spheres may be discussed. The theory in sec. 2.3 shows that computation of the radiation pressure requires the knowledge of the average value of $\cos \theta$.

For large spheres we may separate the effects of geometrical optics (reflection and refraction) and the effect of diffraction. The efficiency factors (cross section/πa^2), including the possibility of absorbing spheres, may be written as indicated in the table. This table defines g and w; w is

	Diffracted Light	Reflected and Refracted Light	Together
Efficiency factor for extinction	1	1	2
Efficiency factor for scattering	1	w	$1 + w$
Average of $\cos \theta$ for scattered light	1	g	$\dfrac{1 + wg}{1 + w}$

the whiteness or albedo, if diffraction is omitted. These values should be put into the general relation

$$Q_{pr} = Q_{ext} - \overline{\cos \theta}\, Q_{sca}.$$

It is seen that the diffracted light gives a zero contribution. This is correct: virtually no momentum is transferred to the sphere (or particle of other form) by the diffraction process. The scattered light, however, gives

$$Q_{pr} = 1 - wg,$$

i.e., for a non-absorbing sphere

$$Q_{pr} = 1 - g.$$

The numerical value of g may be computed on the basis of the intensity distribution derived in sec. 13.2. An obvious but lengthy way would be to derive first the complete intensity distribution, then multiply by $\cos \theta$, and integrate. A shorter method is to consider that any individual ray characterized by τ, p, and its direction of polarization, say 1, carries the amount of incident energy

$$I_0 a^2 \cos \tau \sin \tau \, d\tau \, d\varphi \cdot \varepsilon_1^2$$

to a direction for which

$$\cos \theta = \cos \theta' = \mathrm{Re}\,(e^{2i\tau - 2ip\tau'}).$$

The product of these two expressions has to be integrated over τ and φ, and to be summed over p. On writing ε_1^2 in terms of r_1 we find that the summation from $p = 1$ to ∞ can be made in the form of a geometrical series. The term with $p = 0$ remains separate. After a number of reductions (not reproduced) we obtain the expression

$$g_1 = \int_0^1 \frac{2r_1^4(1 - \cos 2\tau')\cos 2\tau + (1 - r_1^2)\cos(2\tau - 2\tau')}{1 - 2r_1^2 \cos 2\tau' + r_1^4}\, d(\cos^2 \tau)$$

and the same expression for g_2 with r_2 instead of r_1.

For incident natural light

$$g = \frac{g_1 + g_2}{2}.$$

The only published numerical results are in a small graph in Debye's thesis. From this graph we read

$m =$	1	1.1	1.333	1.5	2	3	4	∞
$1 - Q_{pr} = g =$	1.00	0.92	0.74	0.64	0.40	0.10	0.04	0.00
$\cos \theta = \frac{1}{2} + \frac{1}{2}g =$	1.00	0.96	0.87	0.82	0.70	0.55	0.52	0.50

More accurate values may be computed from the formulae given above, if needed. Generalization of these formulae to include any absorption inside the sphere is also quite simple. If the refracted rays are not completely absorbed the ratio between the terms in the geometrical series is changed by a known factor. If the refracted rays are completely absorbed, only the reflected light gives a contribution to g, and we have for one direction of polarization, e.g., direction 1,

$$w_1 = \int_0^1 |r_1|^2 \, d(\cos^2 \tau),$$

$$g_1 w_1 = \int_0^1 |r_1|^2 \cos 2\tau \, d(\cos^2 \tau),$$

and for the two combined, i.e., for incident natural light,

$$w = \tfrac{1}{2}(w_1 + w_2), \qquad gw = \tfrac{1}{2}(g_1 w_1 + g_2 w_2),$$

from which Q_{pr} may be computed as above.

References

The full correspondence between the results from ray optics and an asymptotic form of the Mie formulae was first presented in this form by

H. C. van de Hulst, "Thesis Utrecht," *Recherches astron. Obs. d'Utrecht*, **11**, part 1 (1946).

In its method it does not go beyond the principles outlined by

P. Debye, *Ann. Physik*, **30**, 59 (1909).

Though this is classical theory, occasional errors were made, e.g., by

R. Mecke, *Ann. Physik*, **62**, 623 (1920),

who did not correctly include the change of phase at the focus, and by

H. Blumer, *Z. Physik*, **38**, 920 (1926),

who failed to recognize the diffraction and reflection parts.

A method of greater mathematical rigor (sec. 13.35) was given by

Balth van der Pol and H. Bremmer, *Phil. Mag.* **24**, 141 and 825 (1937);
H. Bremmer, *Terrestrial Radio Waves*, New York, Elsevier (1949).

However, it was not worked out for scattering by a dielectric sphere beyond a result that can also be obtained by ray optics. This holds also for

T. Ljunggrén, *Arkiv Mat., Astron., Fysik*, **36A**, No. 14, (1948).
T. Ljunggrén, *Arkiv Fysik*, **1**, 1 (1949).

A more advanced discussion of the asymptotic behavior was given by

W. Franz, *Z. Naturforsch.*, **9a**, 705 (1954).
P. Beckmann and W. Franz, *Z. angew. Math. und Mechanik* (in press, (1956)).

The results in sec. 12.4 are simple specifications of the results of the preceding sections. In sec. 12.41 we quote

H. Bucerius, *Optik*, **1**, 188 (1946).

Section 12.5 represents the results on radiation pressure from Debye's thesis cited above, but with a simpler derivation and somewhat extended.

13. OPTICS OF A RAINDROP

Fog, rain, mist, and haze are a few examples of scattering media consisting of small spherical particles. The particles are water drops. Water has a refractive index ranging in the visual region from $m = 1.330$ (red, $\lambda = 0.7\ \mu$) to $m = 1.342$ (violet, $\lambda = 0.4\ \mu$). This is why so many computations have been made that refer to the particular values $m = 1.33$, or $m = 4/3$. Some numerical results are compiled in sec. 13.1. Further problems, most often associated with water drops but with wider theoretical implications, are discussed in secs. 13.2, 13.3, and 13.4.

13.1. Some Numerical Results

13.11. Scattering Diagram from Geometrical Optics

In this subsection the results of a computation based on the formulae of the preceding chapter are presented. The computation was made for $m = 4/3$. It gives the scattering diagram of a large water drop. The refinements needed near the angles for which the simple theory gives infinite intensity (rainbow, glory) will be discussed separately.

The first task is to compute the Fresnel coefficients for $m = 4/3$ for various angles of incidence τ. They may be taken from Fig. 39 (sec. 12.41). The numerical values are given in Table 19. The squares r_1^2 and

Table 19. Squares of Fresnel Reflection Coefficients for $m = 4/3$

τ	r_1^2	r_2^2	τ	r_1^2	r_2^2	τ	r_1^2	r_2^2
0°	1.0000	1.0000	10°	0.4572	0.2388	40°	0.0669	0.0005
1°	0.9239	0.8689	15°	0.3143	0.1107	50°	0.0433	0.0059
2°	0.8536	0.7545	20°	0.2200	0.0473	60°	0.0310	0.0120
3°	0.7888	0.6552	25°	0.1573	0.0172	70°	0.0245	0.0166
4°	0.7290	0.5687	30°	0.1153	0.0043	80°	0.0214	0.0195
5°	0.6740	0.4933	35°	0.0865	0.0001	90°	0.0204	0.0204

r_2^2 are the reflected fractions of the intensity of the incident light. This fraction is 0.0204 for perpendicular incidence and 1 for grazing incidence. The Brewster angle, where a wave with polarization 2 is not reflected, is at $\tau = 36.9°$.

228

The next step is to find the direction in which a ray leaves the sphere. This direction follows from the equation

$$\theta' = 2\tau - 2p\tau'$$

of sec. 13.21. Figure 41 gives the values of θ' as ordinate plotted against τ as abscissa for various values of p.

This figure also serves to specify the parameters k, q, and s defined in secs. 12.21 and 12.22. For any $p \geqslant 2$ a maximum of θ' is present, which is sometimes a maximum and sometimes a minimum of θ. These points, where the deflection of the ray is stationary with respect to small changes in τ, define the rainbows. By solution of the equation $d\theta'/d\tau = 0$ we find that these extremes of θ are reached when

$$\sin^2 \tau = \frac{m^2 - 1}{p^2 - 1}.$$

For $m = 4/3$ this formula gives the specification for the two main rainbows:

$$p = 2, \qquad \tau = 30.6°,$$
$$\theta = 138.0° \text{ (minimum)};$$
$$p = 3, \qquad \tau = 18.2°,$$
$$\theta = 128.7° \text{ (maximum)}.$$

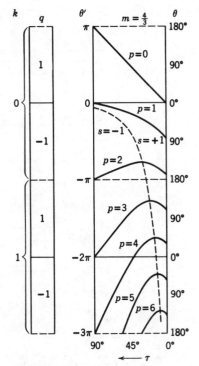

Fig. 41. Angle of emergence θ' versus angle of incidence τ for a geometrical ray with $p-1$ internal reflections passing through a drop with $m = 4/3$. The maxima define the rainbows.

We now have to collect all rays that emerge in the same direction, specified by the scattering angle θ. This is made convenient by Fig. 42, where the data of Fig. 41 are rearranged to show what angles θ occur for each set of parameters. The angle of incidence τ needed to give this θ must be read from the preceding figure. Values of θ for some points of special interest are written in the figure; c means a central ray (perpendicular incidence, $\tau = 90°$) e means an edge ray (grazing incidence, $\tau = 0°$), and r denotes a rainbow.

In order to obtain a survey of the distribution of energy among these rays, the author has divided the θ domain in 6 intervals of 30° and has computed for each interval the total scattered energy in any p. These energies are given in Table 20 as fractions of the total incident energy, for both polarizations separately. They were found by computing numerically $\int \varepsilon^2 \, d(\cos^2 \tau)$ between the proper limits of integration. For instance, for $p = 1$, the $0°$ to $30°$ scattering comes from the interval $90°$ to $39° \; 50'$ of τ For $p = 2$ and $p = 3$ one interval of θ corresponds to different intervals

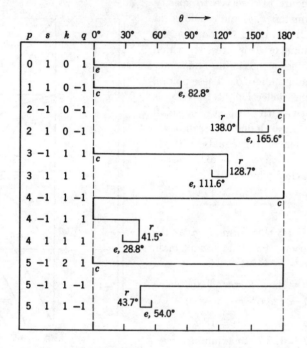

Fig. 42. Data from Fig. 41 in a diagrammatic form:
c = central ray, r = rainbow, e = edge ray.

of τ; the integrals for those intervals are presented separately. The sums of the integrals for a given p correspond to the entire integration interval of τ from $0°$ to $90°$. These sums are identical to those already given with less accuracy in Table 17 of sec. 12.41. Although it is simple to specify the computations needed for constructing a table like Table 20, it takes considerable time to perform them. The accuracy is 1 or 2 units in the last figure given.

The data contained in Table 20 show in a clear manner how the energy scattered by a water drop is distributed. Isotropic scattering would give a distribution proportional to the fractions of the total solid angle 4π that are contained in each interval of θ; these fractions are 0.067, 0.183,

0.250, 0.250, 0.183, and 0.067, respectively. Division of the totals in Table 20 by these numbers thus gives the average "gain" in each interval with respect to an isotropic scatterer. This average "gain" is about 9 in the first interval, less than 2 in the second interval, and 0.04 and 0.01 (for polarizations 1 and 2, respectively) in the interval from 90° to 120°. The enormous range of intensity in the scattered light makes it futile to illustrate the distribution by a polar scattering diagram. The best form is a plot of $\log i_1$ and $\log i_2$ against θ (compare Fig. 44 in sec. 13.12).

Table 20. Fractions of the Incident Energy Scattered by a Large Water Drop in Given Intervals of θ by Different Rays*

Interval of θ	0°–30°		30°–60°		60°–90°		90°–120°		120°–150°		150°–180°		All θ	
$p = 0$	321	184	342	62	194	2	101	18	48	28	14	13	1020	307
$p = 1$	5469	5776	2618	3429	130	251							8217	9456
$p = 2$									523	122	72	45	617	198
											22	31		
$p = 3$	0	0	1	1	2	0	11	0	82	20			98	23
							2	2						
$p = 4$			26	10									26	10
$p \geqslant 5$			15	4					7	2			22	6
All p	5790	5960	3002	3506	326	253	114	20	660	172	108	89	10000	10000

* In each column the first number refers to polarization 1 and the second to polarization 2.

We further want to know the scattered intensity, or "gain" with respect to an isotropic scatterer at each particular value of θ. For this purpose we should make the intervals $\Delta\theta$ in Table 20 ever narrower and find the limiting gain for $\Delta\theta \to 0$. It is doubtful whether this limit exists mathematically, for the intensity scattered by an infinitely large sphere has points of infinity at every rainbow, and it is likely that such points lie in any interval, however narrow. This objection, however, is not a practical one, for the infinities are removed by the finite size of a raindrop and, more important, only the first and second rainbows (with $p = 2$ and 3) contain appreciable energy. Table 20 shows that all further rainbows together contain less than one-half per cent of the incident energy, and the two strongest of those ($p = 4$ and $p = 5$) are located at angles in which the scattering by $p = 1$ is strong.

The narrowing of the intervals of θ and the division by the corresponding fraction of the solid angle were performed analytically in sec. 12.21. Table 21 was computed from the formulae $G_1 = 4\varepsilon_1^2 D$ and $G_2 = 4\varepsilon_2^2 D$, given in sec. 12.21. We thus find the entire scattering diagram for both directions of polarization. The intensities in the two "branches" that coincide for $p = 2$ and $p = 3$, and at the turning point of which the

rainbows occur, have been added in the table.   In those regions we have
to be particularly wary for interference effects; see sec. 13.22 for details.

Table 21. Scattering Pattern of a Very Large Water Drop, $m = 4/3$*

θ	$p = 0$		$p = 1$		$p = 3$ $p = 2$	
0°	1.000	1.000†	15.35	15.35		
5°	0.822	0.701	14.54	14.60		
10°	0.674	0.493	13.27	13.49		
15°	0.553	0.352	10.75	11.12		
20°	0.457	0.239	8.09	8.61		
25°	0.381	0.156	5.85	6.44		
30°	0.314	0.111	4.15	4.77	Very weak	
40°	0.220	0.047	1.951	2.501		
50°	0.157	0.017	0.838	1.257		
60°	0.115	0.004	0.270	0.482		
70°	0.086	0.000	0.051	0.112		
80°	0.067	0.000	0.000	0.001	0.001	0.000
90°	0.053	0.003			0.001	0.000
100°	0.043	0.006			0.002	0.000
110°	0.036	0.009			0.005	0.000‡
120°	0.031	0.012			0.026	0.016‡
130°	0.027	0.015	Empty		Empty‡	
140°	0.025	0.017			1.000	0.090‡
150°	0.023	0.018			0.264	0.183‡
160°	0.021	0.020			0.111	0.093
170°	0.021	0.020			0.082	0.075
180°	0.020	0.020			0.078	0.078

* The numbers give the "gain" relative to isotropic scattering, separately for
polarization 1 and polarization 2.

† Affected by diffraction and not separable from it.

‡ In region of first and second rainbows; these intensities are more strongly
dependent on size than the others.

For all other angles it is *in practice* permitted to add simply the various
intensities, e.g., for $\theta = 60°$:

$$G_1 = 0.115 + 0.270 = 0.385,$$

$$G_2 = 0.004 + 0.482 = 0.486.$$

The values for the central rays ($\tau = 90°$) follow a simple formula found
as a limiting case of the general equation in sec. 12.45, namely,

$$G_1 = G_2 = \varepsilon^2(1 - p/m)^2.$$

The edge rays give zero gain, except for $p = 0$, where we obtain $G_1 = G_2 = 1$ as for a totally reflecting sphere. Unlike the main contribution for $\theta = 0$, arising from $p = 1$, this value cannot be separated, in principle, from the diffracted light, as will be shown later (secs. 17.22 and 17.26).

Table 21 shows the same general features of the intensity distribution that appeared in Table 20. Interpolation between the tabulated values will give the intensity at any angle with an accuracy of a few per cent, except in the region of the rainbows. The work described in this section, and in particular the construction of Table 21, duplicates the work of Wiener. Both Wiener's computations and those of the writer were made mainly by slide rule. Comparison with Wiener's table, which gives the values $1/8\ G_1$ and $1/8\ G_2$, shows differences up to 8 per cent. Table 21 would seem to be more accurate. Wiener has smoothed out the rainbows by a theory about size effects, which has no good theoretical ground. Hence his results for $p = 2$ and $p = 3$ have no value.

13.12. Scattering Diagram from Mie's Formulae; Interpolation Problems

There is now an impressive body of tabulated numerical results based on the rigorous Mie formulae. A bibliography is given at the end of chap. 10. A selection of these data is presented in graphic form in Fig. 25, sec. 10.4. The tables by Lowan and associates and by Gumprecht and his coworkers are the most complete ones to date. They also are the most accurate. The bulk of the computations for the latter tables, terminating in the automatic printing of all Mie coefficients, was made by the electronic ENIAC machine. The Legendre functions $\pi_n(\theta)$ and $\tau_n(\theta)$ were computed with an IBM punched-card machine. The final multiplications and additions giving the scattering diagram, first complex amplitude, then intensity, were also made on a punched-card machine.

What *can* we do with these results and what do we *want* to do with them?

First, it is important to observe that the Mie formulae give the *exact solution* to the scattering problem; e.g., the values of i_1 and i_2 for $m = 1.33$, $x = 35$, $\theta = 130°$ give the intensities scattered by a perfectly smooth sphere of exactly that size at exactly that angle. They include all minute interference effects of rays penetrating or not penetrating into the sphere; in short, they are exact. Any approximate theory which claims to come close to the exact answer may be checked against them. So, where the approximate theory is incomplete, or has doubtful points, it may be possible to complete it by referring to the exact results. This will be done in particular with the extinction formula (sec. 13.42).

Second, these exact results are always *incomplete* because they have been tabulated for certain intervals of x (size) and certain intervals of θ (angle). Undoubtedly, the authors of the tables have hoped that the tabulated values would be representative, i.e., that similar intensities would hold for the non-tabulated values of x and θ. This is not always true, however, and therefore a careful *interpolation* is needed.

This is a problem of great practical value. Intensities and degrees of polarization computed from Mie's formulae have been quoted in the tentative interpretations of the zodiacal light and of the light scattered from the Venus and Mars atmospheres. It is clear that the interest in such cases is not centered on certain precise sizes but on wide ranges of x. In order to compute the scattering diagram from a cloud of particles with a given distribution of sizes an interpolation giving reliable values for all intermediate values of x must be performed first, and then an integration giving the required answer.

The qualitative effect of the interpolation is that the numerous maxima and minima shift in position with changing x. The net effect of the integration then is that these maxima and minima are washed out to a large extent. Only some of them, which are in a sense "bound to position" like the first diffraction rings and the main rainbow peak, may still be visible.

The preceding paragraph describes a quite familiar phenomenon. Interference fringes are lost unless the utmost care is taken to maintain the precise dimensions of the interferometer plate and precisely mono-chromatic light. The Mie formulae give something comparable to the exact interference pattern. The practical user of these formulae wishes to do away with most of this interference and wishes to see something like a true average.

We shall now investigate to what extent a reliable interpolation in the published tables is possible.

Interpolation in θ. The intervals in the Lowan and Gumprecht tables are 10° (except for the range close to 0°). The angular distance between successive maxima or between successive minima in the complete scattering diagram (i_1 or i_2 considered separately) is about $180°/x$. This means that the entire diagram from 0° to 180° contains approximately x maxima and x minima in the i_1 curve and similarly in the i_2 curve. This is a direct consequence of the number of minima and maxima in the functions $\pi_n(\cos \theta)$ and $\tau_n(\cos \theta)$.

For $x = 5$ or lower, the tabulated values define the complete diagram sufficiently, for 19 points suffice for drawing a curve with about 5 maxima. The curves shown in Fig. 25 (sec. 10.4) may serve as an illustration.

For $x = 10$ a free-hand curve drawn through a plot of i_1 (or i_2) against θ at 10° intervals already gives completely erroneous results. A more precise interpolation is obtained by making use of the complex amplitudes. In the tables of Gumprecht and Lowan the notation

$$i_1 = |\, i_1^* \,|^2 = (\mathrm{Re}\, i_1^*)^2 + (\mathrm{Im}\, i_1^*)^2$$

is used, and the real and imaginary parts are tabulated separately. We may plot both of them together in a diagram in the complex domain, or we may plot both separately against θ. Either method gives a more

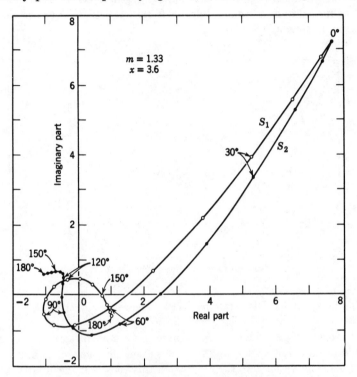

Fig. 43. Example of interpolation curves for the amplitudes of scattered light in the complex domain.

reliable interpolation; preferably they should be combined. The resulting curves in the complex domain are not very fit for illustrative purposes. Figure 43 shows, therefore, a set of curves for a smaller value of x, with fewer turns. One full turn, in which two maxima and two minima usually occur, takes an interval of θ of the order of $360°/x$.

The final scattering diagram for $m = 1.33$, $x = 10$, constructed by means of this method on the basis of Gumprecht's data, is shown in Fig. 44. The values were given at intervals of $10°$; it is clear that a direct interpolation would not suffice.

For $x = 20$ and larger it is likely that this method also fails to reveal all details of the scattering pattern. The only correct method seems to be to repeat the last part of the rigorous computation for more values of θ, choosing an interval of, e.g., 2.5°. Gumprecht and associates have plotted as an illustration the scattering diagram for $m = 1.33$, $x = 40$, but it is quite certain that more than half the maxima and minima are not shown in this diagram. And it is *entirely uncertain* whether the points that happened to be computed coincide fairly well with the average curve between maxima and minima.

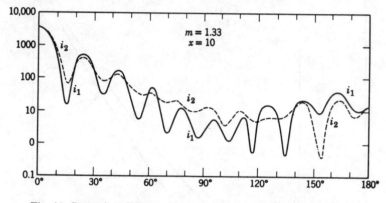

Fig. 44. Scattering diagram for $m = 1.33$, $x = 10$, based on inter-polation between Gumprecht's data.

This makes it hard to draw practical conclusions directly from these data.

Interpolation in x. The interpolation in size is more important but also more difficult. Let us start by the translation equations of Lowan and Gumprecht's notation into the notation of this book. If the notations of Lowan and Gumprecht are temporarily marked by a dagger we have the Mie coefficients:

$$A_n^\dagger = \frac{2n+1}{n(n+1)} \, i^{2n+1} a_n,$$

$$P_n^\dagger = \frac{2n+1}{n(n+1)} \, i^{2n-1} b_n.$$

Amplitude functions:

$$i_1^{*\dagger} = R(i_1^*)\dagger + iI(i_1^*)\dagger = -iS_1(\theta),$$
$$i_2^{*\dagger} = R(i_2^*)\dagger + iI(i_2^*)\dagger = +iS_2(\theta).$$

And our amplitude functions are expressed in terms of the tabulated quantities by means of

$$S_1(\theta) = -I(i_1^*)\dagger + iR(i_1^*)\dagger,$$
$$S_2(\theta) = +I(i_2^*)\dagger - iR(i_2^*)\dagger.$$

Further, the angle used in those tables is the supplement of θ:

$$\gamma\dagger = 180° - \theta.$$

We have seen in secs. 12.21 and 12.33 that $S_1(\theta)/x$ and $S_2(\theta)/x$ approach fixed values if $x \to \infty$, apart from phase effects and interference. So it is wise to make the interpolation between various values of x in terms of these functions.

As a test case the author has taken the light scattered at $\theta = 60°$ in the range $x = 3$ to 40 ($m = 1.33$). This range of x seems adequately covered in Lowan's and Gumprecht's tables. Figure 45 tells the whole story. Each section gives a plot of the amplitude function divided by x in the complex domain. Sections A, C, and E represent $S_1(60°)/x$; sections B, D, and F represent $S_2(60°)/x$. The scale is the same for each section; one small division is 0.1.

Sections A and B show the results taken from Lowan's and Gumprecht's tables in the range $x = 3$ to 10. Up to $x = 6$ it is quite easy to connect the given points by a smooth curve. The curves are considerably more uncertain between $x = 6$ and 10, but the general character of the curve, e.g., the number of loops, is beyond doubt.

Sections C and D show the data from $x = 10$ to 40 by intervals of 5, as tabulated by Gumprecht and associates. They seem to define a fairly smooth curve, and a naive interpolation would, undoubtedly, give smooth curves as shown. However, these curves are *incorrect*, as appears from two lines of reasoning. First, an extrapolation from sections A and B indicates that the curves, in both sections C and D, have to make approximately a full turn between $x = 10$ and 15. Second, the results must resemble the result that may be obtained from ray-optics. A clearer idea of the character of the curve may, therefore, be obtained by using the formulae for ray optics. Referring to Table 21 of the preceding section we find that only the rays with $p = 0$ and $p = 1$ are important. The specifications for $m = 1.33$, taken from the formulae in sec. 12.2, are

$p = 0$ (external reflection): $\tau = 30°$, $\sin \tau = 0.500$, $\cos \tau = 0.866$,

$$\varepsilon_1 = -0.338, \varepsilon_2 = -0.066, D = 0.250, \delta = x.$$

$p = 1$ (refracted ray): $\tau = 12.9°$, $\tau' = 42.9°$,

$$\sin \tau = 0.223, \cos \tau = 0.975,$$

$$\varepsilon_1 = 0.633, \varepsilon_2 = 0.848, D = 0.157, \delta = -1.366x.$$

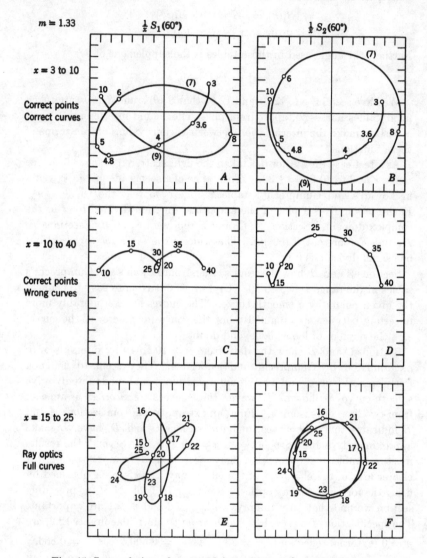

Fig. 45. Interpolation of scattered intensities at $\theta = 60°$ with size.

The final formulae, with proper phase, thus become

$$\frac{S_1(60°)}{x} = -0.169ie^{ix} - 0.252ie^{-1.366ix},$$

$$\frac{S_2(60°)}{x} = -0.033ie^{ix} - 0.338ie^{-1.366ix}.$$

The separate terms represent points in the complex domain moving along separate circles with different speeds and in opposite sense. The full curve can easily be constructed and is shown in sections E and F of Fig. 45 for the range $x = 15$ to 25.

What do these figures show? They show first of all that in this range the ray-optics approximation does *not yet* give a fully correct result, but that the phases are in fair agreement (when comparing C with E, or D with F we find all points in the proper quadrant). They further show that the correct interpolation curves between the points shown in C and D make a full swing around the origin between each two points plotted. They finally indicate that the initial impression that $S_1(60°)/x$ is relatively small throughout the range $x = 20$ to 30 is entirely incorrect. It just happens that in this range the multiples of 5 are at the low part of the swing, i.e., give an unfavorable interference of the reflected and refracted rays. But in this same range of x favorable interferences occur too, just as well as they are seen near $x = 8$, 10, and 40. This is a clear example of a situation in which even a complete inspection of the values tabulated by Gumprecht might still give a wrong impression of the overall orders of magnitude. We believe that this case is not exceptional. The steps of 5 in x are just too big to be sure of the interpolation curves. Generally, a reference to the ray-optics formulae, as sketched here, may be advisable.

The curve shown in section F is more typical for an angle with one dominant ray in the scattering pattern. Virtually all the radiation comes from $p = 1$ (the refracted ray), as is also seen in Table 21 of the preceding section. This means that interference effects are small (not quite absent) and that the value of $S_2(60°)/x$ is relatively stable. This gives us a chance to make a closer comparison of the correct amplitudes (section D) and the ray-optics amplitudes (section F). The first ones are seen to be approximately 1.4 times the last ones; this means that the true intensities are still 2 times the intensities computed from ray optics. A divergence of this amount is not surprising, for the value $\cos\tau = 0.975$ indicates that the geometrical ray is incident very close to the edge and is, for $x = 40$, still in the region where the approximations of sec. 12.33 break down.

13.2. The Rainbow

The rainbow is one of the most beautiful phenomena in nature. It has inspired art and mythology in all peoples, and it has been a pleasure and challenge to the mathematical physicists of four centuries. A person browsing through the old literature receives the impression that a certain affection for this problem pervades even the driest computations. The writer hopes that the following report, though it has to be concise and must leave out most of the history, will to some extent demonstrate this mathematical beauty

13.21. Description; Theory of Descartes

The correct explanation of the rainbow was given by De Dominis (1611). It was developed and experimentally confirmed by Descartes, and Spinoza first computed the position of the rainbow by analytical geometry. The deviation of the once internally reflected ray passes through a minimum when the angle of incidence is varied. Much of the light contained in an incident, parallel beam emerges quite close to this angle of minimum deviation and thus gives a very high intensity near that angle. The beauty of the rainbow is enhanced by the fact that this angle is slightly different for different colors because of the different refractive indices. Thus white sunlight gives a rainbow of intensely colored bands side by side. The refractive index $m = 4/3$ (orange light), to which our further computations refer, gives $\theta = 138.0°$ for the minimum scattering angle. So the rainbow is 42° from the antisolar point.

The maximum of red light is 1.5° further from the antisolar point than the maximum of violet light. We shall see that the maxima are widened for smaller drops. For drops with radii $a < 50\ \mu$ their width is several degrees, so that the colors largely overlap. This gives rise to the "white rainbow." For drops of any size a ray that is not the ray of minimum deviation emerges at larger θ, i.e., closer to the antisolar point, than this ray. Therefore, the inner side (violet side) of the rainbow is always bright.

The explanation of the second rainbow is similar. It is due to rays with two internal reflections. Its geometrical position is at $\theta = 128.7°$, i.e., 51.3° from the antisolar point. Here the bright side is towards smaller θ; the colors are wider apart and in inverted order. The region between the two rainbows is relatively dark. This is also clear from the schematic presentation in Fig. 42.

13.22. Interference; Mascart's Formula

Our further discussion is confined to the main rainbow (rays with $p = 2$). In the bright light adjoining the rainbow at the side of large θ one can often distinguish colored bands, known (by an odd term) as supernumerary bows. They are due to subsequent maxima and minima in the monochromatic intensity distribution. Two geometrical rays coincide at any angle in this bright part, with angles of incidence smaller and larger than the angle giving the minimum deviation. The supernumerary bows are an effect of optical interference of these two. It is not surprising that this explanation was first proposed by Young, who discovered the principle of interference of which this is the finest natural example.

As soon as interference effects come in we leave the domain of geometrical optics. The position of the maxima including the first one, which is the main rainbow, depends on the size of the droplet. Interference maxima and minima occur at any angle where Fig. 42 shows that several rays coincide. Usually, however, they are so close together and shift so rapidly with changing size or wavelength that they are not observed in nature. In the exceptional case of the rainbow they are wider apart and shift less rapidly. We shall illustrate this fact with a numerical example.

A. *Not near the rainbow.* The two rays giving together the intensity for polarization 1 scattered at $\theta = 60°$ are: $p = 0$, $\tau = 30.0°$, and $p = 1$, $\tau = 13.0°$. Their intensities (expressed as the gains with respect to isotropic scattering), taken from Table 21, are $a^2 = 0.115$ and $b^2 = 0.270$. If φ is the phase difference, the total intensity, found as the square of the absolute value of the sum of the complex amplitudes, will be $a^2 + b^2 + 2ab \cos \varphi$.[1] This varies between the maximum $(a + b)^2 = 0.738$ and the minimum $(a - b)^2 = 0.032$. So there is a contrast of a factor 23 between the intensities in maximum and minimum. Applying the formulae given in sec. 13.12 we further find

$$\frac{\partial \varphi}{\partial x} = 2.47, \qquad \frac{\partial \varphi}{\partial \theta} = 1.84x.$$

The first equation means that a change $\Delta x = 1.3$ is sufficient to shift the phase by π, i.e., to convert a maximum into a minimum. Drops with radius $a = 100 \, \mu$ have $x = 1000$. The result just found shows that less than a per cent dispersion in drop sizes or wavelengths obliterates these maxima and minima completely. The observed intensity is the average of maximum and minimum: $a^2 + b^2 = 0.385$. The second formula shows that the maxima are apart by $\Delta \theta = 360°/1.84x = 196°/x$, i.e., by only a fraction of a degree.

B. *Near the rainbow.* The interference maxima of the rainbow are wider apart than ordinary fringes and far less sensitive to changes in the

---

[1] Möbius refers to this expression as Mascart's formula (1892). This formula must have been known from the time of Fresnel.

size. We may anticipate from Fig. 47 that at $x = 1000$ the main rainbow differs by 2.5° from the next maximum and that the size has to be changed by 40 per cent in order to shift the second maximum by 1°; see Fig. 47. The main rainbow shifts less, the higher-order maxima more rapidly. The number of supernumerary bows visible gives a rough estimate of the dispersion of drop sizes.

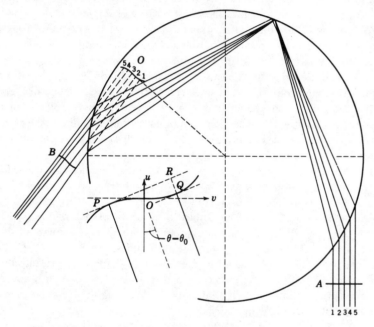

Fig. 46. Drawing in correct proportion of five equidistant rays which contribute to the first rainbow ($m = 4/3$). Insert: schematic drawing of cubic wave front at O.

We may now compute when the phase difference φ between the two rays that emerge at the same θ is $2\pi K$ and thus find the position of the maxima. Here $K = 0$ for the main rainbow, $K = 1, 2$, etc., for the further maxima. The part of the phase difference that is due to the difference in optical path length is

$$\Delta\delta = \delta_a - \delta_b,$$

where a refers to the ray incident closer to the center, b refers to the ray incident closer to the edge, and either δ is found from sec. 12.22:

$$\delta = 2x(\sin \tau - pm \sin \tau').$$

This part corresponds to the path difference RQ in Fig. 46. Yet, this is not the entire story (as Young in his 1804 paper assumed it to be), for we see from the figure intuitively that the phase difference between the "surroundings of Q" and the "surroundings of P" is smaller than $\Delta\delta$. This is indeed true. By the theory of sec. 3.21 there is an additional difference $\frac{1}{4}\lambda$ in the effective path, i.e., a difference $\frac{1}{2}\pi$ in the phase. In the formal representation of sec. 12.22 this additional term arises from the fact that s has a different sign for the two rays. The same section shows that a further reason for the phase difference may be a different sign of the Fresnel reflection coefficients of the two rays. This effect, which occurs only for polarization 2, is rarely important and is discussed in sec. 13.24. We now discuss only the distribution of light in the rainbow in polarization 1, which has by far the stronger intensity.

A maximum occurs in those directions for which the complete phase difference $\Delta\delta - (\pi/2)$ is an integer number, say K times 2π. So

$$\Delta\delta = (K + \tfrac{1}{4})\, 2\pi.$$

The same formula with $K = 1/2$, $3/2$, etc., gives the minima. The position of the maxima and minima for various x has been computed numerically from this formula and is shown in Fig. 47. Beyond $\theta = 165°$ evidently no interference exists because ray b is non-existent.

13.23. Cubic wave front; Airy's theory

So far we have used ray-optics. This means that we have assumed that the wave fronts near any point are sufficiently characterized by their normals and by their local radii of curvature. The approximation breaks down near the rainbow; the next higher approximation is a cubic wave front. The drawing in Fig. 46 shows five rays on scale, the middle one of which is the ray of minimum deflection. We have to apply Huygens' principle to a wave front at B just after these rays emerge. It is convenient but not essential, to replace it by a virtual wave front at O, where the tangent to the point of inflection goes through the center of the sphere. In this front the rays have virtually the same mutual distances as in the incident front at A. We use rectangular coordinates: v along this tangent and u backward along the normal (see inset of Fig. 46).

The equation of the wave front is

$$u = \frac{hv^3}{3a^2},$$

where a is the radius of the drop and h is a constant; $h = 4.89$ for the first rainbow (and $h = 27.86$ for the rainbow of two internal reflections). We may derive h as follows. Let θ_0 be the rainbow angle and τ_0 the angle of incidence belonging to it. Then for $\tau = \tau_0$ we have

$$\theta = \theta_0, \qquad d\theta/d\tau = 0$$

and by further differentiation

$$\frac{d^2\theta}{d\tau^2} = \frac{3}{2 \tan \tau_0},$$

so for neighboring angles

$$\theta - \theta_0 = \frac{3}{4 \tan \tau_0} (\tau - \tau_0)^2.$$

We further have $\qquad du/dv = \theta - \theta_0$

and

$$v = a \cos \tau - a \cos \tau_0 \approx -a(\tau - \tau_0) \sin \tau_0;$$

this gives

$$\frac{du}{dv} = \frac{3v^2}{4a^2 \sin^2 \tau_0 \tan \tau_0},$$

which by integration gives the postulated expression with

$$h = \frac{3}{4 \sin^2 \tau_0 \tan \tau_0} \ (= 4.89 \text{ for } m = 4/3).$$

Airy's theory further assumes a constant amplitude along the effective part of this front. The amplitude function for scattered light at any angle θ is then proportional to

$$\int_{-\infty}^{\infty} e^{-ikv(\theta - \theta_0) + ikhv^3/3a^2} \, dv,$$

where $k = 2\pi/\lambda$. Airy has defined and numerically computed the "rainbow integral":

$$f(z) = \int_0^{\infty} \cos \tfrac{1}{2}\pi(zt - t^3) \, dt.$$

Table 22. Maxima and Minima of the Rainbow Integral

	z	$f^2(z)$	K	$3(K + \tfrac{1}{4})^{2/3}$	$(4/3z)^{1/2}$
Main rainbow	1.08	1.005	0	1.19	1.11
	2.50	0			
First subsidiary maximum	3.47	0.615	1	3.47	0.620
	4.36	0			
Second subsidiary maximum	5.15	0.510	2	5.15	0.510
	5.89	0			
Third subsidiary maximum	6.58	0.450	3	6.58	0.450
	7.24	0			
Fourth subsidiary maximum	7.87	0.412	4	7.87	0.412

The amplitude just found can be brought in this form by combining the integrands for positive and negative v and by the substitutions

$$l = (3\lambda a^2/4h)^{1/3}, \qquad t = v/l,$$

$$z = 4l(\theta - \theta_0)/\lambda = (12/h\pi^2)^{1/3}x^{2/3}(\theta - \theta_0).$$

The geometrical meaning of l is the distance along the tangent from the point of inflection to the point where the front deviates by $\lambda/4$ from this tangent. The intensity in the rainbow is proportional to $f^2(z)$. The heights of the maxima and the values of z for which the maxima and minima are reached are given in Table 22.

To the wave front approximated by a cubic equation we may apply the same principle of interference applied in sec. 13.22 to the precise wave front. The result is that the direction of the normal is $\theta - \theta_0 = hv^2/a^2$ and the path difference (Fig. 46)

$$RQ = 4u = 4hv^3/3a^2.$$

By the same method as in sec. 13.22 this leads to the formula

$$z = 3(K + \tfrac{1}{4})^{2/3},$$

i.e.,

$$\theta - \theta_0 = h^{1/3}\{3\pi(K + \tfrac{1}{4})/2\}^{2/3}x^{-2/3}$$

for the position of the maxima. The comparison in Table 22 shows that this approximation formula is almost exactly correct for all maxima but the main one.

Airy's rainbow integral further gives the following information. The intensity at the geometrical position ($\theta = 138.0°$, $z = 0$) is about 45 per cent of the intensity in the main maximum. At the same distance in the other direction ($z = -1.0$) it has dropped to 5 per cent. The rainbow widens in the same proportion as it is displaced.

The factor in front of the $f^2(z)$ may be found in an unambiguous way by comparing the result from ordinary ray-optics (sec. 13.11) with the mean value of $f^2(z)$ for large z. We confine the argument to polarization 1. Developing near the rainbow ray we find

$$\left|\frac{d\theta'}{d\tau}\right| = 2.54(\tau - \tau_0) = 2.26(\theta - \theta_0)^{1/2}.$$

By taking for the other factors in G_1 the constant values for the rainbow ray and doubling this amount, since two rays are superposed (without interference), we find the gain with respect to an isotropic scatterer:

$$G = 2 \times 4\varepsilon_1^2 D = 0.208(\theta - \theta_0)^{-1/2}.$$

On the other hand the rainbow integral may be computed in the second approximation at each of its points of stationary phase, $t = \pm(z/3)^{1/2}$. The sum of the squares of these integrals is

$$(3z)^{-1/2} = 0.73x^{-1/3}(\theta - \theta_0)^{-1/2}$$

and corresponds again to the intensity of the two rays added without interference. The value of $f^2(z)$ in the maxima (Table 22) is twice this amount.

Comparing the two expressions found we have as the final formula of the complete Airy theory

$$G_1 = 2.85x^{1/3}f^2(z).$$

The corresponding value for polarization 2 is smaller by a factor 25 because ε_2^2 is so much smaller than ε_1^2.

13.24. Derivation from Mie's Formulae; Limitations of Existing Rainbow Theories

The limits of validity of the various theories are not given in the literature. Yet it is clear that *all* theories discussed so far are approximate. Only the Mie solution (chap. 9) is rigorous. Attempts have been made from time to time to give a more exact rainbow theory by deriving an asymptotic expression from Mie's formulae. These attempts have been successful in giving a result identical to the Airy approximation but have not gone beyond. The derivations differ from the derivation in sec. 12.33 (the quadratic approximation) only in going to one further term, i.e., to the cubic term in the Taylor expansion. The resulting formulae are complicated but confirm the Airy approximation in every respect. Also the results of van der Pol and Bremmer (1937) and of Ljunggrén (1948), who apply more exact methods of transformation (sec. 12.35) are reduced by various approximations to the same form. Bucerius (1946) has made an interesting attempt to include fifth-order terms.

In the more conventional way of attack, which was followed in the preceding pages, no drastic improvements have been made or suggested. An exception may be made for the numerical integration carried out by Möbius (1910, 1913), who applied Huygens' principle to the unapproximated wave front by numerical integration. The results of Möbius's computations for glass cylinders show that at least the *positions* of the maxima are not very different than in the simple interference theory. This holds even to some extent for the results for water drops in the range $x = 3$ to 30. Möbius made his integrations in this range of very small x by an unfortunate misconception of the limits of validity of his theory. Many years later one of his students, Rosenberg (1922), completed these computations by adding (for the same range of x) the second rainbow and the externally reflected light. Both authors proudly quote the huge numbers of numerical integrations that were performed, but the results must be regarded as useless.

Let us now review the various rainbow theories (simplified formulae or methods of approximate computation) and estimate their limits of application. This is perhaps the most useful contribution to further research and the only point that is new in this section.

First, let us look at the rays. Figure 46 is a scale drawing with rays incident at distances from the axis given by $\cos \tau = 0.78, 0.82, 0.86, 0.90,$ and 0.94. Ray 3 is the ray of minimum deviation; rays 2 and 4 deviate by 24' and 30', rays 1 and 5 by 1°20' and 2°50', respectively. The differences shown indicate that rays 1 and 5 are far beyond the range of the symmetric cubic approximation. The same follows from the difference of the coefficients ε_1^2, which are 0,134 and 0.062 for rays 1 and 5. We may fairly say that Airy's approximation is useful as a quantitative rainbow theory only if no rays with a deflection of more than about half a degree from the geometrical rainbow are involved. This again means that its use is restricted to those values of x for which the main maximum is shifted by less than say 20' from the geometrical position.

The validity of Airy's theory is thus limited to $x > 5000$, or with light of $\lambda/2\pi = 0.1\ \mu$ to drops with radii $> \frac{1}{2}$ mm. The domain of validity of Airy's theory is indicated by a black rectangle in Fig. 47. We may add that raindrops usually are sufficiently large and that also the experimental verifications (e.g., Möbius) were made with big spheres or cylinders.[2]

Outside the rectangle of Airy's theory we may still apply Huygens' principle; the basic assumption is that the laws of geometrical optics are correct from the incident front A to the emerging front B (in Fig. 46). This is expressed, among other things, by assuming that at incidence and emergence the reflection loss is the same. However, the light arriving at or near B originates from a length on the front A of the order of $(2\lambda a)^{1/2}$. The five rays shown, with their different values of the reflection coefficients, etc., are fairly independent if $(2\lambda a)^{1/2} < 0.08a$, which gives

$$a > 300\lambda \qquad \text{or} \qquad x = 2\pi a/\lambda > 2000.$$

Also this is a very severe limitation. Only for x this large and larger can we find the intensity distribution in the main rainbow maximum by numerical integration along wave front B (Möbius) and the intensities and positions of further maxima and minima by the interference method (Mascart). In all likelihood, however, nearly correct positions of the maxima may be derived for values of x far below this limit.

[2] Möbius further worried about a possible ellipticity and computed the geometrical position of the rainbow for an ellipsoid with arbitrary ratio of the axes and arbitrary direction of incidence.

Further limitations of the preceding theories are the following. The externally reflected light ($p = 0$) becomes stronger than the one branch of the rainbow light for $\theta > 158°$. Here the interference theory breaks down, or rather it has to be completed with the externally reflected light. The outer branch of the rainbow rays ends at $\theta = 165°$, but not abruptly. The asymptotic formulae for the cylinder functions that have to be used

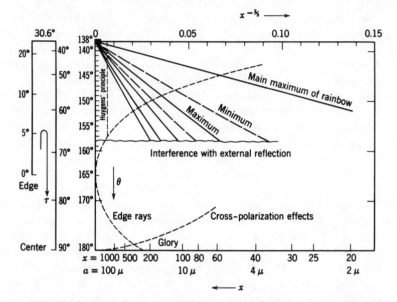

Fig. 47. Diagram for distinguishing the limits of validity of various rainbow theories. Ordinates: scattering angle (linear scale) and angle of incidence τ (reversing non-linear scale). Abscissae: $x^{-2/3}$ (linear scale) and $x = 2\pi a/\lambda$ (non-linear scale).

in order to transform the Mie coefficients into expressions containing the angle of incidence τ (sec. 12.33) break down near the edge. The exceptional region is given roughly by $\tau < x^{-1/3}$, and its border in Fig. 47 is indicated by a parabola. There is good reason to extend this parabola also downward, as we have done in the figure. Inside the lower half of this parabola "edge rays" may have some importance although by geometrical optics they do not penetrate there.

A final effect is the appearance of cross polarization owing to the fact that $\pi_n(\cos \theta)$ cannot be neglected with respect to $\tau_n(\cos \theta)$ in the Mie formulae (compare again sec. 13.3). Taking n of the order of x we find that $\pi_n(\cos \theta)$ is of the order of 10 per cent of $\tau_n(\cos \theta)$ if $\theta = 10x^{-1}$. This limit is also indicated in the figure. Below it cross-polarization effects appear. The region where edge rays *and* cross polarization are important may be interpreted as the region of the glory (sec. 13.32).

The conclusion is as follows. For $x < 30$ (and some points up to $x = 400$) exact computations from Mie's theory are feasible, and for $x > 2000$ Huygens' principle gives reliable results. A quantitative theory of the rainbow for the entire gap $30 < x < 2000$ is lacking, although the position of maxima and minima may be read from Fig. 47, and the distribution of intensity may roughly follow Airy's theory. So far only a guess can be made at the exact scattering pattern for both polarizations in the rainbow region near $x = 100$. A theory for the "gap" might possibly be made by judicious manipulation of further approximations to Mie's formulae. It might be checked by means of data from the tables by Gumprecht and associates.

After this somewhat disappointing conclusion an interesting phenomenon may be mentioned. So far we have only considered polarization 1. In the Airy approximation the intensity in polarization 2 is 0.04 times the intensity in polarization 1 and has the same intensity distribution. However, the sharp variation of r_2 with τ restricts the validity of Airy's approximation for this polarization even more than for the other one. As may be computed from Tables 20 and 21, the ratio of intensities between polarizations 2 and 1 has already climbed to 0.10 at $\theta = 140$ and to 0.23 for the total radiation in the range from 138° to 150°. Beyond $\theta = 150°$ the values of τ are so far from the Brewster angle that both branches have considerable intensity, so that interference maxima in polarization 2 are expected. However, the sign of the reflection coefficient is different for the two interfering rays. So, a maximum in polarization 2 occurs at the angle where we have a minimum in polarization 1, and inversely. This phenomenon was observed by Bricard (1940) in monochromatic light from a searchlight beam scattered by fog. Unfortunately, no precise data are given. He used drops of the order of $2a = 10\mu$, hence $x = 50$, and found the shifted maxima between angles $\theta = 150°$ and 160°, i.e., exactly in the domain of Fig. 47 where they are expected.

It is evident that a theory in which the polarization is entirely neglected and the reflection coefficients for natural light are used cannot give quantitatively correct results. Such theories are found in the work of Wiener (1909) and of Malkus, Bishop, and Briggs (1948).

13.3. The Glory

The diffraction coronae, or colored rings, which surround the sun or moon when covered by a thin veil of cloud, are well known. Their theory has been discussed in secs. 8.31 and 12.32 of this book.

A similar phenomenon can occasionally be observed in the opposite direction. A person standing on a high point observes his shadow projected on low clouds or on a layer of mist. He observes a gradual increase of the intensity of the reflected light towards the shadow's head, and if conditions are favorable some colored rings appear around the head. This phenomenon is called the *glory*. It is often seen in the mist on mountain tops and also frequently from aircraft. According to the position of the observer in the plane, the center of the rings shifts from the shadow of the head of the plane on the clouds to the shadow of its tail. Once or twice the glory has been studied in laboratory experiments.

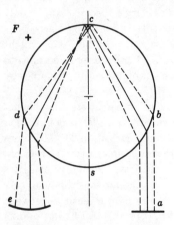

Fig. 48. Schematic diagram of a glory ray. The entire figure has to be rotated around the axis *cs*.

An old explanation of the glory is that it is due to ordinary diffraction of somehow reflected light around the foremost drops of the cloud. This explanation is incorrect, for (*a*) the geometry is unsound, (*b*) the relative diameters of the rings do not agree with ordinary diffraction theory, (*c*) the rings are strongly polarized. The correct explanation is that these peculiar fluctuations in intensity of nearly backward scattered light are already present in the scattering diagram of a single drop. The full theory, then, has to be based on Mie's formulae, but we shall see that the main features can be understood quite simply on the basis of Huygens' principle.

13.31. Radiation from a Torodial Wave Front

We have seen in sec. 12.1 that the ray-optics approximation gives infinite intensity ($D = \infty$) in one circumstance besides the rainbows, namely, for rays which have

$$\sin \theta = 0 \qquad \text{and} \qquad \sin 2\tau \neq 0.$$

The infinity can be removed by considering the actual form of the wave front. The condition just given implies the possibilities of $\theta = 0°$ and $\theta = 180°$. All peculiarities in the direction near $\theta = 0°$ (forward scattering) are blended by the much stronger coronae caused by diffraction of light around the sphere. Therefore, we shall confine our attention to the region near $\theta = 180°$, i.e., to the glory.

Figure 48 shows the simplest example of a light ray *abcde* satisfying the

condition required.   Two adjacent rays, which emerge at slightly different angles, are also drawn.   The linear front of the incident plane wave thus is transformed into the circular front of the emergent wave, which has a virtual focus at F.   The whole figure must be rotated around the axis cs, and all outgoing rays, thus defined, interfere with each other.   They define a *toroidal wave front*, which seems to emerge from the focal circle described by F.   The corresponding interference pattern may be found from Huygens' principle.

The most interesting feature of the present problem is that the two directions of polarization cannot be treated separately.   Let the incident wave be plane-polarized with its electric vector vibrating in the plane of Fig. 48.   Then, for the rays drawn, it possesses a parallel vibration and emerges with an amplitude containing the factor ε_2.   In the perpendicular plane, however, the same wave

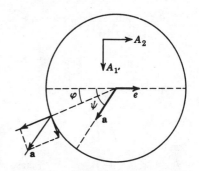

Fig. 49. Decomposition of the electric vectors in the elementary glory theory.

appears as a perpendicularly vibrating wave so that it is transmitted with an amplitude proportional to ε_1. Both transmitted waves are still vibrating in equal directions and are capable of interference.   The interference of rays in different azimuthal planes thus simulates an interference of rays with different directions of polarization.

Figure 49 explains the symbols in our analysis.   The circle represents the focal circle from which the toroidal wave seems to emerge; let its radius be r'.   We compute the intensity radiated in a direction which makes a small angle γ with the axis, and the projection of which on the plane of Fig. 49 is directed towards the right, as shown by the arrow e. The components of the light vector **A** of the emergent wave which vibrate parallel to and perpendicularly to the plane through the axis and e are denoted by A_2 and A_1, respectively.

Let the incident light first be linearly polarized in a direction **a** making an angle ψ with the fixed direction mentioned.   The light emerging from an arbitrary point of the focal circle, as specified by the angle φ, consists of two components which enter with the amplitudes $\cos(\psi - \varphi)$ in the out-ward radial direction and $\sin(\psi - \varphi)$ in the counterclockwise tangential direction.   The amplitudes with which these components emerge are

$$\text{Radial: } C_2 \cos(\psi - \varphi),$$

$$\text{Tangential: } C_1 \sin(\psi - \varphi),$$

where the constants C_1 and C_2 are proportional to ε_1 and ε_2, respectively. To derive the amplitude vectors of the total emergent light wave, we must decompose the emerging vector at each point of the circle into components with directions parallel to A_1 and A_2. These are

$$a_1(\varphi) = (C_1 \cos^2 \varphi + C_2 \sin^2 \varphi) \sin \psi - (C_1 - C_2) \sin \varphi \cos \varphi \cos \psi,$$

$$a_2(\varphi) = (C_1 - C_2) \sin \varphi \cos \varphi \sin \psi - (C_1 \sin^2 \varphi + C_2 \cos^2 \varphi) \cos \psi.$$

Writing $r' \sin \gamma \approx r'\gamma = u$, we now find the total amplitudes of the emergent light to be

$$A_{1,2} = \frac{1}{2\pi} \int_0^{2\pi} e^{-iu \cos \varphi} a_{1,2}(\varphi) \, d\varphi.$$

The definite integrals needed to reduce this expression,

$$\frac{1}{2\pi} \int_0^{2\pi} e^{-iu \cos \varphi} \cos^2 \varphi \, d\varphi = \tfrac{1}{2}\{J_0(u) - J_2(u)\},$$

$$\frac{1}{2\pi} \int_0^{2\pi} e^{-iu \cos \varphi} \sin^2 \varphi \, d\varphi = \tfrac{1}{2}\{J_0(u) + J_2(u)\},$$

$$\frac{1}{2\pi} \int_0^{2\pi} e^{-iu \cos \varphi} \sin \varphi \cos \varphi \, d\varphi = 0,$$

are related to Sommerfeld's integral defining the Bessel functions.

Incident natural light can be considered as a superposition of non-coherent, linearly polarized waves with randomly distributed directions of vibration. Averaging, therefore, the intensities in respect to ψ, we find a factor $\overline{\sin^2 \psi} = \overline{\cos^2 \psi} = \tfrac{1}{2}$ both in A_1^2 and in A_2^2. With omission of the factor $1/8$, the final intensities are

$$I_1 = [C_1\{J_0(u) - J_2(u)\} + C_2\{J_0(u) + J_2(u)\}]^2,$$

$$I_2 = [C_2\{J_0(u) - J_2(u)\} + C_1\{J_0(u) + J_2(u)\}]^2,$$

and in both directions of polarization together

$$I = 2(C_1 + C_2)^2 J_0^2(u) + 2(C_1 - C_2)^2 J_2^2(u).$$

These formulae show the full implications of the interference effect described. If no interference between rays in different azimuthal planes took place, the intensities $I_{1,2}$ would contain only the coefficients $C_{1,2}$ with the same index. This holds for large angles in the present theory, for the factor $J_0(u) + J_2(u)$ decreases like $u^{-3/2}$, leaving

$$J_0(u) - J_2(u) = (8/\pi u)^{1/2} \cos(u - \pi/4)$$

as the dominant term, so that I_1 contains C_1 and I_2 contains C_2. In fact, with fairly large angles γ the "effective area" of the wave front shrinks to two patches on the toroidal surface around the two points at which the normal to this surface has the required direction. A simple theory of the interference of two rays in the same vein as Mascart's theory of the rainbow (sec. 13.22) will then suffice. It would also be possible in this way to find the common intensity factor in front of the expressions just derived.

13.32. Theory Based on Mie's Formulae

The theory of the glory from Mie's formulae is outlined below. Let $\gamma = \pi - \theta$ be small. The asymptotic forms of the Legendre functions then are (compare the corresponding expressions for small θ in sec. 12.32)

$$\pi_n(\cos\theta) = (-1)^{n-1}\tfrac{1}{2}n(n+1)\{J_0(z) + J_2(z)\},$$

$$\tau_n(\cos\theta) = (-1)^n\tfrac{1}{2}n(n+1)\{J_0(z) - J_2(z)\},$$

with $z = (n + \tfrac{1}{2})\gamma$. The amplitude functions $S_1(\theta)$ and $S_2(\theta)$ (sec. 9.31) in the exact backward direction have the form:

$$S_1(180°) = -S_2(180°) = \sum_{n=1}^{\infty} \tfrac{1}{2}(2n+1)(-1)^n(-a_n + b_n).$$

When we are dealing with large values of x, of the order of 100 or 200, most terms in this sum will nearly cancel each other, but we may suppose that near some value N the terms of the order

$$\cdots, N-3, N-2, N-1, N, N+1, N+2, N+3, \cdots$$

have nearly equal phases, so that an appreciable total amplitude results. This "stationary phase" defines a ray in geometrical optics; hence we have just (although not very precisely) translated the basic idea of Fig. 48 into the Mie theory.

With this supposition made, let us define

$$c_1 = \sum_n (2n+1)(-1)^n b_n \qquad \text{and} \qquad c_2 = \sum_n (2n+1)(-1)^{n-1} a_n,$$

where the sums are taken only over the terms in the immediate neighborhood of $n = N$ and further terms are neglected. The total amplitude functions for any small value of γ then become

$$S_1(180° - \gamma) = \tfrac{1}{2}c_2\{J_0(u) + J_2(u)\} + \tfrac{1}{2}c_1\{J_0(u) - J_2(u)\},$$

$$-S_2(180° - \gamma) = \tfrac{1}{2}c_1\{J_0(u) + J_2(u)\} + \tfrac{1}{2}c_2\{J_0(u) - J_2(u)\},$$

where $u = N\gamma$. The intensities in both directions of polarization are proportional to the squares of the absolute values of these functions. The result thus found is in perfect agreement with the formula derived in the preceding section. In particular we can now see that the appearance of a term with the "alien" Fresnel coefficients in the rigorous formulae near $\theta = 0°$ and $180°$ is due to a quite simple interference effect. This solves a question which was first met in sec. 12.33.

It would be possible to work out more precisely the coefficients c_1 and c_2 in this section and C_1 and C_2 in the preceding section and thus to prove the exact correspondence in the limiting case of very large spheres. This derivation will be omitted (there is no reason to expect any difficulties) for we have to face a more severe problem. A ray like that drawn in Fig. 48 does not exist for $m = 1.33$! This ray exists only for m between $\sqrt{2}$ and 2, as indicated by Table 23. The simplest ray satisfying the conditions for

Table 23. Angles of Incidence Needed for Backscatter after a Single Internal Reflection

m	τ	$\cos \tau$
2.0	90°	0
1.9	54°	0.59
1.8	38°	0.78
1.7	26°	0.90
1.6	16°	0.96
1.5	7°	0.999
1.41	0°	1.00
1.33	—	—

$m = 4/3$ is the ray specified by $\tau = 56°$, $p = 5$ (4 internal reflections). It is certain that this ray has nothing to do with the glory observed in nature, for an estimate of the total intensity available in $p = 5$ (compare Table 20) shows that this is very unlikely to give anything observable at all.

A much more probable explanation is that Fig. 48 is still closest to the truth. Although in the limit $x \to \infty$ (ray optics) a ray like this is impossible for $m = 1.33$, its influence may be felt for smaller values of x. The edge region in which ray optics breaks down (secs. 12.33 and 17.2) may be defined by

$$\sin \tau \approx \tau < x^{-1/3}, \qquad 1 - \cos \tau \approx \tfrac{1}{2}\tau^2 < \tfrac{1}{2}x^{-2/3}.$$

The glory ray for $m = 1.5$ (see Table 23) falls in this edge region for $x < 350$. It is quite likely that for similar values of x, say from 100 to 300, the edge rays for $m = 1.33$ also define a glory. The analysis in sec. 17.42 suggests that in this case we may visualize the light path by a ray, as in Fig. 48, but connected with one or more small pieces of a surface wave, as shown by Fig. 86. The status of the theory may be summarized as follows. The parameters in the theory of sec. 13.31 are r' and C_1/C_2. Since very probably rays close to the edge are involved, r' is nearly equal to a, and the argument u of the Bessel functions is defined in the same manner as for ordinary diffraction, viz.:

$$u = x\gamma = 2\pi a\gamma/\lambda,$$

where $a =$ the radius of drop. The value of C_1/C_2 is quite uncertain. The classical theory with the Fresnel coefficients is of no use at all. The Mie formulae may help us out only so far as it would be possible to judge from numerical computations in this range of x what the values of C_1 and C_2 are. Dr. F. Volz wrote the author that the results of Gumprecht agreed with $C_1/C_2 = 0.1$ to 0.3 for $x = 10$ to 30 and with $C_1/C_2 = 0$ for $x = 40$. The theory of the surface wave also gives a definitely stronger intensity for polarization 2; in fact, $C_1/C_2 = 1/n^2 = 0.55$ under conditions that are not directly applicable in the present problem.

13.33. Numerical Results and Comparison with Observation[3]

In the absence of a reliable theoretical determination of the ratio of C_1 to C_2, we may try to estimate this ratio from the observations. The computed results for a few ratios are:

1. $C_1 = C_2$. The glory is wholly unpolarized. The central field is very luminous, and dark rings appear at $u = 2.5$, 5.6, 8.7, 11.8, $\cdots$.

2. $C_1 = -C_2$. The glory is again unpolarized. The antisolar point is dark and is surrounded by a luminous ring at $u = 3.1$. Dark rings are situated at $u = 5.2$, 8.5, 11.6, $\cdots$. Only this ratio was considered in the paper by Bucerius.

3. $C_1 = 0$. The intensities for this case are shown by Fig. 50. . In the central field the "alien" polarization, index 1, is preponderant, except at the antisolar point itself. At $u = 2.3$ we find a fairly dark ring in which the polarization changes its sign. The bright ring at $u = 3.5$ and all further rings are nearly completely polarized in the "proper" direction, index 2. They are separated by dark rings at $u = 5.4$, 8.6, 11.7, $\cdots$.

Data concerning glories on natural clouds have been collected by Pernter and Exner (1910). Measurements on artificial mist, together with qualitative observations of the polarization, have been published by Mierdel (1919). These data, though not obtained with modern observational techniques, indicate clearly that case 3 ($C_1 = 0$) is close to the actual value (see also sec. 20.3).

The glory shows distinct differences from common coronae, all of which can be explained on the basis of the present theory.

1. Their variability: It is obvious that the interference of rays which are refracted by opposite sides of a water drop will be much more sensitive to slight deformations of the droplets than are the rainbows in which only adjacent rays interfere, or the coronae in which non-refracted rays interfere.

2. The outer rings of the glory are much more pronounced than are the outer rings of common coronae. On one occasion as many as 5 minima

[3] See also sec. 20.3.

could be observed. The explanation is that the intensity in the glory decreases proportionally to γ^{-1}, whereas common coronae follow a θ^{-3} law.

3. A final striking feature is the haziness of the first dark ring (which for normal sunlight is observed as the first red ring). Mierdel calls it a slight depression separating the inner and outer parts of the central field. This haziness can be understood if the intensity distribution of Fig. 50 is approximately correct. A further increase of the first bright

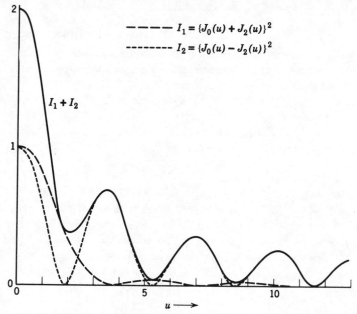

Fig. 50. Intensity and polarization in the glory on the assumption that
$$C_1 = 0.$$

ring is obtained if we approach case 2, where C_1 and C_2 have opposite signs. Estimating from Mierdel's description that

$$0.30 < \frac{\text{brightness of first ring}}{\text{brightness of central field}} < 0.80,$$

we find the value of either C_2/C_1 or C_1/C_2 to lie between 0 and -0.25.

Further information is obtained from Mierdel's observations of the polarization. From the fact that the picture rotates with the analyzing Nicol we infer that circular symmetry existed, so that the glass plates in Mierdel's experiment cannot have had a serious effect. The complementary colors of the central field and the rings indicate different planes of polarization, in agreement with Fig. 50. The preponderant polarization in all bright rings is the one in which the electric vector vibrates in the radial direction. This shows that $C_2 > C_1$, disregarding signs. Still, the tangential component is also visible. Estimating its intensity to exceed 4 per cent of the total intensity of the rings, we find $C_1^2/C_2^2 > 0.04$. Together with the former data this indicates that C_1/C_2 must be about $-\frac{1}{4}$ or $-\frac{1}{5}$. This estimate is based on laboratory data. No observations of the polarization of natural glories seem to have been published. Incidental observations from aircraft, kindly reported to the writer by Dr. O. Struve and Dr. M. Minnaert, confirm this state of polarization. A systematic investigation might give valuable additions.

Finally, the radii of the rings may be examined. As is true for common coronae, the radii of the rings are inversely proportional to the sizes of the droplets. However, it was noticed long ago that the rings of the glory do not give consistent sizes when interpreted with the older theory. Let us denote the radii of the dark rings by γ_1, γ_2, γ_3, etc. For $C_1 = 0$, the present theory gives

$$\gamma_1 : \gamma_2 : \gamma_3 : \gamma_4 = 2.3 : 5.4 : 8.6 : 11.7 = 0.43 : 1.00 : 1.60 : 2.17.$$

For other values of C_1/C_2 the ratios are not very different. Table 24 shows the observed and computed ratios.

Table 24. Ratios of the Radii of the Dark Rings

Observer	Light	Cloud	γ_1/γ_2	γ_3/γ_2
	White	Artificial	0.34 ± 0.05	1.66 ± 0.03
Mierdel	Red	Artificial		1.68 ± 0.04
	Photographic	Artificial		1.59
Average of several observations	White	Natural	0.46 ± 0.05	1.67 ± 0.06
Wegener	Photographic	Natural	0.41	(1.74)*
	$C_1 = 0$		0.43	1.60
Theory	$C_1/C_2 = -0.25$		0.35	1.61
	$C_1/C_2 = -1$		0.61	1.37

* Uncertain.

The agreement between theory and observations is as good as can be expected. No correction has been applied for the size of the sun, and it was assumed that the edges of the red rings, which were measured by the observers, indicate the positions of the dark rings in the monochromatic picture for yellow light.[4] This introduces considerable uncertainty, but the ratio $\theta_2/\theta_1 = 1.80$, valid for the common diffraction coronae, differs strongly enough from the tabulated values to indicate again that that theory does not apply to the glory.

[4] See Pernter-Exner, p. 460.

On the basis of the present theory, measurements of the various dark rings give consistent sizes for the droplets. From the average diameters, which are about $2\gamma_1 = 1° 30'$, $2\gamma_2 = 3° 50'$, $2\gamma_3 = 6° 30'$, we find $x = 160$ for yellow light, i.e., 0.028 mm for the average diameter of the water drops in the clouds on which the glory has been observed.

13.4. Diffraction and Extinction

The scattered amplitude at and near $\theta = 0$ requires special attention and is discussed in this section.

13.41. Anomalous Diffraction by Water Drops

Mecke (1920) was the first to call attention to the fact that there is a range of drop sizes for which the refracted and reflected light in the directions around $\theta = 0$ and the diffracted light have amplitudes of comparable magnitude. He also suggested that the interference of these three components would give rise to the effects of anomalous diffraction.

Since then, a precise discussion of this problem has been given in the limit of m near 1. The externally reflected ray may then be neglected, and only two components, the diffracted and refracted light, interfere. In addition, polarization effects vanish in this case. Full numerical results are given in sec. 11.33. It will be shown below that those results are quite reliable even for $m = 1.33$, if the approximate displacement of maxima and minima or a qualitative idea of their change in amplitude are sought. Having obtained this much, however, it is natural to ask for more and to see whether a graph corresponding to Fig. 34 can be constructed accurately for $m = 1.33$ and for both polarizations.

The simplest way to approach the problem is by Mecke's assumptions. The answer is then ready in sec. 12.2, where amplitudes and phases of the three components are given. Numerical values of the intensities for refraction and reflection have been compiled in Table 21 (sec. 13.11), and those for the amplitude of the diffracted light may be found in Table 6 (sec. 7.4). We just have to put the proper factor in front, add the correct phase, and make the summation.

When Mecke first did this, he made a phase error in neglecting the change of phase at the passage of a focus. Later computations on the same principle but with correct phases were made by Bricard (1946) and by Ljunggrén (1949). Numerical computations in the range $x = 19$ to 30 were made by Ramachandran (1943), but the author has not had the occasion for a detailed comparison.

Bricard (1946) starts his paper with the simple addition of intensities, but in later sections he also computes the phases and adds the three

components with the proper phases without considering the two polarizations separately. Table I of his paper gives the resultant intensity without
phase effects and Table II with phase effects; both tables cover the range of
$x = 2\pi a/\lambda$ from 10 to 37 and of $x \sin \theta$ from 0 to 10.4.

Ljunggrén (1949) has been more ambitious and has not, a priori, been
content with the approximations that may be obtained from geometrical
optics. But after a number of manipulations with the Mie formulae the
refracted component appears to come to precisely this approximation
again with the correct phase (see also sec. 12.35). Ljunggrén states
correctly that a similar transformation of the reflected light is not possible,
as the Debye approximations are different for terms with n near x (nearly
grazing rays). This point had not been made by Mecke or Bricard.
Ljunggrén's own solution for the reflection component is incomplete, and
was unfortunately published without details. This problem is really
quite difficult, as will be shown in chap. 17. Fortunately, the reflection is
numerically insignificant, except in the main diffraction peak, which is
made noticeably higher by it than would follow from the simplistic theory
(secs. 13.42 and 17.26). The even more ambitious approach by Hart and
Montroll has been reviewed in sec. 11.6.

The more recent computations based on the Mie formulae for $x = 15$,
20, 25, 30, 35, 40 can be compared directly with the approximate theories;
this will be done in two figures. They are too widely spaced, both in x and
in θ, to provide, in themselves, sufficient numerical material.

Figure 51 shows the diffraction pattern for three sizes, $x = 30, 35, 40$
(which corresponds to drop radii of 3 μ, 3.5 μ, 4 μ if $\lambda = 0.63 \mu$). They
are given as plots of amplitude vs. $x \sin \theta$, and the actual values of θ are
also indicated. The *square* of the ordinates is proportional to the intensity,
so that the real contrasts are much stronger than suggested. The ordinates
are

$$A_1 = \frac{|S_1|}{x^2} = \frac{\sqrt{i_1}}{x^2}, \quad A_2 = \frac{|S_2|}{x^2} = \frac{\sqrt{i_2}}{x^2}.$$

Points and curves for the separate polarizations have not been drawn, as
their difference was invariably smaller than could be represented in the
figure: $|A_1 - A_2| < 0.01$. Each section of Fig. 51 shows three curves:

a. The curves connecting the rigorously computed points from the
paper by Gumprecht and co-authors. The points are at 1° intervals from
$\theta = 0$ to 10°.

b. The curves of A for $m \to 1$ at the values of $\rho = 2x(m - 1) = 2x/3$.
These amplitudes were defined and computed in sec. 11.3, and the thin
curves in Fig. 51 have been drawn by reading the values directly from the
altitude chart (Fig. 34).

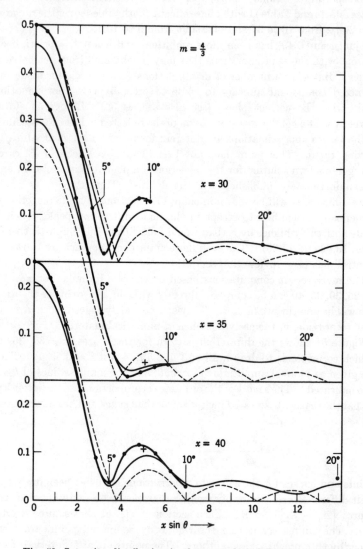

Fig. 51. Intensity distribution in the anomalous-diffraction patterns for water drops with $x = 30$, 35 and 40. The average intensity of both polarizations is given. Thick line and dots: rigorous data for $m = 4/3$. Crosses: approximation formula of Ljunggrén for $m = 4/3$. Thin line: curves for m near 1. Dotted line: Fraunhofer diffraction formula.

c. The curve that represents the scattering pattern of an opaque disk. This is the diffraction component alone, for which, according to sec. 8.31,

$$\frac{S_1}{x^2} = \frac{S_2}{x^2} = \frac{J_1(x \sin \theta)}{x \sin \theta} = \tfrac{1}{2} F(x \sin \theta);$$

the function $F(u)$ is tabulated in sec. 7.4.

d. In addition a few data read from Ljunggrén's graphs and converted to the present units are shown by crosses.

By comparing these curves we see that the central maximum shifts up and down a little (see sec. 13.42 for a more precise discussion), that the second maximum (first bright ring) shows very striking differences, and that the third and further maxima do not yet resemble the classical diffraction pattern for these values of x.

Figure 51 also shows that the theory for $m \rightarrow 1$ comes quite close to rendering the peculiarities in the exact pattern correctly. Yet Ljunggrén's values are even closer to the correct ones.

The values $x = 30, 35, 40$ happen to show extreme cases. This can be seen directly from Fig. 52. This figure gives the positions of the minima in the diffraction pattern without showing the intensities of maxima or minima. The abscissa is $x \sin \theta$, but a few straight lines make it possible to read the actual angles. Four sets of data are represented.

a. Some data from Mie's formulae. These were estimated from the tables of Gumprecht and co-authors, and from Fig. 51 in whole degrees as follows.

$$x = 25 \quad 30 \quad 35 \quad 40$$

First minimum at $x \sin \theta = 10° \quad 6° \quad 7° \quad 5°$

The tables were not helpful for angles $> 10°$.

b. The minima read from Bricard's table and connected by dotted curves. The corresponding graph in Bricard's paper is not reliable.

c. The values read from Ljunggrén's graphs.

d. The minima read from the altitude chart (Fig. 34), for $m \rightarrow 1$. This chart shows conclusively that the separate minima do not preserve their identity as the drops grow in size. At a certain size the second maximum may become so weak that it vanishes, and the two minima that were at either side become one. This happens at $x = 13$, $x = 25$, and again (or very nearly) at $x = 35$. For still larger values of x than shown in the figure the curve indicating the position of the first minimum oscillates with decreasing amplitude about the asymptotic position $x \sin \theta = 3.83$.

All these data are again in very satisfactory agreement. The values for $m = 4/3$ seem to be shifted slightly to the left compared with those for $m \rightarrow 1$. The rigorous data agree quite well both with the theory for $m \rightarrow 1$ and with Bricard's or Ljunggrén's theories for $m = 4/3$.

It has been pointed out by Naik (1954) from less complete material than that used here that the data from the Mie theory also explain why the measurements of diffraction rings of drops with radii $< 5\,\mu$, when interpreted by means of the conventional theory, always give slightly

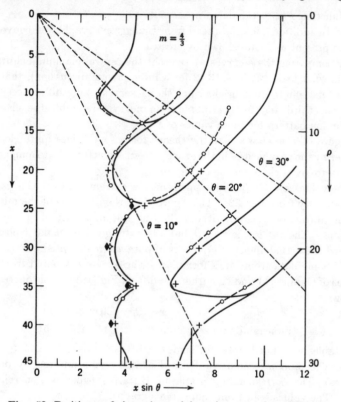

Fig. 52. Positions of intensity minima in anomalous-diffraction patterns: $x = 2\pi a/\lambda$, $\theta =$ scattering angle. Full-drawn curves give positions for $m \to 1$ with ordinate $\rho = 2x(m-1)$ at right. Dotted curves give positions computed by Bricard for water drops, $m = 4/3$ with ordinate x at left. Diamonds give positions from rigorous Mie formulae, and crosses from Ljunggrén's approximation. The positions for diffraction only (large particles) are indicated at the bottom.

higher values of a than the correct values (which may be found from the velocity of fall). This is well illustrated by Fig. 52: If we avoid the sizes for which the first minimum becomes fuzzy we see that it usually lies between $x \sin \theta = 3.0$ and 3.5 instead of the classical value 3.83. Also Paranjpe and Patel (1954) conclude that Ramachandran's theory is better than the uniform-disk formula.

We may finally inquire if it would be possible to construct successfully a complete "altitude chart" for $m = 4/3$, such as Fig. 34 has given for $m \to 1$. At present the direct and reliable way to make such a chart would be to extend the computations of Gumprecht and associates and thus to base the chart directly on Mie's formulae (for two polarizations separately). The other way, by means of approximate theories, as in Bricard's and Ljunggrén's work, can hardly be expected to give a *much* better approximation than the theory for $m \to 1$ gives already. If, however, an attempt at a more complete theory of the Mecke type were to be made, the following points should be kept in mind:

1. For small ρ there is a "remainder" even in the limit of $m \to 1$ (cf. sec. 11.32).

2. For small x the refraction component cannot follow the laws of geometrical optics, as the amplitude (reflection coefficient, etc.) changes over the area over which the integration is taken. (This point is closely related to 1.)

3. For any x the reflection component does not follow the laws of geometrical optics if θ is small. The condition for correctness of the geometrical-optics formula is (compare sec. 17.11)

$$\theta x^{1/3} \gg 1.$$

If $x = 30$ this means $\theta > 20°$, which shows that this effect has to be considered.

4. The "ripple" observed in the numerical results for $\theta = 0$ (sec. 13.42) has probably a quite different cause and may be interpreted as a kind of forward glory (sec. 17.52). This phenomenon will give an additional term for any θ which is within the range of the classical diffraction pattern.

13.42. Forward Radiation; the Extinction Curve

Numerical results for $\theta = 0$ are much more complete than those for other θ. They are available either in the form of the forward scattered amplitude $S(x, 0)$ or in the form of the extinction factor (e.g., sec. 9.32):

$$Q(x) = \frac{4}{x^2} \operatorname{Re}\{S(x, 0)\}.$$

Figure 53 shows the result for $m = 1.33$. It gives the plot of $4S(0)/x^2$ in the complex domain for $x = 0$ to 15 according to computations by Lowan and by Gucker. Most of the dots are based on exact computations; the remaining ones have been put in by interpolation and make the graph complete for intervals of 0.1 in x.

Figure 32, middle curve (sec. 11.22), shows the graph of the real parts. It is the extinction curve computed by Goldberg for $x = 0$ to 30. Both figures may be compared with the corresponding ones for $m \to 1$ (Figs. 31 and 32, lower curve). The striking similarity of the main features has already been brought out in Fig. 32. The question that has intrigued a number of authors is to find a simple and approximate formula which will give nearly correct values for the extinction for various values of m (say

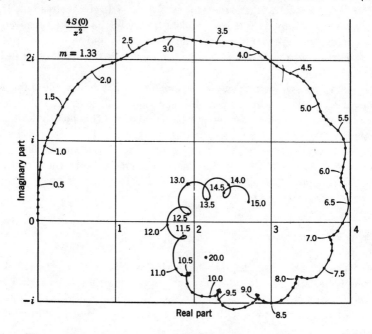

Fig. 53. Plot of $4S(0)/x^2$ for $m = 1.33$ as a function of x in the complex domain. The real part is the extinction factor Q.

1.20 to 1.60) and for all x larger than about 5, where the rigorous computations start to become cumbersome.

The best formula that may be suggested on the basis of the studies made in this book is

$$S(x, 0) = \tfrac{1}{2}x^2 - x \frac{2m^2}{(m + 1)(m^2 - 1)} \, ie^{-2ix(m-1)}$$

$$+(0.46 - 0.80i)x^{4/3} + \text{"ripple."}$$

For $m = 1.33$ this becomes

$$S(x, 0) = \tfrac{1}{2}x^2 - 1.975x(\sin 0.66x + i \cos 0.66x)$$

$$+(0.46 - 0.80i)x^{4/3} + \text{"ripple."}$$

And the corresponding extinction formula is

$$Q(x) = 2 - 7.90x^{-1} \sin 0.66x + 1.84x^{-2/3} + \text{"ripple."}$$

The terms are given in order of importance. The following differences with sec. 12.45, where the formulae for $x \to \infty$ on the basis of the ray-optics approximation were summarized, may be noted.

First term: diffraction, unchanged.

Second term: refracted ray $(p = 1)$, unchanged. The corresponding terms due to central rays that cover the diameter 3, 5, or more times have been omitted, as they are numerically insignificant in the range of m considered.

Third term: grazing reflection, drastically changed. The form given is derived theoretically and checked by numerical results in sec. 17.26; the precise value of the coefficient is somewhat uncertain. It is important that, unlike the ray approximation of the reflection term in sec. 12.45, this term has a real part. This shifts the extinction curve up so that the average curve lies above $Q = 2$. This term decreases less rapidly with growing x than does the refraction term. So, for very large sizes $(x > 100)$ it is the more important one.

Fourth term: the ripple in the extinction curve is characterized by small fluctuations with a period of about 0.8 in x. In the $S(0)$ graph it is shown by a kind of epicycles. The writer has not found an empirical or theoretical formula, but sec. 17.52 gives a tentative explanation as a glory-like phenomenon connected with surface waves.

References

Computations of the scattering diagram based on geometrical optics were first made by

Chr. Wiener, *Abhandl. Kaiser-Leopold-Carol. deut. Akad. Naturforsch.*, **73** (1907) and **91** (1909).

More complete results for $\theta = 0°$ to $25°$ are given by

J. Bricard, *J. phys.*, **4**, 57 (1943).

All the numerical data in sec. 13.11 have been newly computed. Sources of the results based on Mie's formulae (sec. 13.12) are given at the end of chap. 10. The discussion of the interpolation problem is based on a method used in

H. C. van de Hulst, *J. Colloid Sci.*, **4**, 79 (1949).

H. C. van de Hulst, *Mém. Acad. Roy. Sci. Liège*, **15** (*Rep. Intern. Astrophys. Symposium*), 89 (1955).

The section on rainbow theories is largely new. In particular, Fig. 47 is presented for the first time. Airy's formula is derived in many textbooks.

More exact approaches from Mie's theory have been made by

Balth van der Pol and H. Bremmer, *Phil. Mag.*, **24**, 141 and 825 (1937).

H. Bucerius, *Optik*, **1**, 188 (1946).

T. Ljunggrén, *Arkiv Mat., Astron. Fysik*, **36A**, No. 14 (1948).

More complete computations based on Huygens' principle have been made by

W. Möbius, *Ann. Physik*, **33**, 1493 (1910); virtually identical with: *Abhandl. sächs. Ges. Wiss., Math. phys. Kl.*, **30**, 107 (1907).

W. Möbius, *Ann. Physik*, **40**, 736 (1913); virtually identical with: *Preisschr. Jablonski Ges.*, **42** (1912).

J. Rosenberg, *Ann. Physik*, **68**, 414 (1922).

W. V. R. Malkus, R. H. Bishop, and R. O. Briggs, *Nat. Advisory Comm. Aeronaut., Tech. Notes*, No. 1622 (1948.)

The last authors seem to have fallen back into the phase error of Young, explained at the end of sec. 13.22.

The curious effect of alternate maxima in the two polarizations (end of sec. 13.24) was noted in measurements by

J. Bricard, *Ann. phys.*, **14**, 148 (1940).

The discussion of the glory largely follows

H. C. van de Hulst, *J. Opt. Soc. Amer.*, **37**, 16 (1947).

It is believed that the alternate theories as presented by Bricard (1940, above) and Bucerius (above) are not correct, and the present theory is still incomplete.   Observational data are quoted from

J. M. Pernter and F. M. Exner, *Meteorologische Optik*, Vienna, W. Braumüller, 1910.

F. Mierdel, *Beit. Physik freien Atmosphäre*, **8**, 95 (1919).

For more observational data see sec. 20.3.   A review of the rainbow and the glory problem was also given by

R. Meyer, *Himmelswelt*, **56**, 169 (1949).

The theory of the anomalous-diffraction patterns due to a superposition of diffracted, reflected, and refracted light was first proposed by

R. Mecke, *Ann. Physik*, **62**, 623 (1920).

The phase error in this paper was avoided in subsequent discussions of the same problem by

G. N. Ramachandran, *Proc. Indian Acad. Sci.*, **A17**, 202 (1943).

J. Bricard, *Ann. geophys.*, **2**, 231 (1946).

T. Ljunggrén, *Arkiv Fysik*, **1**, 1 (1949).

Further reference is made to

Y. G. Naik, *J. Colloid Sci.*, **9**, 393 (1954).

M. M. Paranjpe and D. T. Patel, *J. Univ. Bombay*, **22**, 1 (1954).

Section 13.42 is a summary of new results, presented more fully in sec. 17.26.

14. ABSORBING SPHERES

This is the final chapter devoted to a discussion of numerical results of the Mie theory (chap. 9) and to approximation methods and their physical interpretation. It deals with complex values of m. Statements on the formulae for complex m have already been made wherever they could be considered as a simple extension of the formulae for real m. Such statements are found in secs. 6.13, 6.4, 7.12, 8.42, 11.23, 12.21, 12.44, and 12.5.

14.1. The Complex Refractive Index

The propagation of electromagnetic waves of circular frequency ω in a homogeneous medium with conductivity σ, dielectric constant ε, and diamagnetic constant 1 can be characterized by the complex refractive index

$$m = \sqrt{(\varepsilon - 4\pi i\sigma/\omega)}.$$

This well-known formula was derived in sec. 9.12. The units are non-rationalized electrostatic units, time factor $\exp(+i\omega t)$. For conversion into practical units of conductivity see sec. 14.41. On writing

$$m = n - in',$$

we find $n^2 - n'^2 = \varepsilon$, $2nn' = 4\pi\sigma/\omega$, from which we may express n and n' in terms of ε, σ, and ω.

The electric and magnetic fields (briefly, the amplitudes) of a wave traveling in the z-direction are proportional to

$$e^{ik(ct-mz)} = e^{i\omega t - imkz},$$

where $k = 2\pi/\lambda = \omega/c$ and λ is the vacuum wavelength. So the Poynting vector determining the intensity of the waves is proportional to

$$e^{-2kn'z} = e^{-\gamma z}.$$

The quantity $\gamma = 2kn' = 4\pi n'/\lambda$ is the absorption coefficient; its dimension is cm^{-1}.

The constants ε and σ depend strongly on ω and approach their static values only for very low frequencies. Measurements at each frequency have to be made in order to find the correct values. Absorption measurements in a very thin sheet or measurements of the intensity and polarization of a wave reflected against a plane surface have been made to this end. Such measurements have yielded reliable values of n and n' for numerous materials at a variety of wavelengths. As expected, the measurements in the far infrared region, started by Hagen and Rubens, have shown a gradual transition of ε and σ to their static values. In this limiting case the penetration depth γ^{-1} is related to the skin depth (sec. 14.41).

There is a wide variety of absorbing materials and a corresponding variety in approximation formulae to the rigorous Mie formulae. Five typical examples are mentioned below, together with the section where details of the Mie computations may be found.

Type 1. $m = 1.27 - 1.37i$ (cf. sec. 14.2).

This is the refractive index of iron at $\lambda = 0.420\ \mu$. For all metals at optical frequencies both n and n' are of the order of unity. They vary considerably with the wavelength, as can be inferred at once from the colors of various metals. Big metal spheres absorb an amount of energy that is of the order of half the radiation striking their surfaces. Very small spheres absorb proportionally to their volumes, and this absorption exceeds the total scattering.

Type 2. $m = 37 - 41i$ (cf. sec. 14.4).

This is the refractive index of platinum at $\lambda = 10\ \mu$ (infrared). For all metals at low frequencies n and n' are very large and almost equal. The material is nearly totally reflecting. Only a small part, which is 4 per cent of the perpendicularly incident radiation in the present example, is refracted into the metal and absorbed.

Type 3. $m = 1.5 - 0.0001i$ (cf. sec. 14.5).

Such a refractive index might be ascribed to black glass. A slide 1 mm thick having this value of m would transmit only 8 per cent of the incident intensity. A dielectric called strongly absorbing by ordinary standards has a value of n' that is still very small. For spheres of such a material the absorption need not be considered, except for very large values of x. Every dielectric, however, will absorb if the frequency approaches the proper frequencies. Near the proper frequencies themselves the refractive index has the type 1 (metallic absorption). Between the extreme types 1 and 3 there is a gradual change.

Type 4. $m = 8.90 - 0.69i$ (cf. sec. 14.3).

The refractive index of water at frequency 3000 Mc/sec, $\lambda = 10$ cm. The value is close to the statical value; $m = \sqrt{\varepsilon} = 9$. The scattering properties of a raindrop at this frequency resemble closely those of a totally reflecting sphere. The large value of n, combined with the small value of n', results in noticeable resonance peaks.   The transition between types 1 and 4 is gradual.

Type 5. $m = 1.01 - 0.01i$ (cf. secs. 11.23 and 14.5).

A sphere having this hypothetical refractive index would behave like an almost black body, provided that x were far greater than 100.   A perfectly black body does not exist in the theory of Maxwell.   It can only be approached by satisfying two conditions:

a.   A negligible fraction is reflected, so $\mid m - 1 \mid \ll 1$.
b.   The sphere is not transparent, so $n'x \gg 1$.

Since condition a implies that $n' \ll 1$, a black body can only be realized in the limit of $x \rightarrow \infty$.

Of course the multitude of absorbing media is not exhausted by these five types.   Very often the description by a complex refractive index is not at all correct, on account of an inhomogeneous structure of the substance.   Yet, as a working outline the division will be useful.

14.2.   General Methods of Computation; Metallic Spheres

Let us recall the rule laid down in sec. 9.4: If the sphere is embedded in a transparent medium with a real refractive index n_0, different from 1, the value of m to be used in all formulae is the actual refractive index of the scattering sphere divided by n_0 and the wavelength in the medium, not in air, should be used in computing $x = 2\pi a/\lambda$.

Example: Gold particles ($m = 0.57 - 2.45i$ at $\lambda = 0.550 \ \mu$) in water ($n_0 = 1.333$).   When yellow light ($\lambda = 0.550 \ \mu$) is used in the experiments the refractive index $0.43 - 1.84i$ and wavelength $0.412 \ \mu$ should be used in the computations.

14.21.   Series Expansion

The numerical computation for complex values of m presents a new problem, since no tables of the Riccati-Bessel functions ψ_n for complex arguments are available.   Nor is the method of phase angles, which proved so successful for real m, a great help, for these angles are themselves complex quantities.   When n or n', or both, have either particularly large or particularly small values, special approximation methods may be devised; they are discussed in subsequent sections.

For ordinary refractive indices (e.g., type 1 and type 4, above) no other method seems available than to start from the definitions and the rigorous formulae.   Such a computation involves the following steps (see chap. 9):

$$\left. \begin{array}{c} x \\ m \end{array} \right\} \rightarrow \left\{ \begin{array}{c} \psi_n \text{ and derivatives} \\ \zeta_n \end{array} \right\} \rightarrow \left\{ \begin{array}{c} a_n \\ b_n \end{array} \right\} \rightarrow \left\{ \begin{array}{c} S_1(\theta) \text{ and } S_2(\theta) \\ Q \text{ values} \end{array} \right. .$$

A number of variations in the actual procedure are possible:

(*A*) Compute $\psi_n(mx)$ and $\psi_n'(mx)$, or any combination needed, from their series expansions or by recurrence formulae. The other functions have real arguments and may be obtained from a table. Proceed from there by numerical substitution. Suitable forms of the recurrence relations have been worked out, e.g., in Aden's method of "logarithmic derivatives" and by Gucker and Cohn.

(*B*) Express all Bessel functions in power series of x, and thus find the Mie coefficients a_n and b_n as power series of x. The coefficients in these power series are functions of m. Find numerical values of a_n and b_n from these series.

(*C*) Go even further and express Q_{ext} and Q_{abs} as a power series of x, and find their values from these series. This is simple after B has been accomplished.

For small x (approximately $x < 0.6$) method C has the decided advantage that it gives the results in an analytical form which leaves no doubt about possible irregularities. For larger values of x methods B and C are more tedious than method A. With any method a great number of points have to be computed before a certain curve, e.g., of Q_{ext} as a function of x, can be drawn.

The series expansions have for a large part already been given. Let us define the functions s, t, u, and w of m as in sec. 10.3; these quantities now have complex values. The main terms of the expansion of a_1, b_1, and a_2 are then as given in sec. 10.3, and the amplitude function for $\theta = 0$ is

$$S(0) = \tfrac{1}{4}\{6isx^3 + is(6t + 6u + 10w)x^5 + 6s^2x^6 + \cdots\},$$

entirely as it was for real m. However, in taking the real part, according to the formula (sec. 9.32),

$$Q_{ext} = \frac{4}{x^2}\,\mathrm{Re}\{S(0)\},$$

the first terms, which dropped out for real m, stay in, with the result that

$$Q_{ext} = -\mathrm{Im}\left\{4x\,\frac{m^2-1}{m^2+2} + \frac{4}{15}\,x^3\left(\frac{m^2-1}{m^2+2}\right)^2\frac{m^4+27m^2+38}{2m^2+3}\right\}$$
$$+\ x^4\,\mathrm{Re}\left\{\frac{8}{3}\left(\frac{m^2-1}{m^2+2}\right)^2\right\} + \cdots.$$

The first term is recognized as the main absorption term, derived in sec. 6.31 from the theory for small particles of arbitrary shape. The scattering is, derived in secs. 6.31 and 10.3:

$$Q_{sca} = x^4\,\frac{8}{3}\left|\frac{m^2-1}{m^2+2}\right|^2 + \cdots.$$

Although this formula is identical with the formula for Rayleigh scattering, it does not necessarily mean that the scattered intensity is proportional to λ^{-4}, for the refractive index of most metals depends very strongly on λ. This is the reason why colloidal metal solutions give colors different from sky blue, even if the particles are very small. This was brought out quite clearly in Mie's original memoir.

Some examples of the coefficients of the main terms and the approximate range of x in which only the first term is needed, are given in Table 25.

Table 25. Main Terms of Absorption and Scattering for Selected Values of m

Material	Gold in Water $\lambda_{air} = 0.60\,\mu$ $\lambda_{water} = 0.45\,\mu$	Iron $\lambda = 0.44\,\mu$	Water $\lambda = 3$ mm	Water $\lambda = 3$ cm	Perfect Conductor
m	$0.28-2.22i$	$1.27-1.37i$	$3.41-1.94i$	$8.18-1.96i$	∞
m^2	$-4.84-1.26i$	$-0.26-3.48i$	$7.87-13.23i$	$63.1-32.1i$	∞
$\dfrac{m^2-1}{m^2+2}$	$1.887-0.393i$	$0.655-0.690i$	$0.891-0.146i$	$0.960-0.018i$	1
Q_{abs}	$1.57x$	$2.76x$	$0.58x$	$0.07x$	0
Q_{sca}	$9.85x^4$	$2.41x^4$	$2.18x^4$	$2.44x^4$	$3.33x^4$*
One term suffices for	$x < 0.3$	$x < 0.3$	$x < 0.12$	$x < 0.06$	$x < 0.6$

* In this result the contribution of the magnetic dipole radiation is included (sec. 10.61).

These series expansions, which form the basis for any computation by means of method C, have been derived by Schoenberg and Jung (1934, 1937) and have also been used by Schalén (see below) as a check on the results computed by methods A and B. In their first paper Schoenberg and Jung give the factors occurring in the main terms of absorption and scattering for over 100 values of m, which correspond to the refractive indices (relative to vacuum) of the metals Pt, Fe, Cu, Ni, Ag, Au, Zn, Mg, Al, Na, K at a number of wavelengths in the visual range. In their second paper they extend the series expansion of Q_{ext} to include a term in x^5 and compute the coefficients of the four terms then needed for 28 values of m. Neglecting higher terms they compute the values of Q_{ext} for values of x ranging to $x = 0.9$. This, however, is stretching method C too far; results with a fair accuracy can be obtained by means of these expansions only for $x < 0.5$ or 0.6.

14.22. Survey of Numerical Results

We shall now describe the results obtained by a numerical summation of the Mie series (method A of the preceding section).

Full computations of the Mie functions for complex values of m have been made by several authors. For those readers who wish to compare their experimental or theoretical data with the available data in the literature a full list of the m values and of the ranges of x is given in Table 26. The reliability of these values of m for the particular metals and wavelengths cannot be discussed in this book. Readers should be warned that some of these values were based on old measurements and that the constants may be different for metals in thin layers and certainly in colloidal solutions that are not strictly lyophobic.

For all values of m mentioned in Table 26 and a number of selected values of x the authors cited give numerical values of numbers equivalent to the coefficients a_n and b_n and to the efficiency factors Q_{ext} and Q_{sca}. Some results from Mie's paper are reproduced in sec. 19.21, Fig. 90. The coefficients of Schalén have been further employed by Schalén and Wernberg (1941) for the computation of Q_{pr}; unfortunately, an error of sign was made so that the tabulated and plotted values are $(1 + \overline{\cos \theta})Q_{ext}$ instead of $Q_{pr} = (1 - \overline{\cos \theta})Q_{ext}$. Schalén (1945) computed the scattering diagram for the iron particles and approximated the phase function by a formula of the type

$$f(\theta) = \frac{1}{4\pi(1 + \frac{1}{3}q)}(1 + p \cos \theta + q \cos^2 \theta).$$

Kerker's computations refer to $\theta = 30°$ to $\theta = 150°$ with $10°$ intervals. Aden has only considered backscatter $(\theta = 180°)$.

A key to the notations used by these authors is given in Table 27. Full results of the efficiency factors for one particular value of m, viz., $m = 1.27 - 1.37i$, are shown in Fig. 54. It was constructed from Schalén's data with the exception of the points for $m = \infty$ that were computed by the method explained in sec. 14.23.

A check on the individual coefficients, as easy as the checks on the phase angles explained in chap. 10, does not exist. However, when the values of a_n or b_n are plotted as points in the complex domain, a smooth curve results. Figure 55 shows for $m = 1.27 - 1.37i$ the locus of a_1 and b_1. The arrowheads correspond to $x = 3.93$. See also sec. 14.23.

It is important to know how many terms have to be computed in order to obtain an accurate result with the series expansion. Suppose we wish to compute the value of Q_{ext}. The tables of Schalén show that the contribution of a_1 is preponderant for $x < 1.0$. At $x = 1.0$ the terms with a_2 and b_1 each contribute 5 to 10 per cent of the total extinction. The contribution to the scattering is smaller, so that Q_{sca} may be computed for x as high as 1.4 from a_1 only. Near $x = 2$ the contributions of a_3 and b_2 to the extinction have become of the order of 10 per cent.

Lowan's tables refer to much larger values of m. Here a_1 and b_1 are about equally important, a_2 and b_2 come in near $x = 1.0$; a_3 and b_3 come

Table 26. Complex Values of m for which Computations have been made

A. Aerosols, visual range

Author	m	Metal	λ (micron)	Upper Limit of x	Upper Limit of $2a$ (micron)
S	1.16–1.27i	Iron	0.395	3.98	0.50
	1.27–1.37i	Iron	0.440	3.57	0.50
	1.34–1.45i	Iron	0.468	3.57	0.53
	1.38–1.50i	Iron	0.508	3.18	0.51
	1.51–1.63i	Iron	0.589	2.78	0.52
	1.70–1.84i	Iron	0.668	2.50	0.53
S	1.36–2.30i	Nickel	0.395	2.39	0.30
	1.46–2.68i	Nickel	0.440	2.14	0.30
	1.44–2.88i	Nickel	0.468	1.00	0.15
	1.50–3.10i	Nickel	0.508	1.00	0.16
	1.54–3.26i	Nickel	0.550	1.71	0.30
	1.58–3.42i	Nickel	0.589	1.00	0.19
	1.74–3.80i	Nickel	0.668	1.00	0.21
S	0.84–2.91i	Zinc	0.395	2.39	0.30
	0.93–3.18i	Zinc	0.440	3.57	0.50
	1.05–4.49i	Zinc	0.468	3.57	0.53
	1.41–4.10i	Zinc	0.508	1.99	0.32
	1.93–4.66i	Zinc	0.589	2.78	0.52
	2.62–5.08i	Zinc	0.668	1.71	0.36
S	1.17–1.76i	Copper	0.395	2.39	0.30
	1.14–2.05i	Copper	0.440	2.14	0.30
	1.10–2.34i	Copper	0.500	1.00	0.16
	1.00–2.28i	Copper	0.535	1.00	0.17
	0.89–2.23i	Copper	0.550	1.00	0.18
	0.65–2.43i	Copper	0.575	1.00	0.18
	0.62–2.63i	Copper	0.589	1.00	0.19
	0.56–3.01i	Copper	0.630	1.00	0.20
S	0.06–1.84i	Sodium	0.435	2.14	0.30
	0.05–2.21i	Sodium	0.546	1.71	0.30
SB	1.59–0.66i	Carbon	0.491	1.12	0.18
K	1.46–4.30i	Mercury	0.546	5.0	0.87

B. Hydrosols, visual range

Author	m	Metal	λ (micron) in Air	λ (micron) in Water	Upper Limit of x	Upper Limit of $2a$ (micron)
M	1.27–1.27i	Gold in water	0.420	0.313	1.58	0.16
	1.30–1.29i	Gold in water	0.450	0.336	1.41	0.15
	0.83–1.51i	Gold in water	0.500	0.374	1.41	0.17
	0.59–1.67i	Gold in water	0.525	0.393	1.41	0.18
	0.43–1.84i	Gold in water	0.550	0.412	1.41	0.19
	0.28–2.22i	Gold in water	0.600	0.450	1.41	0.20
	0.31–2.65i	Gold in water	0.650	0.488	1.41	0.22
F	0.16–1.70i	Silver in water	0.420	0.313	2.00	0.20
	0.21–2.62i	Silver in water	0.525	0.393	1.58	0.20
	0.30–3.66i	Silver in water	0.650	0.488	1.58	0.25
	0.33–4.42i	Silver in water	0.750	0.564	1.58	0.28
F	0.72–2.46i	Mercury in water	0.420	0.313	2.83	0.28
	0.96–2.90i	Mercury in water	0.500	0.374	2.00	0.24
	1.10–3.12i	Mercury in water	0.550	0.412	2.00	0.26
	1.27–3.47i	Mercury in water	0.650	0.488	1.73	0.27

C. Water Drops, infrared

Author	m (practicing example)	Upper Limit of x	m (water)	λ (micron)	Upper Limit of x
JET	1.29–0.064i	18.0	1.14–0.114i	11.0	17.5
	1.29–0.129i	7.0	1.17–0.210i	11.9	17.5
	1.29–0.322i	25.0	1.22–0.061i	10.0	17.5
	1.29–0.645i	6.6	1.28–0.051i	8.5	17.5
			1.28–0.294i	12.6	17.5
	1.29–1.161i	1.0	1.33–0.013i	4.6	17.5
	1.29–1.290i	6.0	1.33–0.040i	7.0	17.5
	1.29–1.419i	1.0	1.33–0.399i	13.5	17.5
	1.29–2.234i	6.0	1.42–0.014i	3.6	17.5
	1.29–2.580i	2.0			
	1.29–2.902i	2.0			
	1.29–5.160i	2.0			
	1.29–129i	2.0			

D. Water Drops, microwaves

Author	m (water at 18°C)	λ(mm)	Upper limit of x	2a(mm)
L	3.41–1.94i	3	5.00	4.78
	4.21–2.51i	5	3.00	4.78
	5.55–2.85i	9	2.00	5.73
	6.41–2.86i	12.5		
	7.20–2.65i	17	1.30	7.03
	8.18–1.96i	30	1.00	9.56
	8.90–0.69i	100	0.60	19.1
A	9.01–0.43i	162	6.0	310

References

S: C. Schalén, *Uppsala astron. Obs. Ann.* **1**, No. 2, 1939; C. Schalén and G. Wernberg, *Meddelande Uppsala astron. Obs.*, No. **83**, 1941; C. Schalén, *Uppsala astron. Obs. Ann.*, **1**, No. 9, 1945.

SB: H. Senftleben and E. Benedict, *Ann. Physik*, **60**, 297 (1919).

K: M. Kerker, *Tech. Rept.* **1**, Division of Research and Industrial Service, Clarkson College of Technology, Potsdam, N. Y., 1953; also *J. Opt. Soc. Amer.*, **45**, 1081 (1955).

M: G. Mie, *Ann. Physik*, **25**, 377 (1908).

F: R. Feick, *Ann. Physik*, **77**, 573 (1925) (Note: the same page numbers occur twice in this volume).

JET: J. C. Johnson, R. G. Eldridge, and J. R. Terrell, *Sci. Rept.* **4**, 1954. M.I.T. Dept. of Meteorology.

L: A. N. Lowan, "*Tables of Scattering Functions for Spherical Particles*," *Natl. Bur. Standards (U.S.)*, *Appl. Math. Series* **4**, 1949, Washington, D. C., Govt. Printing Office.

A: A. L. Aden, *J. Appl. Phys.*, **22**, 601 (1951).

Table 27. Comparison of Notations

Schalén		This Book	Lowan	This Book
(M)	α	x	α	x
(M)	ν	n	C_n^1	a_n
(M)	θ	$180°-\theta$	C_n^2	b_n
(M)	a_ν	$-\frac{1}{2}ix^{-3}(2n+1)a_n$	$K(m^*;\alpha)$	Q_{ext}
(M)	p_ν	$+\frac{1}{2}ix^{-3}(2n+1)b_n$	$\overline{K}(m^*;\alpha)$	Q_{sca}
	Im_λ	$\frac{1}{4}Q_{ext}x^{-1}$		
	ψ_λ	$\frac{1}{4}Q_{ext}(2\pi a)^{-1}$		
	χ_λ	$\frac{1}{8}Q_{sca}(2\pi a)^{-4}$	(M) = notation also used by Mie.	
(D)	$V(\alpha)$	$Q_{ext}(1+\overline{\cos\theta})$	(D) = notation also used by Debye.	

in near $x = 1.8$; a_4 and b_4 come in near $x = 2.6$, etc. Near these values the real and imaginary parts of the coefficients are of the order of 0.01.

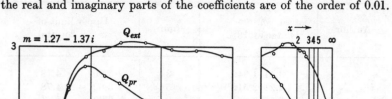

Fig. 54. Efficiency factors for extinction, radiation pressure, absorption, and scattering for $m = 1.27-1.37i$. A tentative interpolation between the computed values and the limiting values for $x = \infty$ is given in the right half of the figure.

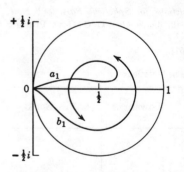

Fig. 55. Locus of the coefficients a_1 and b_1 for $m = 1.27 - 1.37i$.

A characteristic difference in the relative importance of the terms is thus seen. Schalén's finding that a_2 and b_1 (describing the electric quadrupole and magnetic dipole radiation) are of the same order agrees with what is usually found for dielectric spheres (and also in atomic physics). It is explained by the fact that the first terms in the series expansions of these coefficients have the degree x^5 (see sec. 10.3). However, these expansions depend on the expansion of Bessel functions, both with arguments x and with arguments mx. The few terms given suffice only for $|m| \, x < 0.6$, say. Lowan's computations, in which $|m|$ ranges from about 2 to about 9, can therefore be compared with these series expansions up to $x = 0.3$ at the most.

The behavior of the coefficients for *larger* values of x (Lowan) can best be compared with that for $m = \infty$ (sec. 10.62). For $m = \infty$, a_n and b_n are of the same order, and the values of x at which they become important agree well with those mentioned above. The correspondence is also illustrated by Table 28, where some detailed results of the computation for one value of x and several values of m are given as an example. The maximum extinction is reached near $x = 1.5$ for the Schalén values and near $x = 1.0$ or 1.1 for the Lowan values.

Table 28. Example of Computation

	$m = 3.41 - 1.94i$ $x = 1.3$	$m = 7.20 - 2.65i$ $x = 1.3$	$m = \infty$ $x = 1.3$
a_1	$0.5538 + 0.2475i$	$0.5153 + 0.3563i$	$0.474 + 0.499i$
b_1	$0.1780 - 0.2153i$	$0.1730 - 0.2792i$	$0.141 - 0.348i$
a_2	$0.0468 + 0.1050i$	$0.0305 + 0.1055i$	$0.011 + 0.106i$
b_2	$0.0270 - 0.0228i$	$0.0201 - 0.0415i$	$0.003 - 0.053i$
a_3	$0.0008 + 0.0044i$	$0.0004 + 0.0045i$	$0.000 + 0.005i$
b_3	$0.0013 - 0.0003i$	$0.0014 - 0.0019i$	$0.000 - 0.003i$
$S(0°)$	$1.290 + 0.268i$	$1.165 + 0.285i$	$0.957 + 0.366i$
$S_1(180°)$	$0.512 + 0.391i$	$0.484 + 0.608i$	$0.478 + 0.903i$
Q_{ext}	3.053	2.758	2.266
Q_{sca}	1.669	1.860	2.266
Q_{abs}	1.384	0.898	0.0
$i(0°)$	1.734	1.439	1.050
$i(180°)$	0.416	0.604	1.044
$\overline{\cos \theta}$	0.30	0.31	-0.05

An important addition to the available numerical data is depicted in Fig. 56. It represents the extinction cross sections computed by Johnson, Eldridge, and Terrell (1954) for a set of refractive indices, of which the real part remains constant, $= 1.29$, and the imaginary part varies from 0 to ∞. In the first limit the extinction curve is closely similar to the curves in Figs. 24 (sec. 10.4) and 32 (sec. 11.22). In the second limit it is the curve shown in Fig. 28 (sec. 10.62). The refractive index is written in the form $m = 1.29(1 - ik)$. All points were computed with the high-speed digital computer at M.I.T. Although it is clear from a comparison with Fig. 32 that the number of points were insufficient to show all minor wiggles, the gradual change in the character of the curves is very well illustrated. For $k < 0.5$ they may be compared to the curves shown in Fig. 33 (sec. 11.23). For $k > 1$ the peak value is reached between $x = 1$ and 2, as was illustrated for $k \approx 1.1$ in more detail in Fig. 54. The precise behavior for $k > 2$ is not entirely clear. The author has added an enlarged graph for

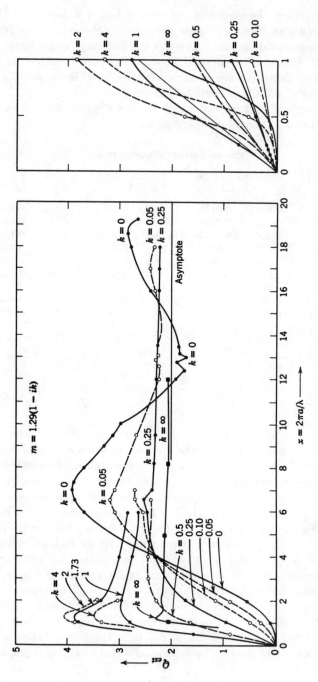

Fig. 56. Gradual variation of the extinction curve if the imaginary part of the refractive index is varied, according to computations by Johnson, Eldridge, and Terrell for $m = 1.29(1 - ik)$.

the part $0 < x < 1$; the linear approximations (see sec. 14.21) have the coefficients

0	0.30	0.75	1.50	2.65	1.50	0.23

for $k = 0$ 0.1   0.25   0.5   1   2   4

The maximum coefficient, 2.72, is reached for $k = 1.23$.

14.23. Asymptotic Formulae for Large Values of x

The form of the scattering pattern and the values of Q_{ext}, Q_{sca}, Q_{abs}, $\overline{\cos \theta}$, and Q_{pr} all approach definite limits if the size of the sphere increases. Of these, only Q_{ext} follows at once from the general theory, for it is equal to 2 (sec. 8.22). The asymptotic behavior of the Mie coefficients is also simple, e.g., the coefficients in Fig. 55 approach a circle with radius 0.26, which is half the absolute value of the Fresnel reflection coefficient for perpendicular incidence.

The scattering pattern of a large metallic sphere consists, like that for a totally reflecting sphere, of one part due to diffraction and one due to reflection. The only difference is that the reflected light contains the factors $|r_1|^2$ and $|r_2|^2$ in the two directions of polarization, respectively; here r_1 and r_2 are the Fresnel reflection coefficients. These are the only components following from the theory of sec. 12.44, which is based on geometrical optics. The additional component due to "spray" of the surface wave (secs. 17.32 and 17.41) will be neglected. The non-reflected part of the incident light is absorbed and gives Q_{abs}.

By putting the reflection coefficients into the formulae for $m = \infty$ (sec. 12.44) we obtain

$$S_1(\theta) = \tfrac{1}{2} i x r_1 e^{2ix \sin \frac{1}{2}\theta}$$

and similarly $S_2(\theta)$, from which, for incident natural light (sec. 4.42):

$$I = \frac{1}{8} \frac{a^2}{r^2} (|r_1|^2 + |r_2|^2) I_0,$$

where a is the radius of the sphere. This formula, with $\theta = 180°$, was applied by Haddock in computing the asymptotic value of the radar cross-section (sec. 14.32). The definitions in secs. 2.1 to 2.4 now give

$$Q_{sca} = 1 + \int_0^{\pi/2} \tfrac{1}{2}(|r_1|^2 + |r_2|^2)\, d(\cos^2 \tau) \equiv 1 + w,$$

$$Q_{abs} = 1 - w,$$

$$Q_{sca} \overline{\cos \theta} = 1 + \int_0^{\pi/2} \tfrac{1}{2}(|r_1|^2 + |r_2|^2) \cos 2\tau \, d(\cos^2 \tau) \equiv 1 + wg.$$

The term 1 is added because of the diffraction part, as explained in sec. 12.5, where the notations w and g are introduced.

Finally, the efficiency factor for radiation pressure is (sec. 12.5)

$$Q_{pr} = 1 - wg.$$

The formulae for r_1 and r_2 are in sec. 12.21. A numerical example, computed for $m = 1.27 - 1.57i$, is shown in Fig. 57, where only the square of the absolute value, not the phase of the complex numbers r_1 and r_2, is shown; $R = \frac{1}{2}(|r_1|^2 + |r_2|^2)$. It is easily seen that in this figure w is the area below the curve R, and the abscissa of the center of gravity of this area is $\frac{1}{2}(1 + g)$.

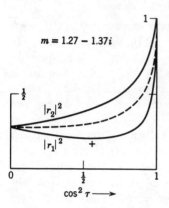

An approximate formula for Q_{abs} found by analytical integration and useful if $|m| > 2$ is given at the end of sec. 14.42.

The results for $m = 1.27 - 1.37i$ are

$$w = 0.34, \quad wg = 0.05, \quad g = 0.14,$$

$$Q_{sca} = 1.34, \quad Q_{abs} = 0.66, \quad Q_{pr} = 0.95.$$

Fig. 57. Squares of the absolute values of the Fresnel reflection coefficients for $m = 1.27 - 1.37i$.

They have been plotted in Fig. 54. The plot against $1/x$ shows that a smooth interpolation between the rigorously computed points ($x \leq 3.5$) and the limiting points at $x = \infty$ is possible.

It is, of course, unsatisfactory to leave it at such a conjectured interpolation. Therefore, several authors have attempted to compute not only the limiting value for $x = \infty$ but an asymptotic formula, i.e., a piece of curve for the very large values of x. Debye in his thesis suggested the problem but did not discuss it. The most complete and basically correct attempt of this kind has been made by Jöbst, who actually derived a (complicated) formula for the function $Q_{ext}(x)$. As a numerical example he computed the curves for gold particles and connected them with the rigorous curves for $x < 1.4$ computed by Mie. The asymptotic curves would hold for any $x > 2.5$. It is certain that the assumption that the asymptotic curve would be joined already at $x = 2.5$ is far too optimistic, and it is doubtful whether Jöbst's formula is correct, as in the simpler case of $m = \infty$ it seems definitely incorrect (secs. 10.62, 17.22).

It seems fairly certain, in analogy with the empirical results for $m = \infty$, sec. 10.62, that the scattering diagram for all angles, *except near the forward and backward direction*, will closely follow the theory for very large spheres for any x larger than about 3 or 5. But it is quite certain that fairly strong changes in the value of $S(0)$, and therefore in the value of Q_{ext} and the form of the diffraction pattern, occur to values of x as high as 10 or 20. This is connected with the fact that the value of $S(0)$ is strongly influenced by the grazing reflection (edge rays). A typical difference with $m = \infty$ is that the reflection coefficients r_1 and r_2 have equal phase for grazing reflection so that the interference effects do not tend to cancel out, as they do for $m = \infty$. Further a sharp turn of the phase of r_2 is seen near the angle for which $|r_2|$ has a minimum (comparable to the Brewster angle for real m). This determines a queer twist in the extinction curve at the high values of m mentioned. This problem is discussed in sec. 17.25.

14.3. Water at Microwave Frequencies

14.31. Resonance Effects

Saxton has given a survey of the complex refractive indices of liquid water at a range of frequencies and temperatures. The values quoted in Table 26 have been selected for numerical computation of the Mie coefficients. They form the transition between the real values $m = 1.33$ in the visual region and $m = 0.9$ for very low frequencies.

The rather large value of $|m|$ opens possibilities of optical resonance (compare sec. 10.5). This resonance is strongly damped, so that the results do not closely resemble those for pure dielectrics. We shall investigate to what extent this damped resonance shows in scattering characteristics.

It was noted in sec. 10.5 that strong resonance effects occur when m is large and real or nearly so. The strongest resonance peak is the peak of b_1 (magnetic dipole radiation), which occurs near $nx = \pi$. Figure 58 shows the locus of a_1 and b_1 in the complex domain, plotted from the tables of Lowan for $m = 8.90 - 0.69i$. The graph of b_1 shows the typical swing in a full circle described in sec. 10.5 for real values of m. The circle is described between $x = 0.30$ and $x = 0.40$, so the peak is fairly sharp. The diameter of the circle is small; the largest real part of b_1, reached at $x = \pi/8.90 = 0.35$ in this example, is 0.050, whereas for any real m it is 1. This is a direct consequence of the imaginary part of m, which causes the standing waves to be damped. The electric dipole resonance near $x = 0.50$ is visible in the curve of a_1.

Figure 59 shows the corresponding graphs for $m = 8.18 - 1.96i$. The stronger damping term has erased the resonance in a_1 and greatly weakened that in b_1.

In the circumstances where the resonance effect is well visible, namely, $m = n - in'$, n large, n' small, its magnitude may be estimated from the formulae as follows. Let β_1 be defined as in sec. 10.21; it is now a complex function of x and m. Near the resonance point x is small but

$y = mx$ is not, so the functions of x may be approximated by their first terms:

$$\psi_1(x) = x^2/3, \qquad \psi'_1(x) = 2x/3,$$
$$\chi_1(x) = 1/x, \qquad \chi'_1(x) = -1/x^2.$$

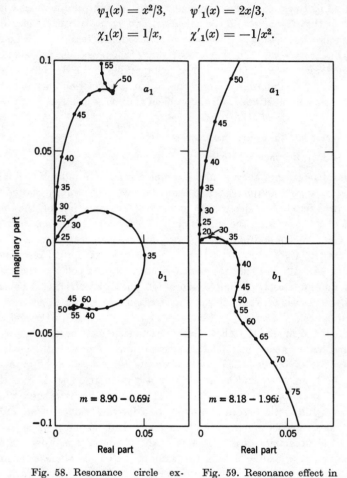

Fig. 58. Resonance circle exhibited by the magnetic dipole coefficient b_1 for $m = 8.90 - 0.69i$.

Fig. 59. Resonance effect in the case of stronger damping than in Fig. 58.

This makes

$$\tan \beta_1 = \frac{2x^3}{3} \frac{\psi_1(y) - \frac{1}{2}y\psi_1'(y)}{\psi_1(y) + y\psi_1'(y)} = \frac{2x^3}{3}\left\{ -\frac{1}{2} + \frac{3}{2y^2} - \frac{3}{2}\frac{\cot y}{y} \right\},$$

which may be approximated by

$$\tan \beta_1 = -\frac{x^3}{3}\left\{ 1 + \frac{3\pi}{y^2(y - \pi)} \right\}$$

near the resonance point.   From this value of $\tan \beta_1$ the value of b_1 and any further quantities may be computed

Numerical example: $m = 8.90 - 0.69i$. The y is closest to π at $x = 0.352$, $y = 3.14 - 0.254i$. The approximation formula gives for this y:

$$\tan \beta_1 = -0.0145(0.40 + 3.76i) = -0.006 - 0.054i.$$

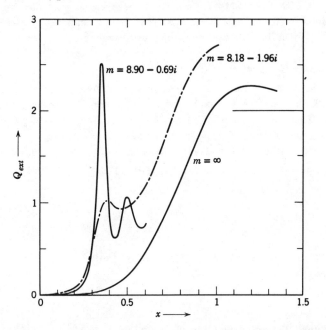

Fig. 60. Extinction curves for three values of the refractive index, showing resonance peaks.

This value is small enough to write

$$b_1 \approx i\beta_1 \approx i \tan \beta_1 = 0.054 - 0.006i.$$

The exact value (interpolated in Lowan's table) is $0.050 - 0.008i$.   The value of a_1 is not negligible compared to this value, so the scattering diagram is not the simple magnetic dipole pattern, as it is when the resonance is more fully developed.   The effect on the extinction curves is clearly shown in Fig. 60.

We may conclude that a rather small value of n' is sufficient to reduce the resonance peak greatly in importance.   Generally, the computation by means of the rigorous formulae (sec. 14.21) is safer and less cumbersome than a special theory for the resonance effect.

Experiments with decimeter waves were made by Schäfer and Wilmsen (1924). Spheres of thin glass (1 or 2 mm) with diameters from 8 to 19 cm were filled with distilled water and the diffraction measured with the source and receiver at a finite distance. The imaginary part of the refractive index is much smaller for these wavelengths. Six resonance values could be demonstrated in the range of x from 0.3 to 0.7 in exact agreement with the theory for $m = 9$. These values may also be read approximately from Fig. 26 (sec. 10.52). Similar spheres of metal, or covered with tin foil, showed no pronounced resonance.

14.32. Backscatter

The efficiency of an object for scattering radiation back to the source ($\theta = 180°$) is conveniently expressed by the *radar cross section* σ of the object. This cross section may be defined as 4π *times the power scattered back per steradian divided by the power incident on a unit area.*

In the notations of this book (e.g., sec. 4.42) this is expressed by

$$\sigma = 4\pi r^2 I(r, 180°)/I_0.$$

When reduced by means of the relations

$$I(r, \theta) = \frac{I_0\{i_1(\theta) + i_2(\theta)\}}{2k^2 r^2}$$

and

$$i_1(180°) = i_2(180°) = \left| S_1(180°) \right|^2$$

this becomes

$$\sigma = \frac{4\pi}{k^2} \left| S_1(180°) \right|^2$$

and divided by the geometrical cross section $G = \pi a^2$ we obtain (with $x = ka$)

$$\frac{\sigma}{G} = \frac{4 \left| S_1(180°) \right|^2}{x^2}.$$

As

$$-\pi_n(180°) = +\tau_n(180°) = (-1)^n \cdot \tfrac{1}{2}n(n+1),$$

we have from sec. 9.31

$$-S_1(180°) = S_2(180°) = \sum_{n=1}^{\infty} (n + \tfrac{1}{2})(-1)^n(a_n - b_n).$$

If a particle scatters the light incident on its geometrical cross section isotropically, its radar cross section equals its geometrical cross section: $\sigma/G = 1$. This follows from the definition and also from substitution of the formula $\left| S_1(\theta) \right| = \tfrac{1}{2}x$ derived in sec. 12.44 for big totally reflecting spheres.

As an example we may compute σ for the data in the first column of Table 28. Here $\lambda = 3$ mm, $m = 3.41 - 1.94i$, $x = 1.30$, $|S(180°)|^2 = 0.416$. These values give $a = 0.62$ mm, $G = \pi a^2 = 1.21$ mm^2, $\sigma = 1.19$ mm^2, $\sigma/4\pi = 0.094$ mm^2, and $\sigma/G = 0.983$. Figure 61 gives a full plot of the values of σ/G computed by Haddock for $\lambda = 3$mm. The curves for the other wavelengths are closely similar, all showing a first maximum near $x = 1$, a very deep minimum near $x = 1.65$, etc. This clearly indicates

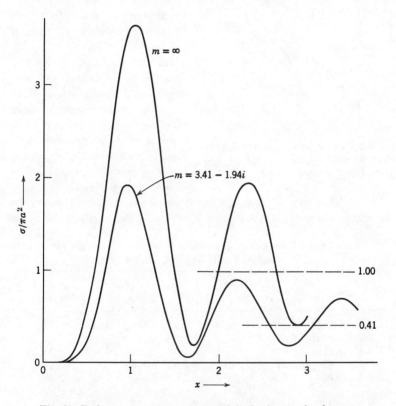

Fig. 61. Radar cross section σ computed for backscattering by water drops at $\lambda = 3$ mm (Haddock) and for totally reflecting spheres.

that these fluctuations are not due to a resonance effect of the kind described in sec. 14.31. Quite the same fluctuations occur also in the curve for totally reflecting spheres ($m = \infty$), which was newly computed for this purpose. Figure 61 and Table 29 show the results. Also given are the values for $x = \infty$, which are simply proportional to the square of the Fresnel reflection coefficients for perpendicular incidence (sec.

12.44). The phase of the backscattered wave was also computed. The angle ψ, defined by

$$-iS_1(180°) = iS_2(180°) = |S| e^{i\psi},$$

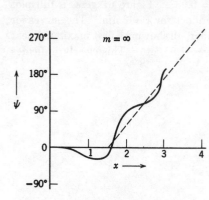

Fig. 62. Phase of the wave scattered back from a totally reflecting sphere (see definition in text).

is given in Table 29 and Fig. 62. The asymptotic formula

$$\psi = 2x - \pi,$$

which is easily found from sec. 12.44, is represented by a straight line.

Aden (1951) has made measurements of the backscatter by water spheres at a wavelength of 16.23 cm. The measurements, which were actually carried out with hemispherical styrofoam shells filled with water and lying against a metal plane, ranged from $x = 0.74$ to 5.90 (diameters 3.7 cm to 30 cm). They give excellent agreement with Aden's own computations of σ/G, which in turn agree with those of Haddock. Also the measured phase

Table 29. Radar Cross Sections of Water Drops at $\lambda = 3$ mm and 18°C
and for Totally Reflecting Spheres

x	σ/G Water	σ/G $m = \infty$	ψ $m = \infty$	x	σ/G Water	σ/G $m = \infty$	ψ $m = \infty$
0.1	0.0003	0.0009	0°	1.6	0.044	0.575	3°
0.2	0.0053	0.014	0°	1.7	0.087		
0.3	0.028	0.072	−1°	1.8	0.223	0.327	57°
0.4	0.087	0.222	−2°	1.9	0.443		
0.5	0.223	0.528	−3°	2.0	0.610	1.007	89°
0.6	0.490	1.076	−6°	2.5	0.511	1.720	107°
0.7	0.935	1.744	−9°	3.0	0.302	0.517	178°
0.8	1.523	2.622	−14°	3.5	0.659		
0.9	1.875	3.315	−18°	4.0	0.248		
1.0	1.956	3.614	−23°	4.5	0.594		
1.1	1.755	3.627	−26°	5.0	0.328		
1.2	1.398	3.154	−27°	∞	0.416	1.000	∞
1.3	0.983	2.465	−28°				
1.4	0.582	1.780	−26°				
1.5	0.267	1.103	−20°				

follows the curve of Fig. 62 very nicely, if due account is taken of the different definition.

Theory and experiment thus check very well, but this does not yet mean that we have an explanation of the extremely strong fluctuations. The correct explanation, discussed in some detail in sec. 17.41, is that a surface wave excited by the grazing incidence near the edge travels around the sphere $\frac{1}{2}$, $1\frac{1}{2}$, $2\frac{1}{2}$ times, etc., and gives off radiation all the time. The radiation in the back direction is then strongly enhanced by a kind of glory effect (sec. 13.31) and has an amplitude which is, for small values of x, comparable to the amplitude reflected back from the center of the exposed surface.

14.4. Metals in the Infrared Region

14.41. The Refractive Index; Law of Hagen-Rubens

The complex refractive index $m = n - in'$ of a metal is much more strongly dependent on the wavelength than the refractive index n of a dielectric substance. In the ultraviolet and visual regions its course is irregular, but beyond the wavelength 2 μ it shows a common behavior for all metals: both n and n' increase with increasing wavelength, and meanwhile they become more and more equal.

The explanation is found from the definition of m^2 (secs. 9.12 and 14.1). We may write this in the form

$$m^2 = \varepsilon - i\eta,$$

where η is a real number, expressed in the classical notation of theoretical physics (gaussian units) as

$$\eta = 4\pi\sigma/\omega. \tag{1}$$

Let the subscript pr denote quantity expressed in practical units (m.k.s. units, Giorgi system), as are now used generally in electricity. Then the conversion formulae

$$\sigma(\text{sec}^{-1}) = 9 \times 10^9 \text{ (ohm meter sec}^{-1}) \times \sigma_{pr}(\text{mho/meter})$$

and

$$\omega(\text{sec}^{-1}) = 2\pi c/\lambda = 2\pi \times 3 \times 10^8(\text{meter sec}^{-1})/\lambda(\text{meter})$$

give

$$\eta = \sigma_{pr}(\text{mho/meter}) \times \lambda_{pr}(\text{meter}) \times 60 \text{ ohm}. \tag{2}$$

A third system of units has been customary in infrared optics. Here w is the specific resistance in ohm mm^2 meter^{-1}, and λ the wavelength in microns. The formula then is

$$\eta = 60\lambda/w. \tag{3}$$

The number η itself is independent of the units used and increases with increasing wavelength. So the imaginary part of m^2 becomes more and more preponderant. The asymptotical approximations, therefore, are

$$m^2 = -i\eta, \qquad m = (\eta/2)^{\frac{1}{2}}(1 - i). \qquad (\eta \gg 1)$$

This formula has been experimentally confirmed by measurements of reflection coefficients. If a beam of light strikes a plane metal surface, a part R of the energy is reflected and the other part $1 - R$ is absorbed. The absorbed part for perpendicular incidence is (sec. 12.21)

$$1 - R_0 = 1 - \left| \frac{m - 1}{m + 1} \right|^2 = \left(\frac{8}{\eta} \right)^{1/2}.$$

This is known as the law of Hagen-Rubens. Beyond the wavelength 5 μ the observations are in good agreement with the law. This formula contains no arbitrary constant.

In sec. 14.1 we have derived and in other secs. (4.3, 11.23, 12.21, 12.44) we have used the absorption coefficient γ. Its inverse value γ^{-1} is the length that a wave can travel through the medium before being reduced to $1/e$ of its initial *intensity*.

In the asymptotic approximation of the present section ($\eta \gg 1$) the length γ^{-1} equals half the "skin depth" δ used in electricity theory, for this is the length that a wave travels before its *amplitude* is reduced to $1/e$. Since

$$\gamma^{-1} = \frac{\lambda}{4\pi n'} = \frac{\lambda}{4\pi} \left(\frac{2}{\eta} \right)^{1/2},$$

we have in practical units

$$\gamma_{pr}^{-1} \text{ (meter)} = \frac{\lambda_{pr}}{4\pi} \frac{1}{(\sigma_{pr} \cdot \lambda_{pr} \cdot 30)^{1/2}} = 0.0145 \left(\frac{\lambda_{pr}}{\sigma_{pr}} \right)^{1/2}.$$

The formula $\frac{1}{2}\delta = (2\omega\mu_1\sigma_{pr})^{-1}$ (Stratton, p. 504) can be reduced to the same expression by the substitutions $\omega = 2\pi \times 3 \times 10^8 \lambda^{-1}$, $\mu_1 = \mu_0 = 4\pi \times 10^{-7}$, for throughout this book we have assumed that the magnetic permeability μ of the materials is equal to that of vacuum.

Numerical example: Silver at room temperature has $\sigma = 5.55 \times 10^{17}$, $\sigma_{pr} = 6.14 \times 10^7$. At $\lambda = 30\mu = 3 \times 10^{-5}$ meter we obtain $\omega = 6.3 \times 10^{13}$ sec^{-1}. So either by formula 1 or 2 we have $\eta = 1.11 \times 10^5$, which gives $m = 236 - 236i$, $1 - R_0 = 0.0085$, $\gamma^{-1} = 1.0 \times 10^{-6}$ cm, and $\delta = 2 \times 10^{-6}$ cm. At this wavelength the approximation is quite sufficient.

The term "skin depth" is more usually applied in the domain of much lower frequencies. Jumping a factor 10^{10} in wavelength, we jump a factor

10^{10} in η and 10^5 in δ, and we arrive at the more familiar values: $\nu = 1$ kc/sec, $\lambda = 300.000$ meters, $\delta = 2$ mm.

We refer to other textbooks for a more complete discussion.

14.42. Absorption by a Sphere

We now consider a metal *sphere* of arbitrary size, exposed to infrared radiation. The value of m just derived is so close to ∞ that the scattering by this sphere is virtually that for $m = \infty$. So the efficiency factor for *absorption* Q_{abs} must be small for any size. A closer examination of the exact formulae of chap. 9 is needed in order to find correct approximation formulae for this small quantity. The reductions are somewhat tedious, hence only a brief enumeration of the results is given.

Let $x = 2\pi a/\lambda$, as usual, and $r = (2\eta)^{1/2}x$; then $y = mx = \frac{1}{2}r(1-i)$. Dependent on the order of magnitude of the real parameters x and r with respect to 1, five size ranges may be distinguished.

(1)	(2)	(3)	(4)	(5)
x small	x small	x small	x arbitrary	x large
r small	r arbitrary	r large	r large	r large

Throughout the regions 1, 2, and 3, x is small and a_1 and b_1 are the only coefficients needed. When the functions of x in the exact formulae (sec. 9.22) are replaced by their main terms, the expressions are

$$a_1 = \frac{2}{3} ix^3 \frac{1 - iG\eta^{-1}}{1 + 2iG\eta^{-1}}, \qquad b_1 = \frac{2}{3} ix^3 \frac{1 - G}{1 + 2G},$$

where $G = \frac{y}{2} \frac{\psi_1'(y)}{\psi_1(y)}$, which can be written exactly in the form $G = \frac{1}{2}(H - 1)$

with $H = \dfrac{y^2}{1 - y \cot y}$. Throughout this region the scattering cross section is small compared to the absorption cross section, and

$$Q_{abs} \approx Q_{ext} = 6x^{-2} \operatorname{Re}\{a_1 + b_1\}.$$

The two terms will be denoted separately as $Q(a_1)$ and $Q(b_1)$. Further reduction gives

$$Q(a_1) = \frac{6x}{\eta} [-1 + \operatorname{Re}\{H\}], \qquad Q(b_1) = 6x \operatorname{Re}\{iH^{-1}\}.$$

The series expansions in region 1 are

$$H = 3 - \frac{ir^2}{10} + \frac{r^4}{700} + \cdots, \qquad H^{-1} = \frac{1}{3} - \frac{ir^2}{90} - \frac{r^4}{1890} + \cdots.$$

They give

$$Q(a_1) = \frac{12x}{\eta}\left(1 + \frac{r^4}{1400} + \cdots\right), \qquad Q(b_1) = \frac{2}{15}\,\eta x^3 \left(1 - \frac{r^4}{420} + \cdots\right).$$

The final result for region 1 thus is

$$Q_{abs} \approx Q_{ext} = \frac{12x}{\eta} + \frac{2}{15}\,\eta x^3 + \cdots.$$

Curiously enough region 1 has to be subdivided into two parts (1a) and (1b). In 1a the first term is largest and in 1b the second term is largest. They have the same order of magnitude near $x = 10\eta^{-1}$.

In region 3 we have, on neglecting terms containing e^{-r}, the approximations:

$$\cot y = i, \qquad H = iy + 1 + \cdots, \qquad H^{-1} = -iy^{-1} + y^{-2} + \cdots,$$

which yield

$$Q(a_1) = \frac{6x^3}{r} + \cdots = \frac{6}{(2\eta)^{1/2}}\,x^2 + \cdots,$$

$$Q(b_1) = \frac{6x}{r}\left(1 - \frac{2}{r}\right) = \frac{6}{(2\eta)^{1/2}} - \frac{6}{\eta x}.$$

The second value is the higher one. So Q_{abs} is virtually a constant in region 3.

In region 2 the full expressions may be found from

$$H^{-1} = \frac{2i}{r^2} + \frac{1}{r}\,\frac{(\text{sh } r - \sin r) - i(\text{sh } r + \sin r)}{\text{ch } r - \cos r}.$$

Here, as in the bordering regions 1b and 3 the term $Q(b_1)$ has a larger order of magnitude than $Q(a_1)$. The final formula for region 2 is

$$Q_{abs} \approx Q_{ext} \approx Q(b_1) = \frac{6x}{r}\left(-\frac{2}{r} + \frac{\text{sh } r - \sin r}{\text{ch } r - \cos r}\right);$$

some numerical values computed from this formula and useful to cover the range between the approximation formulae already given are

$x^{-1}Q(b_1) =$	0.0665	0.256	0.50	0.67	0.71	0.67	0.56	0.48
for $r =$	1	2	3	4	5	6	8	10

We now consider regions 3 to 5, inclusive, where r is very large. Throughout these regions the scattering by the sphere is virtually the same as the scattering by a totally reflecting sphere ($m = \infty$). This is shown e.g., by the main terms in the expansions of a_1 and b_1 in region 3:

$$a_1 = \tfrac{2}{3}ix^3 + \cdots, \qquad b_1 = -\tfrac{1}{3}ix^3 + \cdots.$$

The transition from relatively large absorption (albedo near 0) to relatively large scattering efficiency (albedo near 1) occurs in region 3. On examination of the formulae for this region, absorption and scattering are found to have the same order of magnitude when $x^3 r \approx 1$, i.e., near $x = \eta^{-1/8}$. This means that, unless η is extremely large, i.e., of the order of 10^8 or so, virtually the whole region 3 is characterized by a relatively large absorption.

In region 4 further coefficients besides a_1 and b_1 gradually become important. The entire results for the scattering diagram and the scattering cross section may be taken from sec. 10.62. It would be interesting to compute Q_{abs} as a function of x for this region, but this computation has not been carried out.[1] By inference from the bordering regions 3 and 5 we may suppose that Q_{abs} is of the order of $\eta^{-1/2}$.

Region 5 is the region of large spheres, for which the laws of geometrical optics hold; the formulae of sec. 14.23 are applicable. The approximations for the reflected fractions of the intensity of a ray that makes an angle τ with the surface are for any large m:

$$\left| r_2 \right|^2 = 1 - 4p \, (\sin \tau)^{-1} + 8p^2 \, (\sin \tau)^{-2} + \cdots,$$

$$\left| r_1 \right|^2 = 1 - 4p \sin \tau + 8p^2 \, (\sin \tau)^2 + \cdots,$$

where $p = \mathrm{Re} \, (m^{-1})$. On retaining only the linear terms in p, we obtain by the formula in sec. 14.23

$$Q_{abs} = \tfrac{1}{2} \left\{ \frac{8}{3}p + 8p \right\} = \frac{16}{3} \, p,$$

where the two separate terms refer to polarization 1 and 2. This reduces to the final formula for region 5:

$$Q_{abs} = \frac{8}{3} \left(\frac{2}{\eta} \right)^{1/2} .$$

The approximations made are not valid near grazing angles of incidence, but this does not significantly affect the result of the integration.

It may be remarked that, on the plausible approximation $\sin \tau' = 1$ which is fairly good already for a moderately large $\left| m \right|$, the Fresnel coefficients become

$$r_1 = \frac{\sin \tau - m}{\sin \tau + m}, \qquad r_2 = \frac{\sin \tau - 1/m}{\sin \tau + 1/m},$$

[1] A formula is given by Morse and Feshbach, *Methods of Theoretical Physics*, part II, p. 1885, New York, McGraw-Hill Book Co., derived on the assumption that the tangential component of the magnetic field at the surface, H_t, is the same as for a perfect conductor. The losses (absorption) are proportional to the integral of H_t^2 over the surface provided that the skin depth is $\ll a$. This formula should presumably be identical to the one which may be obtained by reducing the rigorous solution under the mere condition that r is very large.

and the integration can be performed exactly, giving $Q_{abs} = \frac{1}{2}(f_1 + f_2)$, where

$$f_1 = 8q^{-2}\{q - \ln(1 + q + \frac{1}{2}q^2)\} \text{ with } q = (2/\eta)^{1/2},$$

$$f_2 = \text{the same expression} \qquad \text{with } q = (2\eta)^{1/2}.$$

These expressions reduce to those given above if η is very large.

An incidental advantage is that the latter formulae are also useful for much smaller values of m, as occur, e.g., for the metals at visual wavelengths. For an illustration the results computed for $m = 1.32 - 1.32i$, $\eta = 3.50$, by means of the q formula are

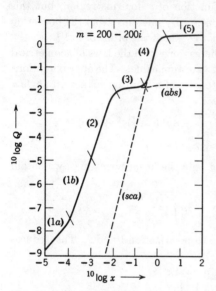

$$f_1 = 0.56; \; f_2 = 0.79; \; Q_{abs} = 0.67$$

as compared with the values

$$f_1 = 0.56; \; f_2 = 0.76; \; Q_{abs} = 0.66,$$

which were found by numerical integration of the correct values for $m = 1.27 - 1.37i$. Also Jöbst's graph for the absorption by gold spheres suspended in water at 0.45μ gives a similar result, $Q_{abs} = 0.65$, at the corresponding refractive index $m = 1.30 - 1.29i$ (Table 26).

The results derived in this section are illustrated and summarized by Table 30 and Fig. 63, showing the resulting efficiency factors for a particular value of m. We have chosen $m = 200 - 200i$, which corresponds to $\eta = 80,000$, $r = 400x$.

Fig. 63. Logarithmic sketch of the extinction curve for spheres with $m = 200 - 200i$.

14.5. Analytical Continuation

The discussions in the preceding sections have had the character of comments on the rigorous Mie solution. Approximate expressions were derived, but in the event of doubt it was a simple procedure to use the exact solution. The interesting range of sizes was at small values of x, at which only a few terms are usually needed.

This is not true for the present section, where values of m of the type $m = 1.5 - in'$ (n' small) are considered. The interesting range of x, in which the extinction curve shows its set of maxima and minima and the scattering diagram changes from the Rayleigh pattern to the pattern for large spheres, is similar for the real value $m = 1.5$ and ranges from

$x = 0.5$ to $x = 15$, say. It is a major job to work through the computations of the Bessel functions for complex arguments,[2] even though the fact that these arguments are nearly real must introduce a great simplification.

Fortunately, an elegant method is available, in which the fact that m is nearly real is used more directly.

Table 30. Efficiency Factors for $m = 200 - 200i$

x	r	Q_{abs}	Q_{sca}	Region
0.00001	0.004	1.5×10^{-9}	2.7×10^{-20}	
0.00002	0.008	3.1×10^{-9}	4.3×10^{-19}	$1a$
0.00005	0.02	8.8×10^{-9}	1.7×10^{-17}	
0.0001	0.04	2.6×10^{-8}	2.7×10^{-16}	
0.0002	0.08	1.1×10^{-7}	4.3×10^{-15}	$1b$
0.0005	0.2	1.4×10^{-6}	1.7×10^{-13}	
0.001	0.4	1.1×10^{-5}	2.7×10^{-12}	
0.002	0.8	1.1×10^{-4}	$5 \quad \times 10^{-11}$	2
0.005	2	1.3×10^{-3}	$2 \quad \times 10^{-9}$	
0.01	4	0.007	3.3×10^{-8}	
0.02	8	0.011	5.3×10^{-7}	
0.05	20	0.013	2.1×10^{-5}	3
0.1	40	0.014	3.3×10^{-4}	
0.2	80	0.015	0.005	
0.5	200	?	0.22	
1	400	?	2.04	4
2	800	?	2.21	
5	2000	0.013	2.12	
10	4000	0.013	2.06	5
20	8000	0.013	2.03	

When in mathematical analysis a function $F(x)$ is known precisely for a range of values of the real argument x, it is often possible to extend this function into the complex domain. By this is meant that it is possible to define uniquely a function $F(z)$ of the *complex* variable z (within a certain area of the complex domain), which has the property that it is regular in this area and that it equals the given function on the real axis. This extension from a real range of variables into a complex domain of variables is called *analytical continuation*. The function $F(z)$ will generally be complex, though it may be real in some places.

This method can be correctly applied if the given function possesses all derivatives. This is true for all functions that we have encountered in this

[2] Johnson, Eldridge, and Terrell (1954) have employed automatic machine computation (see Table 26).

book, except those that are written with the symbols Re, Im, or | |, i.e., those functions in which the real and imaginary parts of a given function have been treated separately. This means that we *cannot* apply the method (directly) to Q_{abs}, Q_{sca}, Q_{ext}, and that we *can* apply the method directly to $S(x, \theta)$, $S(0)$, etc.

The function $S(m, x, \theta)$ is known for real values of m from the preceding chapters. It is found by analytical continuation for complex values of m. We shall discuss only the derivation of the extinction curve. It is based on the continuation of $S(m, x, 0)$ into the complex domain. The actual procedure is quite simple; it is necessary to represent this (complex) function as exactly as possible by an analytical formula. Then this formula, applied to complex values of m can be trusted to give correct values for any complex value of m that is close to the real values from which we started. This book is not the place to discuss further mathematical subtleties.

If a formula follows already from the theory for real m, *this* formula may be used for the analytical continuation. This has, in fact, been done with the series expansion for small x (derived in sec. 10.3, extended in sec. 14.21) and in a more striking form with the formula for m close to 1 (derived in sec. 11.21, extended in sec. 11.23).

However, the method is more of a real help if the given values for real m are not found from a simple formula but have been computed numerically. The choice of approximation formula is then free; a series of powers, of sines and cosines, of exponential functions, etc., might be chosen. Horenstein, who worked out a method for the computations of Lowan's tables, has chosen to use power series.

The function Q_{ext} (in his notation K) is completed by an imaginary part L in order to form an analytical function, which is $(2/x^2)S(0)$ in our notation. Both K and L are computed numerically for fixed value of x and for the real values

$$m = 1.44, 1.45, 1.46, \cdots, 1.55.$$

A Taylor expansion of both functions is then made in terms of powers of $t = m - 1.50$. It is then possible to replace t by $t - in'$, so that

$$m = 1.50 + t - in'.$$

By taking again the real part of the function thus found he obtains the Q_{ext} for this value of m and the same, fixed value of x. For example, the result for $x = 3$ is

$$Q_{ext} = (3.4180 + 6.8978t - 25.68t^2 + 8.2t^3 + 396t^4)$$
$$+ (5.7433 - 15.720t + 312.9t^2)n'$$
$$+ (25.68 - 24.6t - 23.76t^2)n'^2$$
$$- 104.3n'^3 + 386n'^4.$$

For very small n' only the term in the first power of n' is important. The results for $m = 1.50 - in'$ are given in Table 31. This table shows the same trend as that shown in Fig. 33 (sec. 11.23) and in Fig. 56: the maxima are decreased and the minima increased. A similar method applied to cylinders is described in sec. 15.52.

Table 31. Extinction by Spheres with $m = 1.50 - in'$ (n' small)

x	Q_{ext}	x	Q_{ext}
0.5	$0.0146 + 1.13n'$	3.2	$3.532 - 4.26n'$
0.6	$0.0302 + 1.42n'$	3.6	$4.185 - 12.10n'$
1.0	$0.215 + 2.75n'$	4.0	$4.052 - 8.28n'$
1.2	$0.395 + 3.35n'$	4.5	$4.202 - 13.56n'$
1.5	$0.753 + 4.26n'$	4.8	$3.819 - 7.14n'$
1.8	$1.349 + 3.78n'$	5.0	$3.928 - 11.41n'$
2.0	$1.798 + 1.42n'$	5.5	$3.182 - 2.57n'$
2.4	$2.338 + 1.42n'$	6.0	$2.904 - 2.49n'$
2.5	$2.540 + 1.11n'$	6.5	$2.368 + 3.98n'$
3.0	$3.418 + 5.74n'$	7.0	$1.848 + 11.75n'$

References

Metal optics in general (sec. 14.1) is reviewed by

R. Gans, *Handbuch Exp. Phys.*, **19**, 201 (1928),
M. Born, *Optik*, Berlin, Julius Springer, 1933,

and in the infrared region (sec. 14.41) by

C. Schaefer and F. Matossi, *Das ultrarote Spektrum*, Berlin, Julius Springer, 1930.

The first computations for metallic spheres were made by

G. Mie, *Ann. Physik*, **25**, 377 (1908).

Recurrence formulae as used in sec. 14.21A were given by

A. L. Aden, *J. Appl. Phys.*, **22**, 601 (1951).
F. T. Gucker and S. H. Cohn, *J. Colloid Sci.*, **8**, 555 (1953).

The series expansions (sec. 14.21) were derived by

E. Schoenberg and B. Jung, *Astron. Nachr.*, **253**, 261 (1934).
E. Schoenberg and B. Jung, *Mitt. Sternw. Breslau*, **4**, 61 (1937).
C. Schalén, *Meddelande Uppsala astron. Obs.*, **64** (1936).

The numerical results quoted in sec. 14.22 were computed in these papers and in the references given with Table 26.

The problem of the asymptotic forms (sec. 14.23) was discussed by
G. Jöbst, *Ann. Physik*, **76**, 863 (1925).

Computations for water drops at microwave frequencies (sec. 14.31) were
made by Lowan, cited on p. 275.   Backscatter was computed by

F. Haddock, Paper presented at the U.R.S.I.-I.R.E. meeting, Washington, D.C.,
1947, *Report NRL Progress*, June 1956, p. 15.
A. L. Aden, *J. Appl. Phys.*, **22**, 601 (1951),

and also measured by Aden.   The computations for $m = \infty$ in sec. 14.32
are new.   Resonance effects were measured by

C. Schäfer and K. Wilmsen, *Z. Physik*, **24**, 345 (1924).

The formulae for metal spheres in the infrared (sec. 14.42) have been
derived by

H. C. van de Hulst, "Thesis Utrecht," op. cit., chap. 12.

and are given here in a more complete form.   The method of analytical
continuation (sec. 14.5) was described and used in Lowan's tables (cited
in Table 26 references).

15. CIRCULAR CYLINDERS

15.1. Rigorous Solution of Maxwell's Equations

Infinitely long circular cylinders of a homogeneous material exposed to a plane wave of radiation present a somewhat similar scattering problem as homogeneous spheres. The problem is quite comparable to the Mie problem (chap. 9) if the incident light travels perpendicularly to

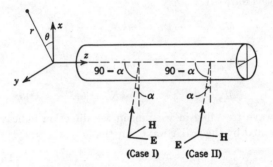

Fig. 64. Coordinates and orientation of the vectors used in the cylinder problem. In subsequent sections we have $\alpha = 0$.

the cylinder axis. This case will be solved in detail (secs. 15.12, 15.13, and 15.23). Oblique incidence gives rise to more complicated formulae, as will be shown first (sec. 15.11).

15.11. Oblique Incidence

The coordinates to be used are illustrated in Fig. 64. The z-axis is along the axis of the cylinder, the x-axis in the plane containing the z-axis and the direction of incidence, and the y-axis perpendicular to this plane. The coordinates r and θ are defined by

$$x = r \cos \theta, \qquad y = r \sin \theta.$$

The Maxwell equations have the form $1a$ and $2a$ of sec. 9.12. Again we shall express $\mathbf{E}$ and $\mathbf{H}$ by means of equation 10 (sec. 9.21) in terms of the four vectors $\mathbf{M}_u$, $\mathbf{N}_u$, $\mathbf{M}_v$, $\mathbf{N}_v$ derived from two solutions, u and v, of the scalar wave equation. These vectors obey the same relations

(8) and (9) (sec. 9.21), but their derivation from the functions u and v is different. The definition to be used here and replacing equation 7 of sec. 9.21 is

$$\mathbf{M}_\psi = \text{curl} \, (\mathbf{a}_z \cdot \psi),$$

where $\mathbf{a}_z$ is the unit vector in the z-direction and ψ is either u or v. The proof that $\mathbf{M}$ and $\mathbf{N}$ have the same properties as in sec. 9.21 is omitted.

The elementary solutions of the scalar wave equation in a homogeneous medium with refractive index m in cylindrical coordinates have the form

$$\psi_n = e^{in\theta} Z_n \left[r \sqrt{(m^2 k^2 - h^2)} \right] e^{-ihz + i\omega t},$$

where h is arbitrary, n is an integer, and Z_n is any Bessel function of order n. The components of the vectors $\mathbf{M}$ and $\mathbf{N}$ for an elementary solution of this form are

$$M_r = -\frac{in}{r} \psi, \qquad mkN_r = -ih \frac{\partial \psi}{\partial r},$$

$$M_\theta = -\frac{\partial \psi}{\partial r}, \qquad mkN_\theta = -\frac{nh}{r} \psi,$$

$$M_z = 0, \qquad mkN_z = (m^2 k^2 - h^2) \, \psi.$$

A plane wave traveling in vacuum in the direction indicated by Fig. 64 is represented by the scalar wave function

$$\psi_0 = e^{i\omega t - ik(x \cos \alpha + z \sin \alpha)},$$

which by substitution of $h = k \sin \alpha, l = k \cos \alpha$, obtains the form

$$\psi_0 = e^{i\omega t - ilx - ihz}.$$

The corresponding vectors $\mathbf{M}$ and $\mathbf{N}$ are derived most easily from the curl equations in rectangular coordinates. We find:

$$M_x = 0, \qquad kN_x = -hl \, \psi_0,$$
$$M_y = il \, \psi_0, \qquad kN_y = 0,$$
$$M_z = 0, \qquad kN_z = l^2 \, \psi_0.$$

This solution may be used to construct a plane electromagnetic wave traveling in the same direction. We shall consider two simple cases separately. (An arbitrary elliptically polarized wave may be formed by linear superposition of these two solutions with complex coefficients.)

Case I: We choose $u = \psi_0$, $v = 0$. Direct substitution of $\mathbf{M}_u$ and $\mathbf{N}_u$ in the equations gives

$$E_x = -\sin \alpha, \qquad E_z = \cos \alpha, \qquad H_y = -1,$$

with the omission of a common factor ilu.

Case II: We choose $u = 0$, $v = \psi_0$. Then

$$H_x = -\sin \alpha, \qquad H_z = \cos \alpha, \qquad E_y = 1,$$

with the omission of a common factor ilv.

The directions of the vectors in each case are shown in Fig. 64.

Before this representation of the incident wave can be utilized for solving the scattering problem, ψ_0 has to be expanded in a series of elementary wave functions. By means of the identity

$$e^{-ilx} = e^{-ilr\cos\theta} = \sum_{n=-\infty}^{\infty} (-i)^n J_n(lr)\, e^{in\theta},$$

we find

$$\psi_0 = e^{i\omega t - ihz} \sum_{n=-\infty}^{\infty} (-i)^n J_n(lr) e^{in\theta},$$

which has the required form, because $l = \sqrt{(k^2 - h^2)}$ and $m = 1$ in vacuum.

The solution has to be completed by assuming similar expansions with undetermined coefficients for the outside scattered wave and for the inside wave. Though we have restricted the incident wave by the assumption $v = 0$ (case I) or $u = 0$ (case II) it appears necessary to introduce u *and* v for the scattered wave and internal wave. So four sets of undetermined coefficients are needed both in case I and in case II.

The boundary conditions require that h is the same for all waves; the radial functions are

$$H_n^{(2)} [r\sqrt{(k^2 - h^2)}]$$

for the scattered wave, and

$$J_n [r\sqrt{(m^2 k^2 - h^2)}]$$

for the internal wave.

The boundary conditions (equations 1b and 2b, sec. 9.13) require that the tangential components of **E** and **H** are continuous at the surface $r = a$. These components contain the following expressions in each elementary wave of u and v:

$$(E_\theta:) \qquad \frac{\partial v}{\partial r} + \frac{inh}{mkr}\, u,$$

$$(E_z:) \qquad \frac{(m^2 k^2 - h^2)u}{mk},$$

$$(H_\theta:) \qquad m\frac{\partial u}{\partial r} - \frac{inh}{kr}\, v,$$

$$(H_z:) \qquad (m^2 k^2 - h^2)v.$$

The boundary conditions for u and v are, therefore, that these four expressions shall be equal at $r = a$ inside and outside the surface. This gives four linear equations connecting the four undetermined coefficients in case I and four other equations connecting the four other coefficients in case II. Either set of four equations may be solved and gives the complete solution of the problem, for the fields are then known at any point in space.

It is important to note that the four equations in either case cannot be separated into two groups of two independent equations as they could be for spheres (sec. 9.22). The expressions of the coefficients are, therefore, more cumbersome. This problem has been pursued further by Wellmann (1937) and Wait (1955).

15.12. Perpendicular Incidence

Perpendicular incidence means that in the preceding equations $\alpha = 0$, so $l = k$ and $h = 0$. A glance at the boundary conditions given above shows that now the functions u and v are no longer mixed. So the assumption that $v = 0$ for the incident wave (case I) entails that it is 0 also for the scattered wave and internal wave. Likewise, $u = 0$ in case II for all waves. Writing

$$F_n = e^{in\theta + i\omega t}(-1)^n,$$

and combining at once the incident and scattered waves in one formula, we have

Case I, $v = 0$. ($\mathbf{E} \,/\!/$ axis)

$$(r > a) \quad u = \sum_{n=-\infty}^{\infty} F_n\{J_n(kr) - b_n H_n(kr)\},$$

$$(r < a) \quad u = \sum_{n=-\infty}^{\infty} F_n\, d_n\, J_n(mkr),$$

$$(r = a) \quad mu \text{ and } m\,\frac{\partial u}{\partial r} \text{ continuous.}$$

Case II, $u = 0$. ($\mathbf{H} \,/\!/$ axis)

$$(r > a) \quad v = \sum_{n=-\infty}^{\infty} F_n\{J_n(kr) - a_n\, H_n(kr)\},$$

$$(r < a) \quad v = \sum_{n=-\infty}^{\infty} F_n\, c_n\, J_n(mkr),$$

$$(r = a) \quad m^2 v \text{ and } \frac{\partial v}{\partial r} \text{ continuous.}$$

The undetermined coefficients b_n, d_n, a_n, and c_n have been denoted in this order for later comparison with the scattering by spherical particles. The function $H_n(kr)$ denotes here and in all further formulae the *second Hankel function*

$$H_n^{(2)}(z) = J_n(z) - iN_n(z).$$

Solutions for the coefficients are easy and give

Case I:
$$b_n = \frac{mJ_n'(y)\,J_n(x) - J_n(y)\,J_n'(x)}{mJ_n'(y)H_n(x) - J_n(y)H_n'(x)},$$

Case II:
$$a_n = \frac{J_n'(y)\,J_n(x) - mJ_n(y)\,J_n'(x)}{J_n'(y)H_n(x) - mJ_n(y)H_n'(x)},$$

where primes denote derivatives and $x = ka$, $y = mka$. We can easily check that

$$b_n = b_{-n} \qquad \text{and} \qquad a_n = a_{-n},$$

so that actual computations involve only the coefficients with $n = 0$, $1, 2, 3, \cdots$.

15.13. The Fields at Infinity

The coefficients are now known, so we proceed to find the fields at large distances by means of the asymptotic expression:

$$H_n(kr) \sim \left(\frac{2}{\pi kr}\right)^{1/2} e^{-ikr + i(2n+1)\pi/4}.$$

Let us take case I as an example. The incident wave was defined by

$$u = e^{-ikx + i\omega t}.$$

The scattered wave is found to follow from a function u which for $kr \gg 1$ has the form:

$$u = \left(\frac{2}{\pi kr}\right)^{1/2} e^{-ikr + i\omega t - i3\pi/4}\, T_1(\theta),$$

where

$$T_1(\theta) = \sum_{n=-\infty}^{\infty} b_n e^{in\theta} = b_0 + 2 \sum_{n=1}^{\infty} b_n \cos n\theta.$$

$T_1(\theta)$ has a meaning similar to the amplitude function $S(\theta)$ in the case of finite particles. By reference to the definitions we find that E_z (the electric-field component parallel to the axis) is proportional to u everywhere. So the intensity is proportional to $|u|^2$. If I_0 denotes the intensity (watt/m²) of the incident light and I the intensity (watt/m²) of light scattered in the direction θ at a large distance r from the cylinder axis, then

$$I = \frac{2}{\pi r k} \mid T_1(\theta) \mid^2 I_0.$$

Since (for infinite cylinders) the scattered light diverges only in one direction, I decreases as $1/r$, not as $1/r^2$.

In case II the magnetic field is parallel to the cylinder axis and proportional to v. The formulae are identical except for the T-factor, which is here:

$$T_2(\theta) = \sum_{n=-\infty}^{\infty} a_n e^{in\theta} = a_0 + 2 \sum_{n=1}^{\infty} a_n \cos n\theta.$$

15.2. Cross Sections and Efficiency Factors

Infinitely long cylinders are not included in the general scheme of chap. 4. They do not give rise to an outgoing spherical wave at large distances, so we cannot define the scattering functions $S_{1,2,3,4}(\theta)$, nor use the general formula for extinction, derived in sec. 4.2.

The remedy may be found in two ways. In sec. 15.21 we give an equally general definition of two functions $T_1(\theta)$ and $T_2(\theta)$ appropriate to infinitely long cylinders of arbitrary shape and size, and in sec. 15.22 we discuss cylinders that are still very long, but not infinitely long, so that the definitions of chap. 4 are applicable. We shall see that both approaches give the same efficiency factors.

15.21. General Expressions for Infinitely Long Cylinders

It is instructive to drop for a while the assumption that the cylinders are circular and homogeneous. This section refers to cylinders in the more general sense. The coordinate system is the same as that used above. The cross section may have an arbitrary form, and the values of m in the x-y plane may be arbitrary but must be independent of z. This includes circular cylinders, elliptical cylinders, tubes, strips, rods, etc. The intention is to derive for such cylinders a theorem analogous to the "general extinction theorem" for finite particles in sec. 4.21.

The amplitude of the incident wave may be written

$$u_0 = e^{-ikx+i\omega t}.$$

We need not specify whether u_0 is the z-component of an electromagnetic wave or the amplitude of a scalar wave. We postulate that the amplitude of the scattered wave for large r is

$$u = \left(\frac{2}{\pi k r}\right)^{1/2} T(\theta) \cdot e^{-3\pi i/4 - ikr + i\omega t}.$$

The dependence on r is such as corresponds with any outgoing cylindrical wave, $T(\theta)$ gives the dependence on θ, and the constant factors have been added for reasons that will become apparent. Integrating the scattered intensity (compare sec. 15.13),

$$I = \frac{2}{\pi r k} |T(\theta)|^2 I_0,$$

over all angles θ and over a length l in the z-direction we find the total energy scattered per unit time by a length l of the cylinder. This amount may be written by definition

$$l \cdot c_{sca} \cdot I_0.$$

Here c_{sca} is the cross section for scattering per unit length. It may also be called the effective width, for the total amount of scattered light equals the amount incident on a strip with width c_{sca}. Performing the integration we obtain

$$c_{sca} = \frac{2}{\pi k} \int_0^{2\pi} |T(\theta)|^2 \, d\theta.$$

The effective width for extinction, c_{ext}, may be defined in a similar way, but its derivation requires a reasoning analogous to that of secs. 4.21 or 4.41. We use the line of reasoning expounded in sec. 4.21, to which we refer the reader for details. The combined amplitude of incident and scattered wave for very small θ and large r is

$$u_0 + u = u_0 \left\{ 1 + \left(\frac{2}{\pi k r} \right)^{1/2} T(0) \, e^{-3\pi i/4 - iky^2/2r} \right\}$$

and the combined intensity is

$$|u_0 + u|^2 = 1 + \left(\frac{8}{\pi k r} \right)^{1/2} \text{Re} \left\{ T(0) \, e^{-3\pi i/4 - iky^2/2r} \right\}.$$

This intensity is integrated over a rectangular objective with side a in the z-direction and side b in the y-direction. The integration over z gives a factor a; the integration over y gives an expression containing a Fresnel integral (sec. 3.12). Some simplifications result, and the total intensity received by the objective is found to be

$$a(b - c_{ext}),$$

where

$$c_{ext} = \frac{4}{k} \text{Re} \{T(0)\}.$$

This is the effective extinction width of an arbitrary cylinder of infinite length.

The simplest application, analogous to sec. 8.22, is found for cylinders whose geometrical width (shadow width), g, is $\gg \lambda$.   Ordinary diffraction theory gives

$$T(0) = \frac{kg}{2} = \frac{\pi g}{\lambda},$$

from which $c_{ext} = 2g$, so that the extinction paradox is confirmed.

15.22. Cylinders of Finite Length

We shall now consider a cylinder which has a large but finite length. It does not matter very much whether the ends are straight or round. The direction of propagation of the incident radiation is still supposed to be perpendicular to the axis.   Such a cylinder is a finite body, so the

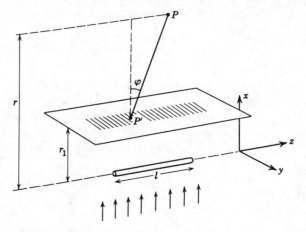

Fig. 65. Determination of the amplitude at infinity of the spherical wave scattered by a very long but finite cylinder.

definitions and theorems of chap. 4 must be applicable. For treating the scattering and extinction by such long needles or strips it is necessary to link up the theory of finite particles [with $S(\varphi, \theta)$ as the amplitude function] with the theory for infinitely long cylinders [with $T(\theta)$ as the amplitude function].   For convenience we discuss at once the field in and close to the forward direction, i.e., the direction of propagation of the incident light.   Let l be the length of the cylinder and a a constant of the order of its width, if the width is larger than λ, or of the order of λ if the width is smaller. Then the waves have a different behaviour in three different zones (Fig. 65).

Zone 1: $r \ll a^2/\lambda$. In this zone the wave motion is complicated.

Wide particles have a shadow, and narrow ones an induction field in this zone.

Zone 2: $a^2/\lambda \ll r \ll l^2/\lambda$. In this zone the scattered wave has the character of a cylindrical wave, except near the ends. So the formulae of sec. 15.21 containing T are applicable. This follows from the physical consideration that the part of the cylinder that effectively influences the field at at point in this zone will have a length of the order $\sqrt{r\lambda}$, which is much smaller than l. So, if the z-coordinate of this point is well within the range $(-\tfrac{1}{2}l, \tfrac{1}{2}l)$ occupied by the cylinder, the field does not differ greatly from that caused by a cylinder of infinite length. If the z-coordinate is well beyond this range the field is undisturbed.

Zone 3: $r \gg l^2/\lambda$. In this zone the scattered wave behaves like a spherical wave, so the formulae of chap. 4, containing S, are applicable. We can derive S from T in the following manner. Let r_1 lie in the second zone. The amplitude in a point P' with coordinates $x' = r_1$, y', z' is

$$u_{P'} = \left(\frac{2}{\pi k r_1}\right)^{1/2} T(0)\, e^{-3\pi i/4 - ikr_1 - iky'^2/2r_1},$$

if $-\tfrac{1}{2}l < z' < \tfrac{1}{2}l$ and 0 if z' is outside this region (see hatched area in Fig. 65). Let P be a point in the third zone with coordinates $x = r\cos\varphi$, $y = 0$, $z = r\sin\varphi$, where φ is a small angle. We apply Huygens' principle (sec. 3.12):

$$u_P = \frac{ik}{2\pi r} \iint u_{P'}\, e^{ik\rho} dz' dy',$$

where ρ is the distance PP'. With legitimate approximations we have

$$\rho = r - r_1 - z' \sin\varphi;$$

it has already been replaced by r in the factor in front of the integrals. Integration over y' gives a complete Fresnel integral (sec. 3.12):

$$\int_{-\infty}^{\infty} e^{-iky'^2/2r_1} dy' = \left(\frac{2\pi r_1}{k}\right)^{1/2} e^{-i\pi/4}.$$

Integration over z' gives an integral of the standard type

$$\int_{-\frac{1}{2}l}^{\frac{1}{2}l} e^{ikz'\varphi} dz = \frac{2}{k\varphi} \sin \tfrac{1}{2}kl\varphi = l \cdot E\left(\frac{kl\varphi}{2}\right),$$

where $E(u) = (\sin u)/u$ is one of the functions tabulated in sec. 7.4. Altogether:

$$u_P = \frac{l}{\pi i r} E\left(\frac{kl\varphi}{2}\right) T(0)\, e^{-ikr}.$$

This is indeed the amplitude of a spherical wave, for it decreases as $1/r$.

The derivation for $\theta \neq 0$ is exactly the same; only $T(0)$ has to be replaced by $T(\theta)$. The extension to electromagnetic waves, i.e., waves with polarization, is simply made by adding an index 1 or 2 to $T(\theta)$, corresponding to case I and case II.

The amplitude function $S(\theta, \varphi)$, defined for arbitrary particles and scalar waves in sec. 4.1, is thus found to be

$$S(\theta, \varphi) = \frac{kl}{\pi} E\left(\frac{kl\varphi}{2}\right) T(\theta);$$

with an index 1 or 2 it holds for electromagnetic waves (sec. 4.41; the cross-polarization terms are virtually zero).

It may be noted that θ and φ do not have the usual meaning; in this formula θ represents geographic longitude and φ geographic latitude, if the cylinder axis is the polar axis and the x-y plane the equatorial plane. Scattering is mainly confined to a narrow belt around the equator. Scattering at high latitudes, e.g., in the positive and negative z-direction, depends on the form of the ends of the cylinder and contributes little to the total amount.

The cross section for extinction of the cylinder is found from the theorem in sec. 4.21 (scalar waves) or sec. 4.41 (electromagnetic waves). For $\theta = 0$, $\varphi = 0$ we have

$$S(0) = \frac{kl}{\pi} T(0)$$

so

$$C_{ext} = \frac{4l}{k} \operatorname{Re}\{T(0)\} = lc_{ext}.$$

This formally confirms the obvious relation that the cross section of a cylinder of length l is l times its effective width. The essential assumption made is that zone 2 exists, i.e., $l \gg a$. In words: the length should be much longer than the width and also much longer than the wavelength. A similar result is easily derived for the scattering cross section.

15.23. Resulting Formulae for Circular Cylinders

We now return to circular cylinders with radius a. Dividing C_{ext} by the geometric cross section $2al$, or c_{ext} by the geometric width $2a$ we obtain in either case the efficiency factor

$$Q_{ext} = \frac{2}{x} \operatorname{Re}\{T(0)\},$$

where $x = ka$ and, likewise,

$$Q_{sca} = \frac{1}{\pi x} \int_0^{2\pi} |T(\theta)|^2 \, d\theta.$$

Upon substitution of $T_1(\theta)$ and $T_2(\theta)$, which specify the T-function for electromagnetic waves (sec. 15.13), these forms reduce to[1]

$$Case\ I: \qquad Q_{ext} = \frac{2}{x} \sum_{n=-\infty}^{\infty} \operatorname{Re} b_n, \qquad Q_{sca} = \frac{2}{x} \sum_{n=-\infty}^{\infty} |b_n|^2,$$

$$Case\ II: \qquad Q_{ext} = \frac{2}{x} \sum_{n=-\infty}^{\infty} \operatorname{Re} a_n, \qquad Q_{sca} = \frac{2}{x} \sum_{n=-\infty}^{\infty} |a_n|^2,$$

analogous to sec. 9.32.

For computing these series it is sometimes useful to define the phase angles β_n and α_n by

$$\tan \beta_n = \frac{mJ_n'(y)J_n(x) - J_n(y)J_n'(x)}{mJ_n'(y)N_n(x) - J_n(y)N_n'(x)},$$

$$\tan \alpha_n = \frac{J_n'(y)J_n(x) - mJ_n(y)J_n'(x)}{J_n'(y)N_n(x) - mJ_n(y)N_n'(x)}.$$

Then, identically with the formulae for spheres (sec. 10.21),

$$b_n = \frac{\tan \beta_n}{\tan \beta_n - i}, \qquad a_n = \frac{\tan \alpha_n}{\tan \alpha_n - 1}.$$

Exclusively if m is real, β_n and α_n are real and

$$\operatorname{Re} b_n = |b_n|^2 = \sin^2 \beta_n = \tfrac{1}{2}(1 - \cos 2\beta_n),$$

$$\operatorname{Re} a_n = |a_n|^2 = \sin^2 \alpha_n = \tfrac{1}{2}(1 - \cos 2\alpha_n),$$

so that

$$Case\ I: \qquad Q_{ext} = Q_{sca} = \frac{1}{x} \sum_{n=-\infty}^{\infty} (1 - \cos 2\beta_n),$$

$$Case\ II: \qquad Q_{ext} = Q_{sca} = \frac{1}{x} \sum_{n=-\infty}^{\infty} (1 - \cos 2\alpha_n).$$

15.3. Some Numerical Results for Real m

15.31. Phase Angles and Nodes

The computations leading from the analytical solution to numerical results are still lengthy. They have so many details in common with the computations for spheres treated in the preceding chapters that no extensive comments are needed. In one respect the present problem is

---

[1] The indices 1 and 2 (referring to cases I and II, respectively) will be written with the efficiency factors only where danger of confusion exists.

even simpler: the angles β_n and α_n refer to different physical problems (case I and case II) and need not appear together in the final formulae. Yet it is wise to make the computation for the two cases alongside each other.

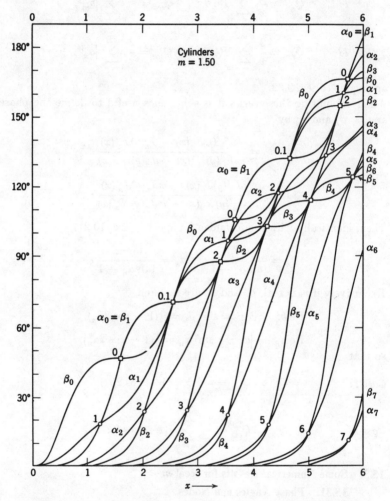

Fig. 66. Phase angles for scattering by cylinders with $m = 1.50$. Squares denote nodes of the first kind, circles, nodes of the second kind.

An example of the phase angles as functions of x is given in Fig. 66 for $m = 1.50$. They might have been computed directly from tables of the Bessel functions and their derivatives, but the labour was considerably reduced by using the tables of Lowan, Morse, Feshbach, and

Lax (LMFL).   The formulae are very similar to those derived for spherical particles (sec. 10.4).   In the section of the tables that refers to circular cylinders five angles are tabulated.   We adopt the notation γ_n, δ_n and δ_n' without change but add an asterisk to α_n and β_n in order to distinguish them from the phase angles.   The definitions are

$$\tan \alpha_n^*(x) = -\frac{xJ_n'(x)}{J_n(x)}, \quad \tan \beta_n^*(x) = -\frac{xN_n'(x)}{N_n(x)},$$

$$\tan \delta_n(x) = -\frac{J_n(x)}{N_n(x)}, \quad \tan \delta_n'(x) = -\frac{J_n'(x)}{N_n'(x)},$$

$$\tan \gamma_n(x) = \tan \delta_n(x) \cos \beta_n^*(x) \sec \alpha_n^*(x).$$

The tangents of the phase angles, $\tan \beta_n$ and $\tan \alpha_n$, can both be written in the form (omitting the common index n and argument x):

$$- \tan \delta \frac{\tan \varphi - \tan \alpha^*}{\tan \varphi - \tan \beta^*} = - \tan \gamma \frac{\sin (\alpha^* - \varphi)}{\sin (\beta^* - \varphi)},$$

where substitution of $\varphi_n(x) = \alpha_n^*(mx)$ gives $\tan \beta_n$ (case I) and of $\tan \varphi_n(x) = (1/m^2) \tan \alpha_n^*(mx)$ gives $\tan \alpha_n$ (case II).

In this way the formulae have been made suitable for logarithmic computation.   The computation was made for thirteen values of x up to $x = 6.4$, including for each x the values of n until the angles fell below $2°$.   As a second step and to ensure better interpolation the *nodes* were computed.   The nodes, defined by $\alpha_n = \beta_n$, are of two kinds (analogous to sec. 10.22).

Node of the first kind: If $\delta_n(mx) = k\pi$, then

$$\alpha_n = \beta_n = \beta_{n-1} = \beta_{n+1} = k\pi - \delta_n(x),$$

$$\frac{d\beta_n}{dx} = 0, \quad \frac{d\beta_{n-1}}{dx} = \frac{d\alpha_n}{dx} = \frac{d\beta_{n+1}}{dx}.$$

Node of second kind: If $\delta_n'(mx) = k\pi$, then

$$\alpha_n = \beta_n = k\pi - \delta_n'(x).$$

The properties of the nodes of the first kind were conjectured from analogy with the spherical problem.   They were confirmed by the plotted angles and also rigorously proved from recursion formulae of the Bessel functions.   A further property,

$$\alpha_0 = \beta_1,$$

is seen in the figure and can easily be shown to be correct for all x and m, including complex values.   In Fig. 66 the nodes of the first kind are shown as squares and the nodes of the second kind as circles.   The values of $mx = y$ for which these nodes occur are the same for all values of m, for they are the zeros of $J_n(y)$ and $J_n'(y)$, respectively. Table 32 may be convenient for reference.

Table 32. Positions of the Nodes for Circular Cylinders

$n =$	0	1	2	3	4	5	6	7
$\delta_n(y) = 180°$	2.405	3.832	5.136	6.380	7.588	8.771	9.94	
$\delta_n(y) = 360°$	5.520	7.016	8.417	9.761	11.065	12.338		
$\delta_n(y) = 540°$	8.654	10.173	11.620	13.015	14.373	15.700		
$\delta_n(y) = 720°$	11.792	13.324	14.796	16.223	17.616	18.980		
$\delta_n(y) = 900°$	14.931	16.471	16.960	19.409	20.827	22.218		
$\delta_n(y) = 1080°$	18.071	19.616				Roots of $J_n(y) = 0$		
$\delta_n'(y) = 0°$	0	1.841	3.054	4.201	5.317	6.42	7.50	8.58
$\delta_n'(y) = 180°$	3.832	5.331	6.706	8.015	9.282	10.51		
$\delta_n'(y) = 360°$	7.016	8.526	9.970	11.346	12.682			
$\delta_n'(y) = 540°$	10.173	11.706	13.171	14.586	15.964			
$\delta_n'(y) = 720°$	13.324	14.864	16.348	17.789	19.196			
$\delta_n'(y) = 900°$	16.471							
$\delta_n'(y) = 1080°$	19.616				Roots of $J_n'(y) = 0$			

In view of a further check and as a preparation of the investigation reported in sec. 15.52, it was thought advisable also to find the expressions of the derivatives of the phase angles in the nodes. A lengthy computation which cannot be reproduced gives the following results. Given are, respectively, the derivative with respect to x at fixed m and the derivative with respect to m at fixed x.

Nodes of first kind:

$$\frac{\partial \beta_n}{\partial x} = 0, \qquad \frac{\partial \beta_n}{\partial m} = \frac{x}{mT},$$

$$\frac{\partial \alpha_n}{\partial x} = \frac{m^2 - 1}{T}, \qquad \frac{\partial \alpha_n}{\partial m} = \frac{mx}{T}.$$

Nodes of second kind:

$$\frac{\partial \beta_n}{\partial x} = \frac{m^2 - 1}{V}, \qquad \frac{\partial \beta_n}{\partial m} = \frac{xm}{V}\left(1 - \frac{n^2}{m^2 x^2}\right),$$

$$\frac{\partial \alpha_n}{\partial x} = \frac{(m^2 - 1)n^2}{Vm^2 x^2}, \qquad \frac{\partial \alpha_n}{\partial m} = \frac{x}{mV}\left(1 - \frac{n^2}{m^2 x^2}\right).$$

Here
$$T = \frac{\pi x}{2}\{J_n^2(x) + N_n^2(x)\} = \frac{\pi x}{2}|H_n(x)|^2,$$

$$V = \frac{\pi x}{2}\{J_n'^2(x) + N_n'^2(x)\} = \frac{\pi x}{2}|H_n'(x)|^2.$$

The function T may be taken directly from the tables computed by Lax and Feshbach (1948).

Numerical example: $m = 1.25.$

A node of the first kind occurs at $n = 3$, $x = 5.10$. Here $\alpha_3 = \beta_3 = 63°$; $T = 1.024$,

$$\frac{\partial \beta_3}{\partial x} = 0, \qquad\qquad \frac{\partial \beta_3}{\partial m} = 3.39 \ (194°),$$

$$\frac{\partial \alpha_3}{\partial x} = 0.47 \ (27°), \qquad\qquad \frac{\partial \alpha_3}{\partial m} = 5.30 \ (304°).$$

A node of the second kind occurs at $n = 6$, $x = 6.0$. Here $\alpha_6 = \beta_6 = 27°$; $V = 0.625$,

$$\frac{\partial \beta_6}{\partial x} = 0.90 \ (52°), \qquad\qquad \frac{\partial \beta_6}{\partial m} = 4.32 \ (248°),$$

$$\frac{\partial \alpha_6}{\partial x} = 0.58 \ (33°), \qquad\qquad \frac{\partial \alpha_6}{\partial m} = 2.77 \ (159°).$$

Table 33 presents an example of the final stage of the computation. The efficiency factors Q are computed separately for case I and case II, using the angles of Fig. 66 and the formulae at the end of sec. 15.23.

Table 33. Example of Computation of Extinction ($m = 1.5$, $x = 2.4$)

n	$2\alpha_n$	$2\beta_n$	$1-\cos 2\alpha_n$	$1-\cos 2\beta_n$
0	122.4°	141.0°	0.768*	0.889*
1	141.0°	119.4°	1.777	1.491
2	116.0°	71.6°	1.438	0.684
3	12.0°	22.4°	0.023	0.075
4	0.8°	2.8°	0.000	0.001
		Sum:	4.006	3.140

(Divided by $\tfrac{1}{2}x$) $Q_2 = 3.338$ $Q_1 = 2.617$

* Factor $\tfrac{1}{2}$ added for $n = 0$.

The result is shown in Fig. 67, together with the Q curves obtained in a similar way for $m = 1.25$. It is seen that the curves show the same general behavior as was found for spheres (Fig. 32 in sec. 11.22). The difference between Q_1 and Q_2 is comparatively small and is smaller for the smaller value of m. The similarity of the Q-curves has been brought out by plotting them on different scales of x in such a way that they have the scale of $\rho = 2(m - 1)x$ in common.

15.32.   Refractive Index Near 1

The limiting form of the curve for $m \to 1$ is also shown in Fig. 67.

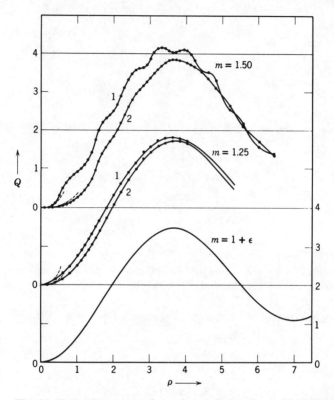

Fig. 67. Extinction curves of very long cylinders for case I ($E \parallel$ axis) and case II ($H \parallel$ axis) computed for three different values of the refractive index.

A very simple modification of the derivation for spheres given in sec. 11.21 gives for cases I and II alike

$$Q(\rho) = 2 \operatorname{Re} \int_0^{\pi/2} (1 - e^{-i\rho \cos \gamma}) \cos \gamma \, d\gamma$$

$$= 2 \int_0^{\pi/2} \{1 - \cos (\rho \cos \gamma)\} \cos \gamma \, d\gamma.$$

By a partial integration this reduces to

$$Q(\rho) = 2\rho \int_0^{\pi/2} \sin (\rho \cos \gamma) \sin^2 \gamma \, d\gamma,$$

which is π times the integral defining the first-order Struve function, so that

$$Q(\rho) = \pi \mathscr{S}_1(\rho).$$

This function was taken directly from the tables of Jahnke-Emde and plotted as the bottom curve in Fig. 67. For large values of ρ the identical expression $2 + \pi\Omega_1(\rho)$ was used; $\Omega_1(\rho)$ is also found in Jahnke-Emde. In the limit of $\rho \to \infty$ we find $Q = 2$ as it should be. For very small ρ the result reduces to

$$Q(\rho) = \frac{2}{3}\rho^2,$$

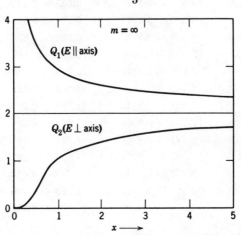

Fig. 68. Extinction curves of very long, totally reflecting cylinders for the two states of linear polarization of the incident light.

a result found also in sec. 7.32. It is the formula for the intermediate case in the sense of sec. 11.1.

15.33. Totally Reflecting Cylinders

The assumption $m = \infty$ (total reflection) leads, as for spheres (sec. 10.62), to some simplification. The phase angles are

$$\alpha_n = -\delta_n'(x), \qquad \beta_n = -\delta_n(x),$$

Both of which may be taken directly from the LMFL tables. Table 34 shows the numerical results for some values of x. For later use, the extinction factor Q has been completed by the imaginary part, denoted by P, of the function from which it was derived (sec. 15.23):

$$Q_1 + iP_1 = \frac{2}{x}T_1(0), \qquad Q_2 + iP_2 = \frac{2}{x}T_2(0).$$

The real parts have been plotted against x in Fig. 68. A graph of the function in the complex domain is found in Fig. 80 (sec. 17.24). Unlike the extinction curves we have met so far, the value of $Q_1(x)$ does not start from 0 for $x = 0$ but from ∞. This is confirmed by a series expansion for small x.

Table 34. Values of $Q + iP$ for Cylinders with $m = \infty$

x	Case I $(\mathbf{E} \| \text{axis})$	Case II $(\mathbf{H} \| \text{axis})$
0.2	$4.572 - 5.574i$	$0.029 + 0.332i$
0.4	$3.696 - 3.347i$	$0.206 + 0.656i$
0.6	$3.319 - 2.528i$	$0.533 + 0.885i$
0.8	$3.101 - 2.060i$	$0.834 + 0.880i$
1.0	$2.957 - 1.761i$	$1.000 + 0.787i$
1.5	$2.740 - 1.340i$	$1.200 + 0.728i$
2.0	$2.621 - 1.101i$	$1.359 + 0.650i$
2.5	$2.530 - 0.950i$	$1.447 + 0.599i$
3.0	$2.471 - 0.836i$	$1.517 + 0.545i$
4.5	$2.361 - 0.637i$	$1.639 + 0.449i$
5.0	$2.337 - 0.594i$	$1.665 + 0.424i$
9.5	$2.221 - 0.386i$	$1.789 + 0.296i$
10.0	$2.213 - 0.373i$	$1.797 + 0.288i$

In case I the dominant coefficient is the one containing β_0. The rigorous expression is

$$\tan \beta_0 = \pi/2l \,,$$

where $l = \gamma + \ln (x/2)$ and $\gamma = 0.5771$ is Euler's constant. The expression resulting if only this term would count is

$$Q_1 + iP_1 = \frac{2}{x}\left\{1 + \frac{2il}{\pi}\right\}\Big/\left\{1 + \left(\frac{2l}{\pi}\right)^2\right\};$$

its real part gives the correct extinction within 5 per cent for $x < 0.4$.

In case II the dominant terms a_0 and a_1 have equal but opposite phase angles for small x, given by

$$-\alpha_0 = \alpha_1 = \pi x^2/4,$$

from which

$$Q_2 + iP_2 = \frac{3\pi^2 x^3}{8} + \frac{\pi x i}{2} \,.$$

The interesting problem of deriving these results at once by a reasoning similar to sec. 10.61 for spheres has not been investigated.

The numerical values for large x may be represented by the empirical formulae

$$Q_1 + iP_1 = 2 + (1.00 - 1.73i)x^{-2/3},$$

$$Q_2 + iP_2 = 2 - (0.86 - 1.49i)x^{-2/3} - (0.36 + 0.73i)x^{-4/3}.$$

These expressions were found on the assumption that the difference with the asymptotical value $Q = 2$, $P = 0$, would be proportional to $x^{-2/3}$. The coefficients were read from graphs in which $x^{2/3}$ times the difference with 2 was plotted against $x^{-2/3}$. The expressions may be expected to give nearly correct values for any $x > 3$. A fuller discussion of the asymptotic behavior is found in secs. 17.21 to 17.24.

15.34. Scattering Diagrams

When the phase angles have been computed, either for finite m or for $m = \infty$, the complete scattering diagrams can be found with little additional work from $T_1(\theta)$ and $T_2(\theta)$ defined in sec. 15.13. The same holds true for the efficiency factor for radiation pressure. The results are very similar to those for spheres. Some sample diagrams, computed by Rayleigh in 1918, are presented in Fig. 25 (sec. 10.4).

15.4. Very Thin Cylinders

15.41. Dielectric Needles

If $x \ll 1$ in the preceding formulae, we have to do with needles that have diameters $\ll \lambda$. For such thin needles expansions in terms of powers of x similar to those given for spheres in sec. 10.31 (non-absorbing) and in sec. 14.21 (absorbing) are appropriate. Figure 66 (sec. 15.31) shows that for small x the phase angles β_0 and α_1 are predominant, followed by α_0 and β_1, which are equal, and still later by the phase angles with $n = 2, 3$, etc.

This is confirmed by series expansion. Expanding the formulae by which α_n and β_n are defined (sec. 15.23) we find

$$\beta_0 = \frac{\pi x^2}{4}(m^2 - 1) + \cdots,$$

$$\alpha_1 = \frac{\pi x^2}{4}\frac{m^2 - 1}{m^2 + 1} + \cdots,$$

$$\alpha_0 = \beta_1 = \frac{\pi x^4}{32}(m^2 - 1) + \cdots.$$

The general formulae for the extinction factors are

Case I: $$Q_1 = \frac{2}{x}(\sin^2 \beta_0 + 2\sin^2 \beta_1 + \cdots),$$

Case II: $$Q_2 = \frac{2}{x}(\sin^2 \alpha_0 + 2\sin^2 \alpha_1 + \cdots).$$

The dominant terms are those containing β_0 and α_1, respectively. So we have in the limit for very thin needles

$$Q_1 = \frac{\pi^2}{8} x^3 (m^2 - 1)^2,$$

$$Q_2 = \frac{\pi^2}{4} x^3 \frac{(m^2 - 1)^2}{(m^2 + 1)^2},$$

with the ratio

$$\frac{Q_1}{Q_2} = \frac{(m^2 + 1)^2}{2}.$$

For instance,

For $m = 1.25$: $Q_1 = 0.39x^3$, $Q_2 = 0.119x^3$, $Q_1/Q_2 = 3.28$;

For $m = 1.50$: $Q_1 = 1.93x^3$, $Q_2 = 0.364x^3$, $Q_1/Q_2 = 5.30$.

These formulae are useful if both x and mx are smaller than 0.4. Beyond this value terms of the order of x^4 in β_0 and α_1 (not given), as well as the terms containing α_0 and β_1, have to be taken into account.

For $m \to 1$ the values of Q_1 and Q_2 do not become equal, as might be surmised, but their ratio goes to 2. This paradoxical result is explained automatically if we try to understand why it is that the term with $n = 0$ dominates in case I and the term with $n = 1$ (more precisely $n = \pm 1$) in case II.

The clue is the formula for Rayleigh scattering by non-spherical particles. From sec. 6.32 we find for a very long circular cylinder the following values of the depolarization factor (L) and the polarizability α.

Case I (electric field $\parallel$ axis):

$$L = 0, \text{ so } \frac{4\pi\alpha}{V} = m^2 - 1.$$

Case II (electric field $\perp$ axis):

$$L = \tfrac{1}{2}, \text{ so } \frac{4\pi\alpha}{V} = \frac{2(m^2 - 1)}{m^2 + 1}.$$

The condition that the cylinder is placed in an effectively homogeneous field is fulfilled in the present section by virtue of the assumptions that the direction of incidence is perpendicular to the axis and that the radius is small. However, the other assumptions needed for Rayleigh scattering, namely, that *all* dimensions are $\ll \lambda$, is not fulfilled, for the formulae derived in the previous sections hold for cylinders that are *very long* compared to the wavelength.

Let us consider a section of the total cylinder that is much longer than the radius, yet shorter than λ. This section gives Rayleigh scattering governed by the formulae quoted above. The total cylinder is made up of many such sections, the scattered light of which interferes. This interference reduces the scattered intensity by a very large factor except for directions that are very closely perpendicular to the axis of the cylinder. The interference cuts, so to say, a two-dimensional slice out of the three-dimensional pattern of Rayleigh scattering. This slice is parallel to the x-y plane (Fig. 64), and any direction in this slice is characterized by one angle, θ, only. Here the difference between case I and case II appears. In case I the radiating dipole is parallel to the axis, so the x-y plane is perpendicular to the dipole. This means that the scattering is constant in all directions in the plane. The scattering function is isotropic (in this plane), and this is why the term with $n = 0$ is the dominant one. In case II the radiating dipole is in the x-y plane. The radiation in the x-y plane now has its full strength in the directions perpendicular to the dipole ($\theta = 0°$ and $180°$) but is zero in the direction of the dipole ($\theta = 90°$). The scattered intensity in any direction is proportional to $\cos^2 \theta$. The fact that the amplitude contains a factor $\cos \theta$ is brought out in the rigorous formulae by the fact that the term with $n = 1$ is the dominant term.

Now let us make m very close to 1. The two values of the polarizability then are equal, as they should be according to sec. 6.22. However, the slice cut from the Rayleigh scattering pattern is more favorable in case I than in case II. The difference is a factor $\cos^2 \theta$, the average value of which over all angles from $0°$ to $180°$ is $\frac{1}{2}$. This is just the factor Q_2/Q_1 for which an explanation was sought.

Further useful formulae for the limiting case of m close to 1 are found in sec. 7.32, which describes the Rayleigh-Gans scattering by cylinders of arbitrary length and width and arbitrary orientation.

15.42. Absorbing Particles

We return to our original problem: very long, very thin cylinders and perpendicular incidence. We now ask for the values of Q_1 and Q_2 in the event that these cylinders are absorbing (metallic). Straight application of the formulae of secs. 15.23 and 15.41 gives, if we retain only the dominant terms:

$$Case\ I:\quad Q_{ext} = \frac{2}{x}\operatorname{Re} b_0 = \frac{2}{x}\operatorname{Re}(i\beta_0) = -\frac{2}{x}\operatorname{Im}(\beta_0) = -\frac{\pi x}{2}\operatorname{Im}(m^2 - 1)$$

and, likewise,

$$Case\ II: Q_{ext} = \frac{2}{x}\operatorname{Re}(a_1 + a_{-1}) = \frac{4}{x}\operatorname{Re}(a_1)^{\cdot} = -\frac{4}{x}\operatorname{Im}(\alpha_1)$$

$$= -\pi x\operatorname{Im}\left(\frac{m^2 - 1}{m^2 + 1}\right).$$

It is curious to compare with the dielectric case (sec. 15.41) and to note that a factor 2 has come in, so that in the limit for $m \to 1$ the values of Q_{ext} for both cases are now equal. This can be easily understood from the physics of the problem. The extinction is now due to absorption *in* the cylinder, and interference effects are irrelevant. It does not matter whether the cylinder is long or short or comparable with λ. In fact, the formulae are identical with those for cylinders which have *all* dimensions $\ll \lambda$, as may be verified from secs. 6.13 and 6.32.

15.5.　Some Results for Complex m

15.51.　Extinction Curves for $m = \sqrt{2}\,(1 - i)$

The lack of tables of the Bessel functions for complex argument makes it a laborious process to compute the extinction by absorbing cylinders as well as by absorbing spheres. If any extensive work were planned it might be worth while to develop series expansions as Schalén has done for spheres (sec. 14.21). Instead, we just wished to have one representative example. The refractive index $m = \sqrt{2}(1 - i)$ was chosen, because it makes the series expansions of

$$J_n(y) = \sum_{r=0}^{\infty} (-1)^r \frac{(\tfrac{1}{2}y)^{n+2r}}{r!\,(n+r)!}$$

particularly simple as $(\tfrac{1}{2}y)^2 = (\tfrac{1}{2}mx)^2 = -ix^2$. This value of m is close to the actual values that hold, e.g., for iron particles in air. Moreover, for this value of m we have

$$J_n(y) = J_n(mx) = (-1)^n\{\operatorname{ber}_n(2x) + i\operatorname{bei}_n(2x)\},$$

where $\operatorname{ber}_n(z)$ and $\operatorname{bei}_n(z)$ are functions tabulated for use in some engineering problems.

It was decided to make the computations for

$$x = 0.2,\ 0.4,\ 0.6,\ 0.8,\ 1.0,\ 1.4,\ 2.0,\ 3.0,\ \text{and}\ 4.0.$$

The first step was to tabulate the Bessel functions that were needed up to $n = 7$. They were obtained from the following sources (for references, see end of chapter):

$J_n(x)$ and $J_n'(x)$, $N_n(x)$ and $N_n'(x)$ from Watson.

ber$_n$ and bei$_n$ for $n = 0$ to 5 and integer values of x from McLachlan.

ber$_n$ and bei$_n$ for $n = 0$ and all values of x from Jahnke-Emde.

$J_1(y)$ for $n \neq 0$ and non-integer values of x from series expansion.

$J_1(y)$ for $n = 6$ and 7 by the relation

$$J_{n+1}(y) = \frac{2n}{y} J_1(y) - J_{n-1}(y).$$

$J_n'(y)$ in all cases by the relations

$$J_n'(y) = \frac{J_{n-1}(y) - J_{n+1}(y)}{2} \, , \; J_0'(y) = -J_1(y).$$

When the functions had been tabulated, straight substitution into the formulae given in sec. 15.12 gave the values for the coefficients a_n and b_n. They are given in Table 35.

Lack of computing time made it impossible to check every step in the calculation independently. A number of checks were applied, and some errors corrected. By means of graphs it was made certain that no gross errors remained. Some values were added by extrapolation. The graphs of a_n and b_n in the complex domain as a function of x are similar to those for spheres (sec. 14.22 and Fig. 55). For $x \to \infty$ the locus approaches a circle around the point $\frac{1}{2}$ with radius

$$\tfrac{1}{2} \left| \frac{m-1}{m+1} \right| = 0.263.$$

The coefficients in Table 35 have to be summed from $-\infty$ to ∞. This means that, e.g., $T_1(0)$ is made up of b_0 plus $2(b_1 + b_2 + \cdots)$. The values of these sums and of $(2/x)T(0) = Q + iP$ (compare sec. 15.23) are given in the table.

As $m^2 - 1 = -1 - 4i$ and $(m^2 - 1)/(m^2 + 1) = (1/17)(15 - 8i)$, we obtain from sec. 15.42 the formulae for small x:

Case I: $Q_{ext} = 2\pi x = 6.28x$,

Case II: $Q_{ext} = 8\pi x/17 = 1.47x$.

Table 35. Coefficients for Metallic Cylinders with $m = \sqrt{2}\,(1 - i)$ for Case I ($E \parallel$ axis) and Case II ($H \parallel$ axis)*

	$x = 0.4$	$x = 0.6$	$x = 0.8$	$x = 1$
Case I				
b_n $n = 0$	$(0.276 - 0.108i)$	$0.4511 - 0.1760i$	$0.5881 - 0.1693i$	$0.6859 - 0.1105i$
$n = 1$	$(0.009 - 0.002i)$	$0.0383 - 0.0196i$	$0.0914 - 0.0612i$	$0.1603 - 0.1250i$
$n = 2$		$(0.0005 - 0.0002i)$	$(0.0030 - 0.0015i)$	$0.0115 - 0.0063i$
$n = 3$				$0.0003 - 0.0001i$
$T_1(0)$	$0.294 - 0.112i$	$0.5287 - 0.2156i$	$0.7769 - 0.2946i$	$1.0301 - 0.3733i$
$(2/x)T_1(0)$	$1.47 \ - 0.56i$	$1.762 \ - 0.719i$	$1.942 \ - 0.736i$	$2.060 \ - 0.747i$
Case II				
a_n $n = 0$	$(0.009 - 0.002i)$	$0.0383 - 0.0196i$	$0.0914 - 0.0611i$	$0.1603 - 0.1251i$
$n = 1$	$(0.074 + 0.093i)$	$0.1732 + 0.1746i$	$0.2798 + 0.2238i$	$0.3554 + 0.2442i$
$n = 2$	$(0.001 + 0.002i)$	$(0.0078 + 0.0130i)$	$(0.0244 + 0.0370i)$	$0.0591 + 0.0718i$
$n = 3$		$(0.0001 + 0.0002i)$	$(0.0005 + 0.0009i)$	$0.0020 + 0.0033i$
$T_2(0)$	$0.159 + 0.188i$	$0.4045 + 0.3560i$	$0.7008 + 0.4623i$	$0.9933 + 0.5135i$
$(2/x)T_2(0)$	$0.80 \ + 0.94i$	$1.348 \ + 1.187i$	$1.752 \ + 1.156i$	$1.987 \ + 1.027i$

	$x = 1.4$	$x = 2$	$x = 3$	$x = 4$
Case I				
b_n $n = 0$	$0.7361 + 0.0809i$	$0.5068 + 0.2588i$	$0.2648 - 0.1118i$	$0.7043 - 0.1637i$
$n = 1$	$0.3321 - 0.2502i$	$0.6338 - 0.2468i$	$0.6902 + 0.1946i$	$0.2543 + 0.1075i$
$n = 2$	$(0.600 \ - 0.0490i)$	$0.1827 - 0.1915i$	$0.5950 - 0.2954i$	$0.7670 + 0.1108i$
$n = 3$	$(0.0030 - 0.0020i)$	$0.0301 - 0.0282i$	$0.1884 - 0.2219i$	$0.5685 - 0.3215i$
$n = 4$		$0.0023 - 0.0015i$	$0.0412 - 0.0507i$	$0.1918 - 0.2401i$
$n = 5$		$0.0001 - 0.0000i$	$0.0054 - 0.0049i$	$0.0481 - 0.0693i$
$n = 6$			$0.0004 - 0.0003i$	$0.0107 - 0.0120i$
$n = 7$				$(0.0012 - 0.0012i)$
$n = 8$				$(0.0001 - 0.0001i)$
$T_1(0)$	$1.5263 - 0.5215i$	$2.2048 - 0.6773i$	$3.3060 - 0.8690i$	$4.3877 - 1.0155i$
$(2/x)T_1(0)$	$2.180 \ - 0.745i$	$2.205 \ - 0.677i$	$2.204 \ - 0.579i$	$2.194 \ - 0.508i$
Case II				
a_n $n = 0$	$0.3322 - 0.2502i$	$0.6337 - 0.2468i$	$0.6902 + 0.1946i$	$0.2543 + 0.1075i$
$n = 1$	$0.3834 + 0.2406i$	$0.2736 + 0.1267i$	$0.3909 - 0.2354i$	$0.7577 - 0.0428i$
$n = 2$	$(0.1780 + 0.1300i)$	$0.4663 + 0.1718i$	$0.3693 + 0.1693i$	$0.3036 - 0.1329i$
$n = 3$	$(0.0160 + 0.0230i)$	$0.1456 + 0.1080i$	$0.5016 + 0.1154i$	$0.4320 + 0.1711i$
$n = 4$	$(0.0010 + 0.0020i)$	$0.0099 + 0.0135i$	$0.2189 + 0.1018i$	$0.5169 + 0.0762i$
$n = 5$		$0.0004 + 0.0007i$	$0.0242 + 0.0258i$	$0.2680 + 0.0791i$
$n = 6$			$0.0016 + 0.0023i$	$0.0374 + 0.0437i$
$n = 7$			$(0.001 \ + 0.0002i)$	$(0.0032 + 0.0060i)$
$n = 8$				$(0.0003 + 0.0004i)$
$T_2(0)$	$1.4890 + 0.5410i$	$2.4253 + 0.5946i$	$3.7034 + 0.5534i$	$4.8925 + 0.5091i$
$(2/x)T_2(0)$	$2.127 \ + 0.773i$	$2.425 \ + 0.595i$	$2.469 \ + 0.369i$	$2.446 \ + 0.255i$

* Q_{ext} is the real part of the numbers given in the bottom lines. Parentheses indicate that the values may be somewhat in error.

From a plot of b_0/x^2 and a_1/x^2 in the complex domain we can see for what values of x this approximation becomes reasonable. It turns out that, even for $x = 0.2$, for which we estimate

$$b_0 = 0.097 \quad - 0.032i, \quad Q_{ext,\,1} = 0.97,$$

$$a_1 = 0.0164 - 0.0260, \quad Q_{ext,\,2} = 0.33,$$

the deviation in case I is over 20 per cent. We have to go to $x < 0.1$ to make fair use of the approximations for very thin cylinders.

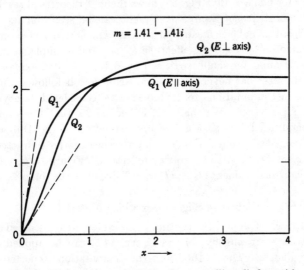

Fig. 69. Extinction curves of very long metallic cylinders with
$m = 1.41 - 1.41i$.

Figure 69 shows the final extinction curves. The straight lines drawn through the origin indicate the linear approximation for thin cylinders. Each curve by itself has a character similar to the extinction curves for absorbing spheres, which was explained in connection with Fig. 54 (sec. 14.22). The most interesting result is that the two curves cross over. The extinction for both directions of polarization is equal at $x = 1.18$, $Q_{ext} = 2.12$. For smaller x the radiation with electric vector parallel to the axis (case I) suffers the stronger extinction. For larger x the radiation with magnetic vector parallel to the axis (case II) suffers the stronger extinction.

The existence of this cross-over point has been known from experiments since the end of the last century. The first effect (small x, i.e., long wavelength compared to the radius) was found in experiments of the scattering of radio waves by wire gratings. It is known as the Hertz effect and is daily used in antenna design. The second effect (large x, i.e., short wavelength compared to the radius) was found in experiments on the transmission of light by wire gratings and is known as the Dubois effect. Such wire gratings have been used extensively as diffraction gratings in front of the objectives of astronomical telescopes. A small but noticeable polarization effect occurs with its sign just opposite to what is familiar in radio. Light with $\mathbf{E} \parallel$ axis comes through with the greater intensity.

The author believes that Fig. 69 shows the first theoretical results from which the existence of a cross-over point can be seen and its position read. This does not yet mean that a clear explanation has been found. The explanation of the Hertz effect (small x) is quite simple, as it follows from the formulae for small particles (sec. 6.32). A good explanation of the Dubois effect has never been given. Nor does it follow simply from the preceding computations, for we have not yet considered the asymptotic behavior for $x \to \infty$. We shall see, in sec. 17.25, that the situation is fairly complicated but that a fair understanding of the Dubois effect may be gained by considering the interference of the diffracted wave with the edge wave caused by grazing reflection. Figure 81, given in that section, shows the values of $Q + iP$ from Table 35 in the complex domain.

15.52. Cylinders of Slightly Absorbing Material

The method of analytical continuation, which was discussed in connection with scattering by spheres in sec. 14.5, can be applied equally well to cylinders. The method and the computations reported in this section were worked out as a separate research problem by Mrs. Elske v. P. Smith.

The starting point is the formulae at the end of sec. 15.23. Analytical continuation into complex values of m is possible if results are available for a range of values of m. It was found useful in the present problem to apply the continuation method to each phase angle separately before the summation is made.

The functions α_n and β_n are real for real m and have complex values for complex m. The first stage of the computation is aimed at finding those values for the refractive index $m - i\mu$ where μ is small. Let, for the moment, the angles and derivatives with argument m be written without prime and those with the argument $m - i\mu$ with a prime. By Taylor expansion we have (for case I)

$$\beta_n' = \beta_n - i\mu \frac{\partial \beta_n}{\partial m} - \frac{\mu^2}{2} \frac{\partial^2 \beta_n}{\partial m^2} + \cdots .$$

Retaining only the first-order term and substituting the value of

$$b_n = \tfrac{1}{2}(1 - e^{-2i\beta_n})$$

into the formulae of sec. 15.23 we obtain, still for case I

$$Q_{ext} = \frac{1}{x} \sum_{n=-\infty}^{\infty} \{1 - \cos 2\beta_n \, e^{-2\mu(\partial\beta_n/\partial m)}\},$$

$$Q_{sca} = \frac{1}{x} \sum_{n=-\infty}^{\infty} \{\tfrac{1}{2} + \tfrac{1}{2}e^{-4\mu(\partial\beta_n/\partial m)} - \cos 2\beta_n e^{-2\mu(\partial\beta_n/\partial m)}\}$$

and by subtraction

$$Q_{abs} = \frac{1}{2x} \sum_{n=-\infty}^{\infty} \{1 - e^{-4\mu(\partial\beta_n/\partial m)}\}.$$

In general, the first-order approximation of the Taylor expansion is adequate only if the exponents are small. This gives a further simplification, viz.,

$$Q_{abs} = \frac{2\mu}{x} \sum_{n=-\infty}^{\infty} \frac{\partial\beta_n}{\delta m}.$$

In all these sums the terms of order n and $-n$ are equal. The formulae for the other polarization (case II, $\mathbf{H} \parallel$ axis) are obtained by replacing β_n by α_n.

The equations may be checked by applying them to very thin cylinders. From the formulae for α_1 and β_0 in sec. 15.41 we obtain

$$\textit{Case I:} \quad Q_{abs} \approx Q_{ext} \approx \pi x m \mu,$$

$$\textit{Case II:} \quad Q_{abs} \approx Q_{ext} \approx \pi x \frac{4m\mu}{(m^2 + 1)^2}.$$

The same results are obtained in the limit of small μ by direct substitution of $m - i\mu$ into the formulae of sec. 15.42.

In order to work out the numerical results for a particular value of the refractive index, e.g., $1.50 - 0.10i$, the following method was chosen. The values of α_n and β_n for $m = 1.50$ and $m = 1.25$ were known from the computations reported in sec. 15.31 (see Fig 66). New values were computed for $m = 2.00$ and $m = 1.10$ for the present purpose. These values were not yet sufficient to obtain full graphs of α_n and β_n against m at a fixed value of x. Additional data were obtained by plotting the nodes, where $\alpha_n = \beta_n$. The formulae for the derivatives $\partial\beta_n/\partial m$ and $\partial\alpha_n/\partial m$ are given in sec. 15.31.

It may also be derived from the definitions of $\tan \alpha_n$ and $\tan \beta_n$ in sec. 15.23 that

$$\lim_{m \to 1} \frac{\beta_n}{m-1} = \frac{\pi x^2}{2}\left\{\left(1 - \frac{n^2}{x^2}\right)J_n{}^2(x) + J_n{}'^2(x)\right\},$$

$$\lim_{m \to 1} \frac{\alpha_n}{m-1} = \frac{\pi x^2}{2}\left\{\left(1 - \frac{n^2}{x^2}\right)J_n{}^2(x) + J_n{}'^2(x) + \frac{2}{x}J_n(x)J_n{}'(x)\right\}.$$

These data together are amply sufficient, to draw for a number of chosen values of x, free-hand curves, from which not only the values of the first derivatives but also of the higher derivatives with respect to m may be estimated.

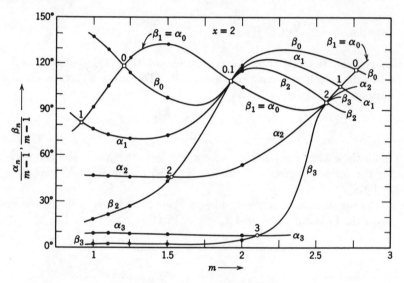

Fig. 70. Plot of the phase angles divided by $m - 1$ versus m for cylinders of a fixed size, $x = 2$. Squares denote nodes of the first kind, circles, nodes of the second kind.

For the practical reason that α_n and β_n increase, on the average, as $m - 1$ it was found more convenient to plot the ratios $\alpha_n/(m - 1)$ and $\beta_n/(m - 1)$. The derivatives of these functions are

$$\frac{\partial}{\partial m}\left(\frac{\alpha_n}{m-1}\right) = \frac{1}{m-1}\left(\frac{\partial \alpha_n}{\partial m} - \frac{\alpha_n}{m-1}\right),$$

and similarly for β_n. The graphs may be continued through $m = 1$ into the domain $m < 1$, and nodes in that domain may be used as well in determining the general character of the curves. Figure 70 gives an example with one node at $m = 0.92$.

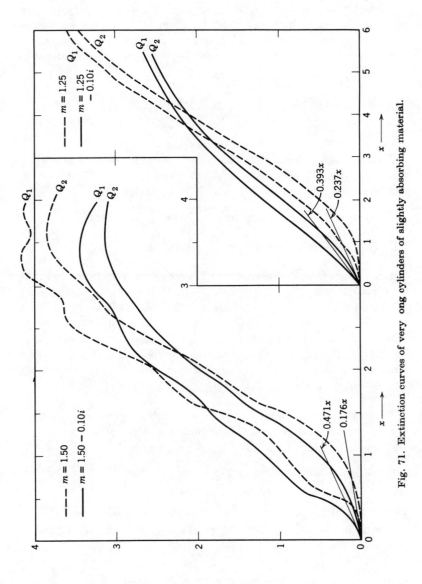

Fig. 71. Extinction curves of very ong cylinders of slightly absorbing material.

These curves, which were constructed for $x = 1, 2, 3, 4, \cdots$ covered the range of m from 1 to 2.5 approximately. From them the values of the first derivatives at $m = 1.25$ and $m = 1.5$ were read. By plotting these derivatives as functions of x they could be interpolated for other values of x. Any errors in this graphical interpolation do not greatly affect the final result since, in converting back to the derivatives of α_n itself by

$$\frac{\partial \alpha_n}{\partial m} = (m - 1) \frac{\partial}{\partial m} \left(\frac{\alpha_n}{m - 1} \right) + \frac{\alpha_n}{m - 1}.$$

(and similarly for β_n), the second term is dominant.

Thus all data entering into the general formulae derived above are known. Figure 71 shows the numerical results for $m = 1.50 - 0.10i$ and for $m = 1.25 - 0.10i$. The tangents for small m are

$$m = 1.25 - 0.10i, \quad Q_1 = 0.393x, \quad Q_2 = 0.237x;$$

$$m = 1.50 - 0.10i, \quad Q_1 = 0.471x, \quad Q_2 = 0.176x.$$

The deviations from the curves for real m are not very large and have the same character as found for spheres (sec. 14.5). They are similar for the two polarizations so that such cylinders can polarize light about as effectively as fully transparent cylinders. This must become quite different if μ gets of the order of $m - 1$ and larger. Then curves, as in Fig. 69 of sec. 15.51, are obtained, and finally for very large μ the curves must become those of totally reflecting cylinders (Fig. 68, sec. 15.32).

References

The complete solution for cylinders of arbitrary radius, arbitrary refractive index, and perpendicular incidence was first given by

Lord Rayleigh, *Phil. Mag.*, **12**, 81 (1881) (*Sci. Papers* 74).

and independently by

W. von Ignatowsky, *Ann. Physik*, **18**, 495 (1905).

The numerical computations in Ignatowsky's paper, and in

W. Seitz, *Ann. Physik*, **16**, 746 (1905); **19**, 554 (1905), and in W. von Ignatowsky, *Ann. Physik*, **23**, 905 (1907),

are devoted to metal wires in the microwave domain ($\lambda \approx 30$ cm). An example for an optical wavelength (gold wire with $2\pi a/\lambda = 3$) was worked out by

W. Seitz, *Ann. Physik*, **21**, 1013 (1906).

The complete solution was rederived, computations for the real value $m = 9$ were made and compared with experiments with water cylinders at $\lambda = 24$ cm by

C. Schaefer, *Sitzber. Königlich Preuss. Akad. Wiss. Berlin*, **11**, 326 (1909).
C. Schaefer and Q. Grossmann, *Ann. Physik*, **31**, 455 (1910).

The computations have been extended to include radiation pressure by

G. Thilo, *Ann. Physik*, **62**, 531 (1920).

Numerical results for $x = 0.4(0.4)2.4$; $\theta = 0°(30°)180°$ (see Fig. 25) were computed by

Lord Rayleigh, *Phil. Mag.*, **36**, 365 (1918) (*Sci. Papers* 434).

The approximation formulae for very small values of $2\pi a/\lambda$ are included in many of these papers and in earlier papers by J. J. Thomson and others. The method for finding the asymptotic behavior for large sizes was first clearly explained by

P. Debye, *Physik. Z.*, **9**, 775 (1908).

In spite of the extensive later studies of this problem by

H. Spohn, *Physik. Z.*, **21**, 444, 469, 501, and 518 (1920),
H. Pfenninger, *Ann. Physik*, **83**, 753 (1927),

no useful numerical results have been obtained.

A modern textbook giving the detailed solution for perpendicular incidences is, for instance,

J. A. Stratton, *Electromagnetic Theory*, New York, McGraw-Hill Book Co. 1941,

to whom we refer for the proofs of the auxiliary formulae quoted in secs. 15.11 and 15.12. The full solution for oblique incidence (sec. 15.11) was given by

P. Wellmann, *Z. Astrophys.*, **14**, 195 (1937),
J. R. Wait, *Can. J. Phys.*, **33**, 189 (1955).
R. Burberg, *Z. Naturforsch.*, **11a**, 300 (1956).

Wellmann's result appears to be incorrect. Wait's paper also contains the simpler formulae for the coefficients arising from special assumptions about x and m, all for an arbitrary angle of incidence, thus supplementing the results reported in secs. 15.32, 15.33, and 15.41.

Experimental investigations and computations on the scattering of waves generated by a source at a finite distance were made by

F. Oberhettinger, *Ann. Physik*, **43**, 136 (1943),
R. D. Kodis, *J. Appl. Phys.*, **23**, 249 (1952).

The form in which the solution for perpendicular incidence is presented and the introduction of the functions $T_1(\theta)$ and $T_2(\theta)$ (sec. 15.13) is new. Also the discussion for cylinders of arbitrary cross section in secs. 15.21 and 15.22 is presented here for the first time.   The same problem has, however, been treated before by

V. Twersky, *J. Appl. Phys.*, **25**, 859 (1954).

The asymptotic value $Q = 2$ for cylinders was also derived by

C. H. Papas, *J. Appl. Phys.*, **21**, 318 (1950),

F. E. Borgnis and C. H. Papas, *Randwertprobleme der Mikrowellenphysik*, Berlin, Springer Verlag, 1955.

In secs. 15.31 and 15.52 reference is made to the tables of auxiliary functions, appearing in

A. N. Lowan, P. M. Morse, H. Feshbach, and M. Lax, *Scattering and Radiation from Circular Cylinders and Spheres, Tables of Amplitudes and Phase Angles*, Washington, D.C., U.S. Navy, 1946.

M. Lax and H. Feshbach, *J. Acoust. Soc. Amer.*, **20**, 108 (1948).

E. Jahnke and F. Emde, *Tables of Functions*, New York, Dover Publications, 1945.

The results in Fig. 67 were first published in

H. C. van de Hulst, *Astrophys. J.*, **112**, 1 (1950).

The further results in secs. 15.3 and 15.4 were newly derived and computed.   Undoubtedly, they are partly contained in the older literature. The computations for $m = \sqrt{2}(1 - i)$ (sec. 15.51) were performed by Mr. A. W. Grootendorst and the writer in 1950, and are based on data from

G. N. Watson, *A Treatise on the Theory of Bessel Functions*, Cambridge, Cambridge Univ. Press, 1944.

A. Fletcher, J. C. P. Miller, and L. Rosenhead, *An Index of Mathematical Tables*, London, Scientific Comp. Service, Ltd., 1946.

N. W. McLachlan, *Bessel Functions for Engineers*, Oxford, Oxford Univ. Press (2nd ed.), 1941.

A good review of the polarization by long cylinders (Hertz effect and Dubois effect) is found in

C. Schaefer and F. Matossi, *Das ultrarote Spektrum*, Berlin, Julius Springer, 1930.

The computations in sec. 15.52 were performed by Mrs. Elske v. P. Smith in 1951.

16. PARTICLES OF OTHER FORMS

16.1. Methods for Solving the Scattering Problem

The general problem posed in part II of this book (chaps. 6–17, inclusive) is to solve the scattering problem for a plane wave incident on a particle of given form, size, and composition. This amounts to the problem of solving Maxwell's equations with given boundary conditions.

Surveying these scattering problems Epstein (1919) wrote:

In order to find new solutions of the fundamental equations that are adequate to the special forms of the boundary surfaces, two methods have been employed. A. Sommerfeld has given a method for finding by function theory branched solutions of the wave equation. The further procedure, by which he satisfies the boundary conditions with these solutions, is a generalization of Thomson's method of the mirror images. Another method that has proved successful is to transform the fundamental equations into such curvilinear coordinates that the surfaces of discontinuity of the parameters correspond to constant values of one parameter. Unfortunately, the applicability of both methods is limited and only a few cases have so far been solved.

Sommerfeld's method is limited to certain particles with plane boundaries (half-plane, wedge), and in its physically useful form it is also limited to perfect conductors.[1] The related problem of a perfectly conducting semiinfinite cone has also been solved rigorously (Siegel and Alperin, 1952; Siegel, Alperin, Crispin, Hunter, Kleinman, Orthwein, and Schensted, 1953). We shall not discuss these solutions but further review the work done on bodies without sharp edges or points.

Somewhat remote from the problem of this book also are the scattering by two spheres close together (Trinks, 1935), the scattering by a hemispherical or semicylindrical boss on a conducting half-plane (Twersky, 1951 and 1954), the diffraction by wire gratings with proper account of the mutual influence of the wires, the scattering by atmospheric inhomogeneities or by wavy surfaces with statistically specified irregularities (cf. sec. 21.1), and finally the radiation from antennas of various forms. All these problems bear some resemblance to those enumerated below. References to papers on some of these problems are given in Bouwkamp's review (1954) on diffraction theory.

---

[1] Sommerfeld also gives the solution for a black screen, in which the energy is assumed to vanish into the second branch of the Riemann surface. The present writer fails to see a physical interpretation of this assumption.

From the extensive literature on wire gratings and directly related subjects we cite only some of the more recent papers: Wessel (1939), MacFarlane (1946), Hönerjäger (1948), Twersky (1952a, b), Lewis and Casey (1952), Groves (1953), Storer and Sevick (1954), Wait (1954).

16.11. Separation of the Wave Equation, Homogeneous Bodies

The method used for spheres (chap. 9) and cylinders (chap. 15) was to separate the vector wave equation in curvilinear coordinates which have been chosen in such a way that the surface of the particle coincides with one of the coordinate surfaces. The same method may be applied to a number of other forms. For a derivation of all coordinate systems in which the vector wave equation is separable, see, e.g., Spencer (1951), Morse and Feshbach (1953).

The separation technique has been applied to:

a. *Circular cylinders.* The solution has been given in chap. 15.

b. *Elliptic cylinders.* The separation by means of Mathieu functions first was made by Sieger (1908) and Aichi (1908). The resultant equations are much more complicated than those for the circular cylinder, because of the extra parameter, but their general structure is the same. The solution is in the form of a series with an infinite number of coefficients, as for spheres and circular cylinders. It has been worked out in Epstein's paper for totally reflecting elliptic cylinders of arbitrary size and for radiation incident in a direction perpendicular to the cylinder axis and falling on the "flat side," i.e., perpendicular to the long axis of the generating ellipse. The numerical result is given in the limit of a flat strip with width $\ll \lambda$. Since then the numerical work on the flat strip has been greatly extended (sec. 16.23). The problem treated by Sinclair (1951), namely, the derivation of the patterns of antennas located near cylinders of elliptical cross section, is reciprocal to that of finding the fields on such cylinders due to an incident plane wave.

c. *Parabolic cylinders.* The subject of Epstein's thesis (1914). This again gives much simpler equations. The rigorous analytic solution can be obtained as easily as for spheres or circular cylinders, but the problem for numerical work is the slow convergence of the series. The relevant parameter is, apart from the refractive index of the parabolic cylinder, the constant

$$kp = 2\pi p/\lambda,$$

where p is the parameter of the generating parabola (equation: $y^2 = 2px$). Sommerfeld's solution for the reflecting half-plane is obtained if this parameter goes to 0 and the refractive index to ∞. Epstein finds that for perpendicular incidence the ratio of the amplitudes scattered by the edge in the case that the electric vector is parallel or perpendicular to the edge (case I and case II) is $\tan(\pi/4 + \theta/2)$ for very thin screens. A finite thickness decreases the amplitude in case II and increases it in case I. If the parabolic cylinder has an imperfect conductivity, the radiation in case II is hardly influenced at all, the radiation in case I strongly. Further work on the parabolic cylinder has been done by Kay (1953). Rice (1954) and Keller (1956) derived the asymptotic expressions for large kp that may be obtained by means of the Watson transformation.

d. *Spheres.* See chaps. 9 to 14.

e. *Spheroids.* Epstein states that the attempts to find a rigorous solution have not been successful. Reviewing the later literature, we find a further attempt by Möglich (1927) for a perfectly conducting oblate spheriod. His solution was later proved to give a wrong answer in the limiting case of a flat circular disk. A full solution for totally conducting prolate spheroids with nose-on incidence of a plane was given by Schultz (1950) and worked out numerically for half-axes $2.01 \, \lambda/2\pi$ and $0.201 \, \lambda/2\pi$.

Fuller numerical results for the nose-on radar cross sections of a prolate spheroid with ratio 10 to 1 of the semiaxes have been computed from these formulae by Siegel, Gere, Marx, and Sleator (1953). The curve is fairly complete for $2\pi a/\lambda = 0$ to 2, and several points between $2\pi a/\lambda = 2$ and 6 have been computed; $a = $ long semiaxis. In the first maximum, which occurs at $2\pi a/\lambda = 1.4$, the radar cross section is 4.4 times the value $\pi b^4/a^2$, which is the small wavelength limit. The formulae for a perfectly conducting oblate spheroid and radiation incident along the axis have been worked out by Rauch (1953), who refers to earlier work by Leitner and Spence (1950).

The result for any ellipsoid is simple in the limit for very small size, or large wavelength, if it is not a perfect conductor (sec. 6.32). It is slightly more complicated for a perfectly conducting ellipsoid, where also the magnetic dipole radiation is important (sec. 6.4).

Further advances in problems b and e will depend largely on the developments in the theory and tabulation of Mathieu functions and spheroidal functions, which are in full progress.

f. *Paraboloids of revolution.* The fields due to the reflection of a plane wave incident in a paraboloid of revolution from the concave side have been studied in some detail in connection with the use of these paraboloids as radar mirrors or radio telescopes (Pinney, 1946, 1947; Bucholz, 1948). The diffraction of a wave incident on the paraboloid on the convex side has been studied by Horton and Karal (1951).

g. *Flat disks and strips.* These form degenerate cases of the preceding particles and have been studied extensively (secs. 16.22 and 16.23).

h. More complicated problems which may be solved by the separation technique. These include the scattering by two spheres close together (Trinks, 1935), by two parallel conducting circular cylinders (Row, 1953a), and by a conducting plane terminated at its edge by a conducting circular cylinder of arbitrary radius (S. N. Karp, cited by Row, 1953b).

16.12. Compound Bodies

The formulae become more complicated but not intrinsically more difficult if, instead of one boundary surface separating the "particle" from the "medium," we have two or more boundaries each separating regions of space in which the refractive index is a constant.

The simplest problem of this kind is the well-known problem of reflection from a plane-coated surface. The equivalent problem for spheres is scattering and extinction by a homogeneous sphere coated with a concentric spherical shell of different material. The formulae have been worked out by Güttler (1952) and by Aden and Kerker (1953). Güttler's treatment includes the electric and magnetic dipole and the electric quadrapole terms and is, therefore, comparable in scope with the formulae for homogeneous spheres reported in sec. 10.3.

Aden's and Kerker's equations were evaluated in one example with a view to a meteorological application (sec. 20.43) and later by Scharfman (1954) in greater numerical detail. In Scharfman's computations the inner sphere, radius a, is a perfect conductor, and the outer shell, between radii a and $a + \delta$, is a dielectric with ε varying from 2.56 to ∞. The radar cross section is found to pass through a maximum if a/λ and δ/λ remain fixed and ε varies. We may note that Scharfman's explanation of this resonance effect, in which the conducting core plays an essential role, is not necessarily correct as resonance occurs also for homogeneous dielectric spheres with such high values of ε. The propagation of surface waves on a cylindrical conductor covered with a cylindrical dielectric shell has been discussed by Horiuchi (1951, 1953).

16.13. Integral Equations and Variational Methods

Another important method is the variational technique. This technique has been worked out by Schwinger and others and has proved very successful in giving approximate but fairly accurate answers to many problems. The principle of the method, as applied to the scalar problem of diffraction by a hole in a plane sheet, is as follows.

By Huygens' principle the diffracted radiation in any direction may be expressed in terms of an integral containing φ, the amplitude in the hole. This function φ is itself the solution of an integral equation, or of an

integro-differential equation, which is difficult to solve.   It is, however,
possible, with the help of this equation, to transform the first expression
for the diffracted field into a slightly more complicated one, which has the
advantage of being stationary with respect to small deviations from the
correct solution for φ. This means that a *somewhat* wrong function
for φ, when substituted into the new expression, still gives virtually
correct values for the diffracted light.   It is then possible to make a
good guess at a simple expression for φ with undetermined coefficients
and to find the coefficients from the condition of stationarity.   Sometimes
it has even been possible to find the completely rigorous solution by using
an infinite number of coefficients.

Space does not permit a more complete exposition. It may be remarked,
however, that this method has proved very workable and that similar
methods may certainly be devised for solving the scattering by three-
dimensional particles.

One such problem is the diffraction by a thick edge, with radius of
curvature larger than the wavelength.   In this case the unknown
function is the current distribution on the surface.   It has been possible
to solve the integral equation numerically, and thus it has become
unnecessary to apply a variational principle, as the distant field may be
found by direct integration (sec. 17.23).

The integral equation for the surface currents has been solved for thin
wires and strips of finite length by van Vleck, Bloch, and Hamermesh
(1947), and additional data have been given by Tai (1952).   It is found
that the scattering, including backscatter, by conducting wires of length
$2l$ and diameter $2a$ depends on the ratios $2l/a$ and $2l/\lambda$. Resonance occurs
at $2l/\lambda = 1/2,\ 3/2,\ 5/2 \cdots$ for very thin wires and at slightly smaller
values if $a/2l$ is not negligible.   Angular distributions for resonant
and non-resonant response are given.   The non-resonant response
increases greatly, the resonant response only slightly, with increasing
diameter.

Approximate solutions valid if the wavelength is small with respect
to the characteristic dimensions of the body may be obtained on the
assumption that the surface currents at any point of a smoothly curved
body are the same as if the wave were incident on a flat reflecting surface
with the same direction of the normal.   The radiation scattered in an
arbitrary direction and the radar cross section may then be computed by
a form of Huygens' principle.   This is called (by a not very specific term)
the physical-optics approximation. Results for a large variety of
bodies have been tabulated by Siegel, Alperin, Bonkowski, Crispin,
Maffett, Schensted, and Schensted (1953).   The most serious defect of
this method is that it neglects the surface waves that may travel around

the body (sec. 17.32), so it may be expected to give better radar cross sections for infinite bodies (paraboloid, semiinfinite cone) than for finite bodies (sphere, spheroid).

16.14. Expansion in Powers of the Ratio Dimension/Wavelength

We have seen in earlier sections that power series expansions are quite useful for particles that are small but not sufficiently small to admit of a solution in which the applied fields can be thought to be homogeneous (chap. 6).   Such expansions have been given in secs. 10.3 (dielectric spheres), 10.61 (totally reflecting spheres), 14.21 (absorbing spheres), and 15.4 (circular cylinders).   All expansions were in powers of $x = ka$, where $k = 2\pi a/\lambda$ and $a =$ the radius, and the expansions were always made starting from a rigorous solution for arbitrary x.

The question may be asked, whether a number of terms of such series expansions can be computed even if a rigorous solution for arbitrary size is not available.   This important problem has been successfully solved by Stevenson (1953a).   Stevenson considers a homogeneous body of arbitrary form and having the complex dielectric constant $\varepsilon + 4\pi i\sigma/\omega$ (corresponding to our m^2 with a different convention about the sign of i) and a magnetic permeability μ.   This means that his treatment covers both dielectric particles ($\sigma = 0$, $\mu = 1$) and totally reflecting particles ($\varepsilon = \infty$, $\mu = 0$).   It is shown that, if $\mathbf{E}$ and $\mathbf{H}$ are assumed to be power series in k, then the coefficients may be found as solutions of certain well-defined problems in potential theory.   The first of these is, of course, the polarization of the body in stationary and homogeneous electric and magnetic fields (secs. 6.1 and 6.4).

The further terms of the series are solutions of successively more complicated problems in potential theory.   This means that, in principle, the problem is solved and just requires the evaluation of certain integrals which can be written down explicitly.   The change from the near-field to the distant-field solution involves certain mathematical finesses which cannot be discussed here.

Logically, Stevenson's solution should have been treated in chap. 6 of this book, as it refers to particles of arbitrary form and is a natural extension of Rayleigh scattering.   In practice, only ellipsoids would seem amenable to a full solution of this kind.   In a second paper Stevenson (1953b) works out the ellipsoid problem in three terms of the series, in which the fields go as k^2 (Rayleigh scattering), k^3 (this term vanishes), and k^4.   The resulting formulae for spheres contain those given in secs. 10.3 and 10.61 as special cases.   The resulting formulae for a perfectly conducting elliptical disk and arbitrary direction of incidence agree in the case of perpendicular incidence with the first terms of

Bouwkamp's expansion (sec. 16.22). Tai's study (1952) of the near and far fields scattered by an oblate spheroid at normal incidence is based on a similar principle.

16.2. Flat, Totally Reflecting Particles

16.21. Babinet's Principle

A great deal of time has been devoted by the mathematical physicists to the so-called planar diffraction problems (compare sec. 3.3). In these problems the only obstacle in the way of the incident plane wave is a plane sheet of a given form. It may extend to infinity and have holes of a given form, etc. A perfectly thin and perfectly opaque sheet must be totally reflecting. In practical experiments, where metal sheets and microwaves are used, the conditions to make the sheet "thin" and "opaque" are (sec. 14.41)

$$d \ll \lambda, \quad d \gg \text{skin depth},$$

where d is the thickness of the sheet. We shall further consider perfectly thin, totally reflecting sheets.

A new problem arises in this connection. We know that, both on the illuminated side and on the shadow side, the ordinary boundary conditions for a perfect conductor hold (sec. 9.14). The physical reality further is that the sheet has a non-zero thickness and is rounded off at the edge so that all along the "surface of the edge" also the same conditions hold. However in the limit of vanishing thickness there is no surface of the edge, and a new set of boundary conditions must hold that dictates a definite behavior of the fields in free space near the edge. These "edge conditions" have presented great difficulties in the past. An authoritative account is given by Bouwkamp (1954), who writes:

The component of the electric fields tangential to the edge vanishes at the edge as $D^{1/2}$, where D is the distance from the edge. The tangential component of the magnetic field remains also finite at the edge, but all other components of the electromagnetic field become infinitely large at the edge as $D^{-1/2}$. Despite these singularities, the edge does not radiate energy: the vectors **E** and **H** are quadratically integrable over any domain of three-dimensional space.

It is curious that these conditions have not been proved conclusively. They have been derived from the rigorous solutions for Sommerfeld's half-plane problem and occur also in the rigorous solution of the circular-disk problem. Bouwkamp therefore suggests that "it may be conjectured that the same type of singularities occur at any sharp edge, whether the obstacle is plane or curved, and independent of the form of the edge."

It follows from the same conditions that the current density on the screen has a component normal to the edge that vanishes as $D^{1/2}$, whereas its component tangential to the edge becomes infinite as $D^{-1/2}$. Also the charge density becomes infinite as $D^{-1/2}$.

Another new problem is the correct formulation of Babinet's principle. We have used Babinet's principle in its simplest and original form in sec. 8.21. It is given there as an approximation valid in the diffraction theory, which itself is an approximate theory holding asymptotically for large disks and apertures, and small angles (meaning a of the term diffraction in sec. 3.3). When the diffraction was made into a rigorous theory for totally reflecting sheets (meaning d of sec. 3.3) also Babinet's principle had to be given a rigorous formulation. This was found some years ago and is also given in Boukwamp's review paper:

Let $(\mathbf{f}, \mathbf{g})$ denote any incident field, where $\mathbf{f}$ stands for the electric vector and $\mathbf{g}$ for the magnetic vector. The "complementary incident field" is defined by $(-\mathbf{g}, \mathbf{f})$ in the order (electric, magnetic) vector. Both fields satisfy Maxwell's equations. First we consider the diffraction of the field $(\mathbf{f}, \mathbf{g})$ by a perfectly conducting plane screen S of zero thickness. Secondly we consider the diffraction of the complementary field $(-\mathbf{g}, \mathbf{f})$ by an aperture A in a perfectly conducting screen such that the aperture in the second problem is of the same size and shape as the screen in the first problem $(A = S)$. For simplicity we call the second diffraction problem the "complementary diffraction problem." The rigorous form of Babinet's principle asserts that the solution of one of these problems gives at once the solution of the other problem. In the first problem the total field everywhere in space is given by $(\mathbf{f} + \mathbf{E^s}, \mathbf{g} + \mathbf{H^s})$. where the scattered field $(\mathbf{E^s}, \mathbf{H^s})$ is due to the electric currents induced in the screen by the incident field. In the complementary problem we can distinguish between the fields in front of and behind the aperture. Let $(\mathbf{E_0}, \mathbf{H_0})$ denote the total field in the illuminated half-space $(z \leqslant 0)$ if there is no hole in the screen, and let $(\mathbf{E^d}, \mathbf{H^d})$ denote the diffracted field in the presence of the aperture. The latter field constitutes the total field behind the aperture, but in front of the aperture the total field is $(\mathbf{E_0} + \mathbf{E^d}, \mathbf{H_0} + \mathbf{H^d})$. Now according to the rigorous form of Babinet's principle we have the simple relations

$$\mathbf{E^d} = \mp \mathbf{H^s} \qquad \mathbf{H^d} = \pm \mathbf{E^s} \qquad (z \lessgtr 0).$$

The functions $E_x^{s,d}$, $E_y^{s,d}$, and $H_z^{s,d}$ are even functions of z, and $H_x^{s,d}$, $H_y^{s,d}$, and $E_z^{s,d}$ are odd functions of z.

The practical value of Babinet's principle is that the solution for the straight slit (in a conducting screen) at once gives the solution for the parallel strip, and the solution for the circular hole at once gives the solution for the circular disk. Since in this book we are dealing only with isolated particles, we are interested primarily in the "disk" and "strip" problems and shall quote the results accordingly.

16.22. Results for a Circular Disk

Let a totally reflecting circular disk of radius a be exposed to a plane incident electromagnetic wave of wavelength λ with direction of propagation perpendicular to the plane of the disk. The solution of this scattering problem depends only on the parameter

$$x = ka = 2\pi a/\lambda,$$

and the scattered fields may be expressed in terms of two amplitude functions $S_1(\theta)$ and $S_2(\theta)$, exactly as for spheres.

After this problem had been analytically solved by Meixner and Andrejewski (1950), a thorough numerical investigation was made by Andrejewski (1953). One important result refers to the total scattering cross section. When divided by the area πa^2 it gives $Q(x)$, the efficiency factor for scattering or extinction.

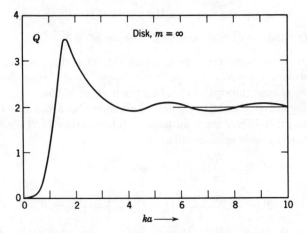

Fig. 72. Extinction by a flat circular disk with radius a and consisting of perfectly conducting material; abscissa $x = 2\pi a/\lambda$.

The formula for small x (useful for $x < 0.8$) is

$$Q(x) = \frac{128}{27\pi^2} x^4 \left(1 + \frac{22}{25} x^2 + \frac{7321}{18375} x^4 + \dots \right).$$

The first term represents the radiation from an electric dipole $\mathbf{p} = (4a^3/3\pi) \mathbf{E_0}$ at the center of the disk. This value of the polarizability is also found if in the formula for oblate spheroids we go to the limit of extreme flatness (sec. 6.4). The value for $x \to \infty$ is $Q \to 2$, as it should be. The intermediate part has been computed by Andrejewski and is presented in Fig. 72. The maximum $Q = 3.4$ is reached near $x = 1.6$.

Andrejewski further gives diagrams of the distribution of scattered amplitude and intensity in both principal planes for $x = 10$. The scattered intensity in both planes comes mainly from the imaginary parts of the amplitudes, which corresponds to the real parts of our $S_1(\theta)$ and $S_2(\theta)$ and is very similar to that found from ordinary Fraunhofer diffraction. This is not surprising and corresponds exactly to our previous findings for a totally conducting sphere (sec. 10.62, Fig. 29). Further diagrams include the distribution of fields on both sides of the disk and on the axis.

16.23. Results for a Parallel Strip

Let a totally reflecting strip of width $2a$ and infinite length be exposed to a plane incident electromagnetic wave of wavelength λ with direction of propagation perpendicular to the strip. As in chap. 15 we distinguish:

Case I: linear polarization with $\mathbf{E} \parallel$ length of strip,

Case II: linear polarization with $\mathbf{E} \perp$ length of strip.

The complementary problem of a straight slit is discussed in extenso in Bouwkamp's report. In order to change to the time factor exp $(-i\omega t)$ we change the sign of i, and on the basis of Babinet's principle we interchange cases I and II and give the amplitudes a factor -1. The results may then be written in the form postulated in sec. 15.21 for cylinders of an arbitrary cross section and give

$$T_1(\theta) = \frac{\pi i}{2p} [1 + \tfrac{1}{4} x^2 \cos^2\theta + \dots].$$

$$T_2(\theta) = \frac{\pi i x^2 \cos \theta}{4} [1 - \tfrac{1}{4}(p - \tfrac{3}{4} + \tfrac{1}{2}\sin^2\theta)x^2 + \dots].$$

Here $x = ka$, and $p = \log(\tfrac{1}{4}\gamma x) + \tfrac{1}{2}\pi i$, where $\gamma = 1.781$ and $\log \gamma = 0.57721$ is Euler's constant. Equivalent formulae are given by Müller and Westphal (1953). The total cross section (for scattering or extinction) may be found either by integration or by using the formula

$$c_{ext} = \frac{4}{k} \{\mathrm{Re} \ T(0)\},$$

derived in sec. 15.21. Using the latter formula we find the efficiency factors

$$Q_1 = \frac{c_1}{2a} = \frac{2}{x} \cdot \frac{\pi^2}{\pi^2 + \{2 \log(\gamma x/4)\}^2} \left[1 + \frac{1}{4} x^2 + \dots \right]$$

$$Q_2 = \frac{c_2}{2a} = \frac{\pi^2 x^3}{16} \left[1 + \frac{5}{16} x^2 \left\{ 1 - \frac{8}{5} \log(\gamma x/4) \right\} + \dots \right].$$

These pseudopower expansions are useful approximations for small x. One further term is given in Bouwkamp's paper.

The numerical values for $x > 0.6$ have to be found by a more powerful method. The best way is a rigorous solution of the diffraction problem in terms of Mathieu functions. Figure 73 is based on the values computed by Skavlem and quoted in Bouwkamp's report. A check with the formulae given above gives good agreement for $x < 0.5$. The general behavior of these curves is similar to that shown in Fig. 68 (sec. 15.33) for totally reflecting cylinders; Q_2 reaches the maximum value 2.48

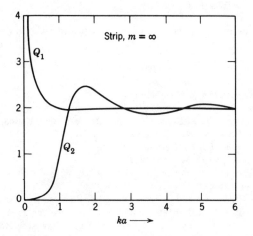

Fig. 73. Efficiency factor for extinction by a flat strip of infinite length and width $2a$ if the radiation is incident perpendicularly to the flat side and is linearly polarized with $E \, // \,$ length (curve 1) or $H \, // \,$ length (curve 2).

near $x = 1.70$. Asymptotic formulae for large x may be derived by a variety of methods (Burger, 1954; Imai, 1956; Clemmov, 1956).

References

The papers on miscellaneous topics cited in sec. 16.1 are:

P. S. Epstein, *Enzyklopädie math. Wiss.*, **5**, Teil 3, 488 (1919).

K. M. Siegel and H. A. Alperin, *Studies in Radar Cross Sections—III*, Willow Run Research Center, Univ. of Michigan Press, 1952.

K. M. Siegel, H. A. Alperin, J. W. Crispin. H. E. Hunter, R. E. Kleinman, W. C. Orthwein, and C. E. Schensted, *Studies in Radar Cross Sections—IV*, Willow Run Research Center, Univ. of Michigan Press, 1953.

V. Twersky, *J. Appl. Phys.*, **22**, 825 (1951); **25**, 859 (1954).

C. J. Bouwkamp, *Repts. Progr. in Phys.*, **17**, 35 (1954).

W. Wessel, *Hochfrequenztechnik und Elektrotechnik*, **54**, 62 (1939).

C. G. MacFarlane, *J. Inst. Elec. Engrs. London*, **93**, 1523 (1946).

R. Hönerjäger, *Ann. Physik*, **4**, 25 (1948).

V. Twersky, *J. Appl. Phys.*, **23**, 407 and 1099 (1952).

E. A. Lewis and J. P. Casey, *J. Appl. Phys.*, **23**, 605 (1952).

W. E. Groves, *J. Appl. Phys.*, **24**, 845 (1953).

J. E. Storer and J. Sevick, *J. Appl. Phys.*, **25**, 369 (1954).

J. R. Wait, *Can. J. Phys.*, **32**, 571 (1954).

The separability conditions are discussed by:

D. E. Spencer, *J. Appl. Phys.*, **22**, 386 (1951).

P. M. Morse and H. Feshbach, *Methods of Theoretical Physics*, New York, McGraw-Hill Book Co., 1953.

Further papers on the scattering by homogeneous bodies (sec. 16.11) are:

B. Sieger, *Ann. Physik*, **27**, 626 (1908).

K. Aichi, *Proc. Tokyo Math. Phys. Soc.*, **4**, 966 (1908).

G. Sinclair, *Proc. I.R.E.*, **39**, 660 (1951).

P. S. Epstein, "*Thesis München*," 1914.

I. Kay, N. Y. Univ. Math. Research Group, *Research Rept.* EM-53, 1953.

S. O. Rice, *Bell System Tech. J.*, **33**, 417 (1954).

J. B. Keller, *Trans. I.R.E.*, **AP-4**, 312 (1956).

F. Möglich, *Ann. Physik*, **83**, 609 (1927).

F. V. Schultz, *Studies in Radar Cross Sections—I*, Willow Run Research Center, Univ. of Michigan Press, 1950.

K. M. Siegel, B. H. Gere, I. Marx, and F. B. Sleator, *Studies in Radar Cross Sections—XI*, Willow Run Research Center, Univ. of Michigan Press, 1953.

L. M. Rauch, *Studies in Radar Cross Sections—IX*, Willow Run Research Center, Univ. of Michigan Press, 1953.

A. Leitner and R. D. Spence, *J. Franklin Inst.*, **249**, 299 (1950).

E. Pinney, *J. Math. and Phys.*, **25**, 49 (1946); **26**, 42 (1947).

H. Bucholz, *Ann. Physik*, **2**, 185 (1948).

C. W. Horton and F. C. Karal, *J. Appl. Phys.*, **22**, 575 (1951).

W. Trinks, *Ann. Physik*, **22**, 561 (1935).

R. V. Row, *Cruft Laboratory Techn. Rept.* 170, Cambridge, Harvard Univ. Press, (1953a).

R. V. Row, *J. Appl. Phys.*, **24**, 1448 (1953b).

Theory and tables of Mathieu functions and spheroidal wave functions are found in:

N. W. McLachlan, *Theory and Application of Mathieu Functions*, Oxford, Oxford Univ. Press, 1947.

Natl. Bur. Standards Mathematical Tables Project, *Tables Relating to Mathieu Functions*, New York, Columbia Univ. Press, 1951.

J. Meixner and F. W. Schäfke, *Mathieu Functions and Spheroidal Functions with Applications to Physical and Technical Problems*, Berlin, Julius Springer, 1954.

C. Flammer, *J. Appl. Phys.*, **24**, 1218 (1953a).

J. A. Stratton, P. M. Morse, L. J. Chu, J. D. C. Little, and F. J. Corbató, *Spheroidal Wave Functions*, New York, Technology Press of M.I.T. and John Wiley & Sons, 1956.

The problem of the two concentric spheres (sec. 16.12) has been treated by

A. Güttler, *Ann. Physik*, **11**, 65 (1952),
A. L. Aden and M. L. Kerker, *J. Appl. Phys.*, **22**, 1242 (1951),
H. Scharfman, *J. Appl. Phys.*, **25**, 1053 (1954),

and a related problem by

K. Horiuchi, *J. Appl. Phys.*, **22**, 504 (1951); **24**, 961 (1953).

The variational method (sec. 16.13) is used in a great number of papers. Useful surveys available at present are Bouwkamp's report and the book of Morse and Feshbach, both cited above, as well as

F. E. Borgnis and C. H. Papas, *Randwertprobleme der Mikrowellenphysik*, Berlin, Springer Verlag, 1955.

Wires of finite length have been treated by

J. H. van Vleck, F. Bloch, and M. Hamermesh, *J. Appl Phys.*, **18**, 274 (1947).
C. T. Tai, *J. Appl. Phys.*, **23**, 909 (1952).

The physical-optics approximation was applied to bodies of various shapes by

K. M. Siegel, H. A. Alperin, R. R. Bonkowski, J. W. Crispin, A. L. Maffett, C. E. Schensted, and I. V. Schensted, *Studies in Radar Cross Sections—VIII*, Willow Run Research Center, Univ. of Michigan Press, 1953.

Power expansions (sec. 16.14) have been used by

A. F. Stevenson, *J. Appl. Phys.*, **24**, 1134 (1953a); **24**, 1143 (1953b).
C. T. Tai, *Trans. I.R.E.*, PGAP-1, February, 1952.

The key reference to sec. 16.2 is again Bouwkamp's review cited above, where many references and further details may be found. Derivations and results for the circular disk (sec. 16.22) may be found in

J. Meixner and W. Andrejewski, *Ann. Physik*, **7**, 157 (1950).
W. Andrejewksi, *Z. angew. Physik*, **5**, 178 (1953).
C. Flammer, *J. Appl. Phys.*, **24**, 1224 (1953b).

The extension to a dipole source at a finite distance has been made by

J. Meixner, *Ann. Physik*, **12**, 227 (1953).

The problem of the flat strip (sec. 16.23) is discussed by

S. Skavlem, *Arch. Math. Naturvidenskab*, **51**, 61 (1951).
R. Müller and K. Westphal, *Z. Physik*, **134**, 245 (1953).
A. P. Burger, "Thesis Delft," Nationaal luchtvaart Laboratorium (Amsterdam), *Report* F.157 (1954).
I. Imai, *Trans. I.R.E.*, **AP-4**, 233 (1956).
P. C. Clemmov, *Trans. I.R.E.*, **AP-4**, 282 (1956).

17. EDGE PHENOMENA AND SURFACE WAVES

The scattering particles considered in this chapter do not have a prescribed form, except that they are rounded with radii of curvature much larger than the wavelength. The refractive index is arbitrary, in principle, but most results are for $m = \infty$. The plane wave, incident from infinity, strikes the body at grazing angles near its "edges." In this way two effects are set up:

A. A particular distribution of fields and currents near the edge, which is to a large extent independent of the further shape of the particle and gives rise to a particular term in the amplitude of forward scattered light (*edge phenomena*).

B. A wave motion which continues along the surface of the particle into the shadow region. If the particle is not too big, the wave may creep entirely around it (*surface wave*).

These effects occur in addition to the more familiar diffraction, reflection, and refraction effects for large particles studied in chaps. 8 and 12. Only a few studies of these effects have appeared in the literature.

In this chapter the problem is approached partly from the theoretical side, partly in an empirical fashion on the basis of the numerical results of rigorous computations.

17.1. Forward Scattering by an Optical Edge

This section deals with the distant fields only. The fields on and near the surface are discussed in secs. 17.23, 17.31, and 17.32.

An opaque screen used in diffraction experiments may be idealized in two totally different ways if a simple theory for these experiments is sought. The criterion is whether the wavelength is large or small compared to the thickness of the screen.

The specifications for a thin, opaque screen in experiments with *microwaves* are that its thickness is $\ll \lambda$, yet large compared to the skin depth. The screen is then totally reflecting, and the common type of diffraction theory, summarized in sec. 16.2, is adequate. By "edge" we understand the separation curve (and its immediate surroundings) between the illuminated and non-illuminated parts of the screen's surface. By "profile" we understand the boundary curve between the material of the screen and the outside medium in a cross section perpendicular to the edge. This profile may be a round curve, a rectangle, or it may have a sharp point (wedge) but the profile of the edge is presumably indifferent in microwave experiments.

The jaws of a spectrograph slit or similar *optical* screens are thicker than a wavelength. A little absorption inside the material of the screen suffices to make it opaque, so it need not be totally reflecting. Although an attempt is often made to make it wedge-shaped, the profile is actually rounded with a large radius of curvature. For instance, a sharp razor blade may have a radius of curvature of 1 or 2 μ. The specification of a radius of curvature *large* compared to the wavelength will be said to define the *optical edge*.[1] It presents a special diffraction problem that will be reviewed below.

We propose to show that the grazing reflection against the rounded edges in the optical case gives a component of forward scattered light that modifies the ordinary diffraction wave. This suggestion had been made already by Young and Fresnel, but its practical consequences are minute and show up only in a refined theory. It is physically obvious that the fields near the edge will not be influenced by the shape of the particle far away from the edge (except for the effects of surface waves). This notion has been formalized by Fock as "the principle of the local field."

17.11. Amplitude Function of a Cylindrical Edge; Cylinders with Two Edges

A smoothly rounded surface has two principal planes through the normal at any point E. These planes are perpendicular to each other. The radii of curvature of the curves by which these planes intersect the body are called the main radii of curvature of the surface at that point. Let E be a point on the geometrical edge of the scattering body, i.e. at the curve that separates the illuminated and dark sides of the body if we neglect any diffraction effects. We shall assume for simplicity that the *direction of propagation of the incident light is in one of the principal planes*. The intersection of this plane with the surface is the "profile" of the edge, and R is the radius of curvature of this curve at the point E. The other principal plane is parallel to the wave front of the incident wave. The radius of curvature of the body in this plane, S, is also the radius of curvature of the shadow cast by the body on a screen perpendicular to the wave.

We may assume that the edge phenomena depend only on R, the curvature of the profile, and not on S, so S will not further be mentioned.

---

[1] The problem of propagation of radio waves around the earth and around gently curved hills is evidently related to that of the optical edge in diffraction experiments.

The condition is that both R and S are $\gg \lambda$; it is not necessary that $S > R$. The plausibility of these statements will appear later (sec. 17.21).

Our first example, which furnishes also the definition of the amplitude function, is a *cylinder with arbitrary but finite profile* and radiation incident perpendicular to the axis. We then have two edges with radii of curvature R and R' (Fig. 74). We assume the incident light to be

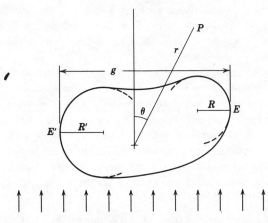

Fig. 74. Cross section through arbitrary cylinder perpendicular to axis.

plane-polarized, either along the edges or perpendicular to them. The time factor will be $e^{i\omega t}$ as in the earlier chapters but contrary to the most common convention in diffraction theory.

Let in case I ($\mathbf{E} \parallel$ edge) v represent the electric field and in case II ($\mathbf{H} \parallel$ edge) v represent the magnetic field. By the general formulation in sec. 15.21 the distant field scattered by the cylinder may be represented by

$$v_{sca}(P) = v_{inc}(O) \left(\frac{2}{\pi k r} \right)^{1/2} T_{1,2}(\theta)\, e^{-3\pi i/4 - ikr}.$$

Here O is a point on the (arbitrarily chosen) axis, P is a distant point, θ is the scattering angle, and r is the distance of P from the axis. We shall avoid duplication by writing the further formulae only for case I; those for case II are found by replacing the index 1 by 2.

If the cylinder is non-transparent and θ is not near zero [more precisely $\theta \gg (kR)^{-1/3}$, sec. 17.21], then $T(\theta)$ is made up of the contributions from one or more places where the external reflection occurs in the direction θ.

The theory, based on geometrical optics, is as simple as that of secs. 12.21 and 12.22 and gives a value

$$T_1^{refl}(\theta) = \left(\frac{\pi k R \sin \frac{1}{2}\theta}{4}\right)^{1/2} r_1 e^{3\pi i/4 + 2ikR \sin \frac{1}{2}\theta}$$

for each of these places. Here r_1 is the Fresnel reflection coefficient, and R is the local radius of curvature of the profile. The phase in this formula is referred to the local axis of curvature. When the phase is referred to the edge point E the exponent becomes $3\pi i/4 + ikR(2 \sin \frac{1}{2}\theta - \sin \theta)$, which is $3\pi i/4 + ix\theta^3/8$ for $\theta \ll 1$.

The nearly grazing rays, which would be reflected in a direction near 0, give rise to the edge effects. If certain odd shapes are excluded, we may assume that for θ near 0 $T_1^{refl}(\theta)$ changes into another function, which will be denoted by $T_1^{edge}(R, \theta)$. Here R is the radius of curvature at the edge. If the cylinder with its two edges is non-transparent, it may be characterized by the three parameters g, the geometrical shadow width, and R and R', the radii of curvature of the two edges (Fig. 74). Its entire amplitude function, as defined in sec. 15.21, then has the form:

$$T_1(\theta) = \frac{kg}{2}\frac{\sin(\frac{1}{2}kg \sin \theta)}{\frac{1}{2}kg \sin \theta} + T_1^{edge}(R,\theta) + T_1^{edge}(R',-\theta).$$

The first term follows from the ordinary (Fraunhofer) diffraction theory; the result $kg/2$ was derived in sec. 15.21, the second factor in sec. 8.32. The additional terms are the edge terms. They depend on the local radii of curvature (R or R'), the angle θ, the material of the cylinder, the wavelength, and the state of polarization of the incident light (case I, index 1, or case II, index 2), and are subject to phase changes due to the choice of reference point.

It is plausible that under these assumptions the edge terms do not depend on other parameters. A formal proof may be found for totally reflecting bodies ($m = \infty$) in Fock's papers.

The result for $\theta = 0$ is, still in case I:

$$T_1(0) = \frac{kg}{2} + T_1^{edge}(R,0) + T_1^{edge}(R',0),$$

from which the effective width for extinction c is found (sec. 15.21) by multiplying the real part with $4/k$. Thus

$$c = 2g + c_1^{edge}(R) + c_1^{edge}(R'),$$

where

$$c_1^{edge}(R) = \frac{4}{k} \operatorname{Re}\{T_1^{edge}(R,0)\}.$$

Similarly, for case II:

$$c_2^{edge}(R) = \frac{4}{k} \, \mathrm{Re} \, \{T_2^{edge}(R, 0)\}.$$

These lengths depend on the radius of curvature, the wavelength, the state of polarization of the incident light, and possibly also on the refractive index of the cylinder.

By these definitions, which depend on a postulated not yet proven form of the asymptotic solutions, our problem is reduced to finding the functions $T_1^{edge}(R,0)$ and $T_2^{edge}(R,0)$ and the derived functions $c_1^{edge}(R)$ and $c_2^{edge}(R)$. Before proceeding with a determination of these functions we shall show their use for bodies of different forms.

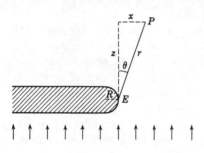

17.12. Screen with One Cylindrical Edge

The diffraction by a screen that covers a half-plane and has one straight cylindrical edge with radius of curvature R is again simple. It cannot be subordinated to one of the earlier chapters as the diffraction pattern does not approach, for large distances, a fixed distribution in the angles θ.

Fig. 75. Cross section through opaque screen.

Let, in Fig. 75 which represents a perpendicular cross section, E be the edge-point separating the illuminated and dark side. Let the amplitude of the undisturbed wave on the front through E be v_0, and let

$$F(u) = C(u) - iS(u) = \int_0^u e^{-\pi i t^2/2} \, dt$$

represent Fresnel's integral; $F(\infty) = \frac{1}{2}(1 - i)$. If the time factor is $\exp(+i\omega t)$, as used throughout this book, the total amplitude in a distant point P, which has the coordinates $z = r \cos \theta$, $x = r \sin \theta$, where θ is small (Fig. 75), is given by the formula

$$v(P) = v_0 \frac{e^{i\pi/4}}{\sqrt{2}} \{F(\infty) - F(-u)\} e^{-ikz}$$
$$+ v_0 \left(\frac{2}{\pi kr}\right)^{1/2} T_{1,2}^{edge}(R,\theta) \, e^{-3\pi i/4 - ikr}.$$

The first term, in which $u = (k/\pi z)^{1/2}x$, follows from the application of the principle of Huygens-Fresnel (sec. 3.12) to the open part of the wave front; it describes the Fresnel diffraction pattern discussed in many textbooks. The second term, in which 1 and 2 denote the two possible polarizations, is copied directly from the preceding section.

The only condition for the preceding formula is $\theta \ll 1$. When θ sweeps from the shadow region into the illuminated region, the first term changes from 0 to the undisturbed wave, and the second term changes from 0 to the reflected wave (sec. 17.11). It is of interest to inquire which change occurs fastest. The "interesting angles" of the Fresnel pattern are by order of magnitude given by

$$u \approx 1, \theta \approx (kz)^{-1/2};$$

those for the edge waves are by order of magnitude given by $\theta \approx (kR)^{-1/3}$. So the "edge pattern" will be relatively wide compared to the Fresnel pattern if

$$(kR)^{-1/3} \gg (kz)^{-1/2},$$

i.e., if

$$z \gg (R^2/k)^{1/3}.$$

In most experiments z is much larger than R, and R much larger than λ, so this condition is amply fulfilled. For instance, an experiment with $1/k = \lambda/2\pi = 0.1\mu$, $R = 10\mu$, $z = 1$ cm makes the Fresnel pattern about 0.1 mm wide, in total, while the edge wave extends several mm at either side.

These order-of-magnitude considerations show that in the range of θ in which the Fresnel pattern shows its typical fluctuations we may write

$$v(P) = v_0 \frac{e^{i\pi/4 - ikz}}{\sqrt{2}} \left\{ F(\infty) - F(-u) - \left(\frac{4}{\pi k r}\right)^{1/2} e^{-\pi i u^2/2} T_{1,2}{}^{edge}(R,0) \right\}.$$

If $T_1{}^{edge}(R_1, 0)$ and $T_2{}^{edge}(R,0)$ had been real numbers, the additional term would have meant just a small change in the argument of $F(-u)$, i.e., a shift of the Fresnel diffraction pattern. Actually these numbers are complex. The practical way to see the consequences is, as always in such problems, by means of Cornu's spiral, which is a plot of $F(u)$ in the complex domain. The pattern is not only displaced but also changes its form slightly.

Analytically, we may proceed as follows. Let u' be defined by the condition that $\left| F(\infty) - F(-u') \right|$ equals the absolute value of the form within braces above. Following Artmann (1950) we may compute the value of u' in the range of u near 0. We obtain

$$\left| \tfrac{1}{2}(1 - i) + u' \right| = \left| \tfrac{1}{2}(1 - i) + u - (4/\pi k r)^{1/2} T_{1, 2}{}^{edge}(R,0) \right|,$$

which gives, after a brief reduction, making use of the fact that u is small

$$u' = u - \left(\frac{4}{\pi kr}\right)^{1/2} \text{Re}\{(1+i)\,T_{1,\,2}{}^{edge}(R,0)\}.$$

The second term describes the displacement of the Fresnel pattern near $u = 0$ in agreement with Artmann.

17.13. Spheroids and Spheres

Let a prolate or oblate spheroid, with semiaxes a, a, and b be illuminated by a plane, plane-polarized wave with direction of propagation parallel to the axis of rotation. The radius of the shadow circle is a, and the radius of curvature at the edge is $R = b^2/a$; we assume that R and a are $\gg \lambda$. For simplicity we consider only the forward scattered wave. Its amplitude and phase are fully described by an amplitude function $S(0)$ independent of polarization, for the formulae of sec. 4.42 are applicable. We may postulate the form

$$S(0) = \frac{k^2 a^2}{2} + S^{edge}(R,0).$$

The first term follows from sec. 8.22 with the geometrical cross section $G = \pi a^2$. The form of the second term may be derived as follows. A length $l = a\,d\varphi$ of the circumference scatters light as a cylindrical edge. If the electric field of the incident wave is parallel to a particular stretch of the circumference, it contributes to $S(0)$ the amount

$$\frac{ka\,d\varphi}{\pi}\,T_1{}^{edge}(R,0).$$

This result is based on the last-but-one formula of sec. 15.22; it holds for zone 3 in the terminology of that section. If the part of the circumference is parallel to the magnetic field, the same formula with index 2 holds. If the electric field makes an arbitrary angle φ with the edge, we have to decompose it in the parallel and perpendicular vibrations, apply the above formulae, and add. It is evident for symmetry reasons that only the components vibrating in the original direction give a non-zero result upon integration over the entire edge. These components are

$$\frac{ka\,d\varphi}{\pi}\{T_1{}^{edge}(R,0)\cos^2\varphi + T_2{}^{edge}(R,0)\sin^2\varphi\},$$

which upon integration along the entire edge gives

$$S^{edge}(R,0) = ka\{T_1{}^{edge}(R,0) + T_2{}^{edge}(R,0)\}.$$

The extinction cross section that follows from $S(0)$ by the general formula (sec. 4.21) is

$$C_{ext} = 2\pi a^2 + \pi a \{c_1^{edge}(R) + c_2^{edge}(R)\},$$

where $c_1^{edge}(R)$ and $c_2^{edge}(R)$ are defined in the preceding section.

In sec. 17.24 and further we shall compare the numerical results for circular cylinders and spheres. Here $R = a$, $x = ka$. The efficiency factors Q_{ext}, found by dividing by the geometrical cross sections, are

For circular cylinders:

$$\text{Case I,} \quad Q_{ext} = 2 + Q_1^{edge}(x);$$
$$\text{Case II,} \quad Q_{ext} = 2 + Q_2^{edge}(x).$$

For a sphere (independent of polarization):

$$Q_{ext} = 2 + Q_1^{edge}(x) + Q_2^{edge}(x).$$

In these formulae we have written

$$Q_{1,2}^{edge}(x) = \frac{c_{1,2}^{edge}(a)}{a} = \frac{4}{x} \, \text{Re} \, \{T_{1,2}^{edge}(a,0)\}.$$

The fact that the edge term for a sphere is not the average but the sum of the edge terms in the two cases of a cylinder might seem strange. It is due to the fact that the sphere's shadow has a twice longer ratio of circumference to area than the cylinder's shadow. Many other bodies, including general ellipsoids, paraboloids, parabolical cylinders, may be treated in the same way. The resulting formulae may be somewhat more complicated but will contain only the functions defined above, provided that all dimensions are $\gg \lambda$.

17.2. The Precise Form of the Edge Functions

17.21. Heuristic Arguments

When this chapter was written no correct formulae for the edge functions $T_1^{edge}(R,\theta)$ and $T_2^{edge}(R,\theta)$ were found in the literature, not even for the simplest case $m = \infty$, $\theta = 0$. It is instructive to approach the problem first by arguments of a heuristic character. They do not lead to numerically correct results and may not even carry much conviction. Yet it is illuminating to see how the complicated dependence on the radius, the polarization, and the refractive index may be understood in a simple way.

Let us first take the—too naive—point of view that all nearly grazing waves are initially reflected as follows from Fresnel's laws for a plane boundary. When the main front of the incident wave has reached the edge, the reflected front is then curved, as shown in Fig. 76. The incident light is from below. E is an edge point on the separation between the "illuminated" and "dark" parts; $OE = R$ is the radius of the curvature The undisturbed front is ET. A ray that would have traveled along PQ is diverted to the point S by the time it would have reached Q. Such reflected rays together form the front ES. Simple geometry gives

$$PS = PQ = R \sin \tau,$$
$$ET = QT - QE = R \sin \tau \cdot \sin 2\tau - R(1 - \cos \tau),$$
$$ST = R \sin \tau \, (1 - \cos 2\tau) = 2R \sin^3 \tau.$$

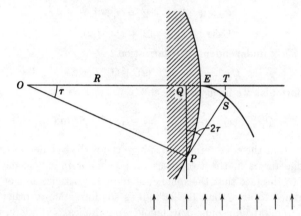

Fig. 76. Diagram illustrating the possibility of interference between the undisturbed wave (wave front ET) and the wave formed by nearly grazing reflection (wave front ES).

The approximation for small τ, which alone is important in the present connection, is

$$ET = 3QE = (3/2)\tau^2 R, \quad TS = 2\tau^3 R.$$

Also, as long as the approximation by ray-optics is permitted, the reflected radiation is spread over a front, ES, 3 times as wide as the original front QE, so that the amplitude is $1/\sqrt{3}$ times the original amplitude, which prevails in the front ET.

The close coincidence between the fronts ET and ES gives rise to interference, which is effective if $TS < \lambda/4$, say. This means that

$$2R\tau^3 < \pi/2k,$$

so

$$\tau < \left(\frac{\pi}{4kR}\right)^{1/3}$$

and for the limiting ray:

$$QE = \tfrac{1}{3} ET = \frac{R}{2} \left(\frac{\pi}{4kR}\right)^{2/3}.$$

This shows that grazing rays are effective in determining a peculiar interference effect when they are incident within a fraction $(kR)^{-2/3}$ of the full radius from the edge. This is exactly the condition for edge rays that was used before in the discussion of rainbow and glory (secs. 13.24 and 13.32).

In order to find the sign of the amplitude of the reflected wave, let us first suppose that the body consists of totally reflecting material ($m = \infty$). If, then, the electric field is parallel to the edge (case I), the reflection coefficient is $r_1 = -1$ (sec. 12.44), which means that the passing wave ET is weakened by the reflected wave ES. The diffraction pattern is then such *as if* the edge cuts away a relatively large part of the incident wave. In the terminology of sec. 17.11, $c_1(R)$ is positive. In case II, however, $\mathbf{H}$ is parallel to the edge, $r_2 = 1$, the passing front is strengthened, and $c_2(R)$ is negative. Totally reflecting spheres show the combined effect of $c_1(R)$ and $c_2(R)$, which is nearly nil (sec. 17.24).

Quite different results are obtained for dielectrics or metals. If the refractive index is real and not too large, the relevant rays (see above) have more nearly grazing angles of incidence than the Brewster angle. So r_1 and r_2 are both negative, though r_2 is somewhat smaller. This means that in both cases a larger part of the front is cut away, and $c_1(R)$ and $c_2(R)$ are positive. Also the combined effect, which shows up in spheres, is positive. *This is the reason why the extinction is systematically higher than 2 and approaches $Q = 2$ only for very high x.* This is true for spheres as well as for cylinders and for dielectrics as well as for metals.

If, however, m is fairly large and x is not very large, the change of phase at the Brewster angle or the related change for metallic reflection may occur right in the range of angles which contribute to the edge functions. This may lead to very complicated phenomena. (sec. 17.25).

Returning to the simpler assumption of $m = \infty$ we may note that, of course, the integration over the curved wave front ES does not give an amplitude that is exactly in phase (for $r = +1$) with the undisturbed wave. The integral is of the type

$$\int_0^\infty e^{-iby^n}\, dy = b^{-1/n} I_n = b^{-1/n} e^{-i\pi/2n}\, \Gamma\left(1 + \frac{1}{n}\right),$$

where $n = 3/2$, $I_{3/2} = 0.4514 - 0.7819i$. The application of the principle of Huygens-Fresnel to the wave front ES thus gives

$$T_{1,2}{}^{edge}(R,0) = \frac{r_{1,2}\sqrt{3}}{8}\,(2kR)^{1/3}\, I_{3/2} = (kR)^{1/3}\, r_{1,2}\,(0.123 - 0.213i).$$

This result does not have the correct numerical factor (it is too low by a factor about 2.0 for polarization 1 and about 1.75 for polarization 2, sec. 17.24). It does, however, show the correct functional dependence on kR and also the correct phase. The exponent $-i\pi/3$ means that the curves of $T_1^{edge}(R,0)$ and $T_2^{edge}(R,0)$, or the curves of $Q + iP$ in the complex domain, all approach the asymptotic point $(kR = \infty)$ from a direction that makes an angle of 60° with the real axis. The 60°

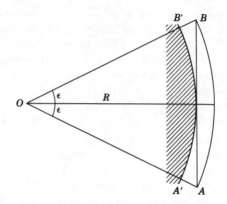

Fig. 77. Diagram illustrating path difference between grazing ray and curved path AB.

phase difference corresponds to a path difference $\lambda/6$, which is the "effective value" of the path ST (Fig. 76) upon integration. This may be compared to the path difference $\lambda/8$ upon integration along a parabolical front in one dimension (secs. 3.12 and 3.21).

We finally note why this reasoning is too naive. It was entirely arbitrary to consider only reflection of the directly incident light and to take only one wave front, namely the one through E. The "elementary waves" emanating from Huygens' "secondary sources" may again strike the surface and be reflected. In fact we might apply such reasoning to a whole succession of wave fronts. Artmann describes this as "multiple diffraction." Approximating the curved profile by a polygon he again computes that the scattered amplitude is proportional to $(kR)^{1/3}$.

Even more interesting is the fact that such reasoning may still be applied in the absence of a plane wave as incident wave. We then obtain a graphic picture of the *surface wave* that travels along the curved surface. The order-of-magnitude estimates follow in the simplest way (but perhaps not the most convincing way!) from Fig. 77. Here between A and B two optical paths are shown; a curved one at a fixed distance from the surface and a straight one including a grazing reflection. If it is asked how large the span AB may be in order to allow a difference of the order of λ between these two paths, the answer follows from simple geometry.

The angle $BOA = 2\varepsilon$, $BB' = AA' = R\varepsilon^2/2$, $AB = 2R\varepsilon$, and the path difference is $R\varepsilon^3/3$. Thus by order of magnitude, $\varepsilon = (kR)^{-1/3}$, which means that the surface wave occupies the region up to a distance of the order of $(kR)^{-2/3} \cdot R = k^{-2/3} \cdot R^{1/3}$ from the surface and includes spans of the order of $(kR)^{-1/3} \cdot R = k^{-1/3} \cdot R^{2/3}$. The same results follow from the exact theory (sec. 17.32).

17.22. Derivation from the Asymptotic Form of a Complete Solution

The most direct though not most elegant way of arriving at the precise values of the edge functions $T_1^{edge}(R,\theta)$ and $T_2^{edge}(R,\theta)$ is to start from a complete analytical solution of the scattering problem for a body of simple geometrical form. This body may, for instance, be a circular cylinder or a parabolic cylinder, or a sphere. The edge functions may be found by asymptotic expansion of the scattering amplitudes for large R.[2] A less ambitious version of this method is to use an empirical formula that represents the numerical results gained from the exact solution for fairly large values of kR (sec. 17.24).

A direct analytical attack was made by Artmann (1950). After heuristic arguments similar to those in sec. 17.21 he gives a complete solution of the scattering by a perfectly conducting circular cylinder. This solution is, apart from differences in notation, identical to the one in sec. 15.33. The approximations for large distances and small diffraction angles are made, and the summation over n replaced by an integration. The integral is separated into a part giving the ordinary diffraction and a part giving the edge wave. The major job then is to evaluate the latter part of the integral, using asymptotic expansions of the Hankel functions for large kR and for values of $|kR - n|$ that remain smaller than $5(kR)^{1/3}$.

Unfortunately, there is reason to suspect that the lengthy derivation contains errors. One reason is that the asymptotic form of $T_{1,2}^{edge}(R,\theta)$ for large values of $\theta/(kR)^{1/3}$ does not quite agree with the form following from geometrical optics (sec. 17.11), which is

$$T_{1,2}^{edge}(R,\theta) = \left(\frac{\pi kR \sin \frac{1}{2}\theta}{4}\right)^{1/2} r_{1,2} e^{3\pi i/4 + ix\theta^3/8}.$$

When Artmann's notation is translated into ours, the same formula is found, except for the last term in the exponent, for which he finds $+ix\theta^3/24$.

[2] Asymptotic solutions may be obtained directly by means of the Luneburg-Kline method. For application to various particle forms and for the most general asymptotic expansions, see F. G. Friedländer and J. B. Keller, *Commun. Pure and Applied Math.*, **8**, 387 (1955); J. B. Keller, R. M. Lewis, and B. D. Seckler, *Commun. Pure and Applied Math.*, **9**, No. 2 (June, 1956).

The second reason for suspecting an error is that the numerical values Artmann gives for $\theta = 0$ do not agree with those found in two other ways (sec. 17.24). His result in our notation is

$$4T_1^{edge}(R,0) = 1.90e^{-i\pi/3}(kR)^{1/3} = (0.95 - 1.64i)\,(kR)^{1/3},$$

$$4T_2^{edge}(R,0) = -1.02e^{-i\pi/3}(kR)^{1/3} = (-0.51 + 0.88i)\,(kR)^{1/3}.$$

An investigation of this kind for spheres was made in 1925 by Jöbst. He considers totally reflecting as well as metallic spheres. If we assume[3] that his $2D$ corresponds to our Q_{ext}, the final formula for $m = \infty$ is

$$Q_{ext} = 2\left\{1 + \frac{1}{x} + \frac{(0.468)^2}{x^{4/3}} + \ldots\right\}.$$

It is evident from the numerical results that this formula is not correct. The final result for gold particles is more complex and cannot so easily be checked.

A related study for cylinders based on earlier work by Spohn was made by Pfenninger. His work is more ambitious than Artmann's, in that the cylinders are assumed to be metallic. The final formulae are complicated and are not worked out to give numerical results.

Fock mentions that he has made a computation of this kind for a paraboloid of revolution and arbitrary angle of incidence, which gives him the opportunity to check the asymptotic forms for arbitrary ratios of the radii of curvature. No details of this highly interesting computation seem to have been published.

17.23. Derivation from an Integral Equation for the Surface Currents

A quite different approach to the problem is to apply Huygens' principle in its rigorous form to a cylindrical body whose cross section (profile) has a form such that $kR \gg 1$ but otherwise arbitrary. Figure 78 represents a plane perpendicular to the cylinder axis. By Huygens' principle the field at an arbitrary point P in space may be expressed in terms of those on two curves: a contour Γ that encloses the cylinder and a curve S at a very large distance around it. If P is between these two curves and the illuminating source is effectively at infinity, i.e. outside S, then the integral over S gives the field of the incident radiation, the integral over Γ the field of the radiation scattered by the cylinder. It is possible to include in the same formula the possibility that P is

---

[3] Some authors have criticized Jöbst for omitting a factor 2. This must be a matter of definition, as the same factor is absent from the very well-known formula for Rayleigh scattering in another paper by Jöbst [*Ann. Physik*, **78**, 15 (1925)].

located on Γ, as has been discussed in detail by Maue (1949). We shall first discuss case II only, magnetic field parallel to axis. The equation is

$$Cv(P) = v_{inc}(P) - \int_\Gamma \frac{\partial L}{\partial \nu}\, v(Q)\, ds.$$

Here $C = 1$ if P is outside Γ, and $C = 1/2$ if P is on Γ. The total magnetic field at any point is v, the incident field is v_{inc}, a point on Γ

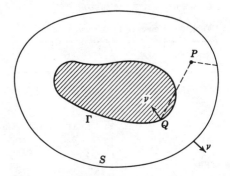

Fig. 78. Contours of integration in the application of Huygens' principle.

is Q, the inward normal of Γ is ν, and the line element of Γ is ds; the distance PQ is r_1, and

$$L = -\frac{i}{4}\, H_0^{(2)}(kr_1).$$

We now make the assumption that Γ is the boundary curve between conductor and vacuum, i.e., the "profile" of the cylinder. A first application of the formula above, with P on Γ, gives an integral equation of the magnetic field on the surface of the cylinder, which may be solved. A second application with P at a very large distance from Γ gives (in the second term) the distant scattered field, which we wish to find. The first step was made by Fock for the asymptotic case $kR \gg 1$. The second and simplest step will be reported in some detail.[4]

Fock's solution is given in the form:

$$v(Q) = v_{inc}(Q) \cdot G^*(\xi).$$

[4] The frequent use of variational methods has sometimes created the impression that the correct cross section can be derived only from a stationary representation (F. E. Borgnis and C. H. Papas, *Randwertprobleme der Mikrowellenphysik*, Berlin, Springer Verlag, 1955; note on p. 59). This is not true: the simple representation suffices if the correct current densities are known.

Here $\xi = (z - z_1)(2R^2/k)^{-1/3}$, where R is the radius of curvature, $k = 2\pi/\lambda$, and z is the rectangular coordinate measured inversely along the direction of propagation of the incident light, while z_1 is the coordinate for the edge point E. Further, $G^*(\xi)$ is a complex function that approaches 0 for $\xi \to -\infty$ (far on the dark side) and 2 for $\xi \to \infty$ far on the illuminated side). The latter value arises from the fact that the reflection coefficient in case II is $r_2 = 1$, so that the incident field is doubled. The asterisk is a reminder that we have to change the sign of i in Fock's table as we use the time factor $e^{+i\omega t}$.

Table 36. Values of Fock's Function $G(\xi)$

ξ	$|G|$	γ	ξ	G^*
-6	0.0093	4405°	-0.5	$1.029 - 0.252i$
-5	0.0223	2536°	0	1.399
-4	0.0537	1339°	0.5	$1.678 + 0.115i$
-3	0.1300	603°	1	$1.857 + 0.119i$
-2	0.315	211°	2	$1.981 + 0.050i$
-1	0.738	45.7°	3	$1.998 + 0.018i$
0	1.399	0°	4	$2.000 + 0.008i$
1	1.861	$-3.7°$	5	$2.000 + 0.004i$

Table 36 gives some values of $G^* = |G| e^{-i\gamma}$ quoted from Fock's paper. Now, let P be a very distant point in the forward direction, at a distance r from the origin O (somewhere in the body). The field in this point is

$$v(P) = v_{inc}(P) + v_{sca}(P).$$

With permitted approximations we have $r_1 = r + z$,

$$L = \frac{1}{(8\pi kr)^{1/2}} e^{-i\pi/4 - ikr_1},$$

$$\frac{\partial L}{\partial \nu} = \frac{k \sin \tau}{(8\pi kr)^{1/2}} e^{+i\pi/4 - ikr_1},$$

$$v_{inc}(Q) = v_{inc}(O) \cdot e^{ikz},$$

so that Huygens' principle in the rigorous form quoted above gives

$$v_{sca}(P) = -v_{inc}(O) e^{i\pi/4 - ikr} \int_\Gamma \frac{k \sin \tau}{(8\pi kr)^{1/2}} G^*(\xi) \, ds.$$

This assumes the form postulated in the first formula of sec. 17.11 if we write

$$T(0) = \frac{k}{4} \int_\Gamma \sin \tau \, G^*(\xi) \, ds.$$

The separation into three terms:

$$T(0) = \frac{kg}{2} + T_2{}^{edge}(R,0) + T_2{}^{edge}(R',0)$$

is effected by separating the integration path as follows (Fig. 79). The letters S and L denote points chosen sufficiently far on the shadow side and on the light side.

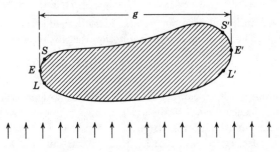

Fig. 79. Separation of the integration contour in order to find the edge terms.

From S to E with factor G^* $\left.\begin{array}{c} \\ \\ \end{array}\right\}$ edge term
From E to L with factor $G^* - 2$

From E to L to L' to E' with factor 2: classical diffraction

From L' to E' with factor $G^* - 2$ $\left.\begin{array}{c} \\ \\ \end{array}\right\}$ edge term.
From E' to S' with factor G^*

The middle part gives the classical diffraction by a strip of width

$$g = \int_{E\,LL'E'} \sin \tau \, ds.$$

Only one edge term has to be discussed. In it we write

$$\sin \tau = \frac{z - z_1}{R} = (2/kR)^{1/3} \cdot \xi,$$

$$ds = dz = (2/kR)^{1/3} R \, d\xi,$$

so that

$$T^{edge}(0) = \left(\frac{kR}{16}\right)^{1/3} (I + J),$$

with

$$I = \int_{-\infty}^{0} G^*(\xi)\xi \, d\xi, \qquad J = \int_{0}^{\infty} [G^*(\xi) - 2]\xi \, d\xi.$$

Numerical integration from the table of $G(\xi)$ given by Fock yielded:

$$I = -0.299 + 0.527i$$
$$J = -0.239 + 0.416i,$$
$$I + J = -0.538 + 0.943i,$$
$$4^{1/3}(I + J) = -0.855 + 1.500i,$$

so that

$$4T_2{}^{edge}(R,0) = (kR)^{1/3}(-0.855 + 1.500i).$$

The corresponding solution for case I (electric field $\parallel$ axis) has not yet been made.[5] The simplest integral equation for this case is

$$\tfrac{1}{2}v(P) = v_{inc}(P) + \int_\Gamma \frac{\partial L}{\partial \nu}\, v(Q)\, ds,$$

where v represents the φ-component of **H** (i.e., the tangential component of **H**, which is perpendicular to the axis) or the z-component of the surface current. The equation differs from that for case II not only by the plus sign but also by the fact that $|v_{inc}|$ was a constant in case II and is proportional to $\sin \tau$ in case I, where τ is defined as the angle between the normal in Q and the normal at the edge point E.

Both equations have been solved analytically for circular cylinders and arbitrary kR by Maue. The solution made by Fourier expansion leads back to the classical solution by separation of the variables (sec. 15.33) and thus gives no advantage for numerical work. Franz and Deppermann have discussed both equations in the limit of large kR. Numerical results are confined to the running wave, i.e., to the asymptotic form of $v(P)$ in the shadow part of the surface far from the edge (sec. 17.32) and to a solution by iteration of the value at the edge point. The agreement with Fock for case II is excellent, namely, $v(E) = 1.399v_{inc}(E)$. Kodis (1956) by means of a variational principle obtains for case I $\mathrm{Re}\,[2T_1{}^{edge}(R,0)] = 0.523x^{1/3}$ in fair agreement with the empirical coefficient which is 0.50.

It may be remarked that the problem of computing the field on a circular cylinder due to a wave incident from infinity is reciprocal to the problem of computing the distant fields radiated by antennas on the surface or slots in the surface of the cylinder. Sensiper (1953) cites dozens of papers and reports on the latter problem; we may quote the most accessible ones: Carter (1943), Silver and Saunders (1950a, b) Wait and Kahana, 1955. A comparison of detailed results might be useful.

---

[5] Since this was written the important study by S. O. Rice [*Bell System Tech. J.*, **33**, 417 (1954)] based on a Watson transformation of the solution for parabolic cylinders has been brought to the author's attention. This work confirms in detail Fock's results for case II and also gives the corresponding solution for case I.

17.24. Empirical Formulae for $m = \infty$

A simple manner of obtaining an estimate of the edge terms is to start from the rigorous solutions for cylinders or spheres, to evaluate the distant fields numerically for large $x = ka$, and to derive an empirical formula for the asymptotic behavior. This the author has done in the following cases.

a. Circular cylinders, $m = \infty$. Formulae and numerical values have been given in sec. 15.33. Also the empirical formulae for large x are

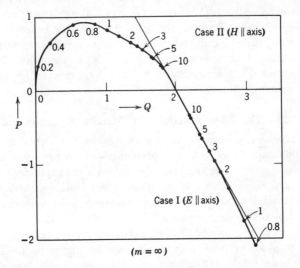

Fig. 80. Plot of $Q + iP$ for perfectly conducting cylinders of infinite length. The real part Q is the efficiency factor for extinction. The numbers with the curves are the values of $x = 2\pi a/\lambda$.

found in that section. Acccording to sec. 17.11 the full formula for $(2/x)T(0)$ is $2 + (4/x)T_{1,2}{}^{edge}(R,0)$. Thus

$$4T_1{}^{edge}(R,0) = (1.00 - 1.73i)x^{1/3} + \dots$$

$$4T_2{}^{edge}(R,0) = (-0.86 + 1.49i)x^{1/3} + \dots$$

The latter result checks extremely well with the value found in sec. 17.23 from Fock's table. The functions $(2/x)T_{1,2}(0) = Q + iP$ are plotted in the complex domain in Fig. 80. It is seen, in agreement with the heuristic argument (sec. 17.21), that these functions approach the asymptotic value 2 from opposite sides and from directions making an angle of $60°$ with the real axis. However, the rates of approach are different.

b. Spheres, $m = \infty$. In accordance with sec. 17.13, the asymptotical formula for $(4/x^2)S(0)$ should be $2 + \{4T_1^{edge}(R,0) + 4T_2^{edge}(R,0)\}/x$. This expression may be denoted by $Q + iP$, as its real part is again the efficiency factor for extinction, Q. When the values found under *a* are inserted this becomes

$$Q + iP = 2 + (0.14 - 0.24i)x^{-2/3} + \ldots$$

There is not a very convincing agreement with the numerical results or empirical formulae given in sec. 10.62. However, the main feature, the existence of only a small excess of Q over 2, is confirmed; with empirical coefficients as small as this the relative errors may be large. Nevertheless, it is disappointing that the correct asymptotic formula for spheres with $m = \infty$ is still in the dark. Deppermann and Franz did not go beyond the dominant term 2. The form derived by Jöbst (sec. 17.22) is evidently in error.

17.25. The Edge Functions for Metallic Cylinders and Spheres

As in previous sections we must warn against a tendency to take too lightly the results for $m = \infty$ as representative for a variety of materials. The attempts at accurate analytical derivation of the asymptotic formulae for arbitrary m, as in sec. 17.22, lead to bewildering formulae. It is, therefore, useful to try an empirical approach from numerical results supported by heuristic arguments.

As soon as $ka = x$ exceeds 1, any wave that travels through a metallic particle is absorbed and may practically be disregarded. The only difference with totally reflecting particles is, then, in the boundary conditions. For plane waves they lead to the Fresnel reflection coefficients, and we have already noted (sec. 17.21) that the sign of r_2 for grazing incidence and finite m differs from that for $m = \infty$. Accordingly, we expect that the values of $Q + iP$ for cylinders in the two polarizations do not approach the limit 2 from opposite sides but from the same side, viz., the side of $Q_1 + iP_1$ in the case $m = \infty$.

The full-drawn line in Fig. 81 is the complete graph of data given in Table 35 (sec. 15.51) for $m = \sqrt{2}\,(1 - i)$. The value $x = 4$, that is, the upper limit of these computations, is too small to permit a guess at an asymptotic formula. However, qualitatively, these data confirm the expectations: at $x = 4$ the curve for case I is already on its way to the limit along the correct asymptote, but the curve for case II still has to make a big swing. The dotted extension suggests how the further curve might run.

Little more can be done in an exact manner, short of solving a full integral equation with new boundary conditions. We may, however,

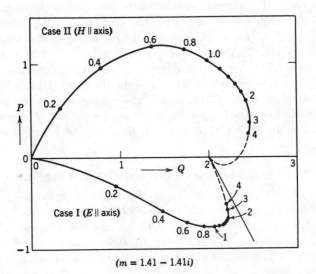

Fig. 81. Plot of $Q + iP$ for cylinders of infinite length with $m = \sqrt{2}(1 - i)$. The real part Q is the efficiency factor for extinction. The numbers with the curves are the values of $x = 2\pi a/\lambda$.

check the preceding interpretation as follows. The angles of incidence that are involved in the edge phenomenon go to $\tau = x^{-1/3}$, approximately (sec. 17.21). Let us take $\tau = (2x)^{-1/3}$ as an effective angle. For $x = 4$

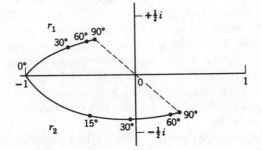

Fig. 82. Fresnel reflection coefficients for external reflection against a body with $m = \sqrt{2}(1 - i)$ plotted in the complex domain as a function of the angle with the surface.

the effective angle is $\frac{1}{2}$ radian $\approx 30°$. Fig. 82 shows a sketch of the reflection coefficients for $m = \sqrt{2}(1 - i)$ computed from the equations in sec. 12.21. If we remember that the asymptote in Fig. 81 corresponds to $r_1 = r_2 = -1$, we see that both functions deviate in the proper sense and by about the proper angle from the asymptote. In particular the point on curve II for $x = 4$ is about at right angles to the asymptote; this corresponds with the value of r_2 for $\tau = 30°$.

The form of the extinction curves for metallic *spheres* (sec. 14.22, Fig. 54) can also be understood a little better now. If, following the general formula in sec. 17.13, we add the ordinates of curves I and II in Fig. 69 and substract 2, we obtain indeed a rapid rise of the real part for small x, an approximate standstill for $x = 2$ to 4 and, thereafter, a slow decline.

17.26. Edge Functions for Dielectric Cylinders and Spheres

The form of the extinction curves for dielectric cylinders (Fig. 67) and spheres (Figs. 24 and 32) is dominated by the interference effects between the centrally transmitted radiation and the radiation diffracted around the particle. This effect was explained in secs. 11.22 and 13.42; it accounts fully for the big fluctuations in Q with a maximum at $2x(m - 1) = 4.0$, etc.

The next striking feature is that the average curve through these fluctuations lies above the asymptote $Q = 2$. The first maxima reach $Q = 3.5$ to 4, the first minima hardly come below $Q = 2$. This effect may be ascribed to the "edge reflection," as will be made plausible below.

A third feature is the existence of small wiggles. They will be neglected in the present section; a tentative explanation by means of a modification of the surface wave is given in sec. 17.52.

At $x = 4$ the "effective angle of incidence" is near $\tau = 30°$ (sec. 17.25). The Brewster angle for $m = 1.33$ is at $\tau = 36.9°$ (sec. 13.11). This means that already at $x = 4$ the edge terms in the two polarizations will have the same sign, and they will more and more approach each other for increasing x. This prediction is in qualitative agreement with the few results for dielectric cylinders shown in Fig. 67 (sec. 15.31). We shall further consider spheres, for which more extensive numerical material is available.

Starting from the data given in the work by Gumprecht and associates for $\theta = 0$, the present writer has first transformed these data into our $S(0)$, so that the real and imaginary parts of $S(0)$ are known for quite a number of values of x up to $x = 400$. The transformation formulae are in sec. 13.12. In sec. 12.45 we have compiled the formulae for the predicted terms in $S(0)$ for large x.

After the main term, $\frac{1}{2}x^2$, due to diffraction, was subtracted, the difference showed the familiar large fluctuations. When the refracted light

$(p = 1)$ was subtracted also, the large differences vanished, and a set of points in the complex plane was left that showed small fluctuations about a fairly well-defined law. The small fluctuations are the "ripple" shown in Fig. 32. The remaining term may be interpreted as the edge term, $S^{edge}(a,0)$, for the term with $p = 3$ is numerically unimportant. So the tentative formula is

$$S(0) = \tfrac{1}{2}x^2 - x \frac{2m^2}{(m+1)(m^2-1)} ie^{2ix(m-1)} + S^{edge}(a,0) + \text{``ripple.''}$$

On the basis of this formula the author plotted for $m = 1.33$ the function

$$\frac{1}{x} S(0) - \tfrac{1}{2}x + 1.975\,(\sin 0.66x + i \cos 0.66x)$$

against $x^{1/3}$ and derived the following empirical law:

$$\frac{1}{x} S^{edge}(a,0) = (0.46 - 0.62i)x^{1/3},$$

or if a constant term is permitted:

$$\frac{1}{x} S^{edge}(a,0) = 1 + i + (0.30 - 0.78i)x^{1/3}.$$

The imaginary coefficient is less certain than the real one, as the number of values for which the imaginary part of $S(0)$ has been published ($x = \cdots, 15, 20, 25, 30, 35, 40, 100, 200, 400$) is low. The real parts of $S(0)$ also may be derived from the values of Q_{ext}, which have been published for many more values of x. Some small inconsistencies were found in the tables themselves, but the major cause of uncertainty is undoubtedly the "ripple" that has not yet been represented by a (theoretical or empirical) formula. Also used were the data of Q_{ext} for $m = 1.20$. By plotting

$$\frac{x}{4} Q_{ext} - \tfrac{1}{2}x + 2.975 \sin 0.40x$$

against $x^{1/3}$ the empirical relation for $m = 1.20$ was found to be

$$\text{Re}\left\{ \frac{1}{x} S^{edge}(a,0) \right\} = 0.51x^{1/3} \text{ (or } 1 + 0.31x^{1/3}).$$

Let us now see what coefficients we should expect. In sec. 17.13 we derived

$$S^{edge}(a,0) = x\{T_1{}^{edge}(a,0) + T_2{}^{edge}(a,0)\}.$$

We may use the formulae for totally reflecting bodies (secs. 17.23 and 17.24), if we reverse the sign of T_2^{edge} $(a,0)$, as was explained in sec. 17.21. Thus

$$\frac{1}{x} S^{edge}(a,0) = \tfrac{1}{4}\{(1.00 - 1.73i)x^{1/3} + (0.86 - 1.49i)x^{1/3}\}$$

$$= (0.46 - 0.80i)x^{1/3}.$$

The coefficient agrees reasonably well with those in the empirical relations for $m = 1.33$ and $m = 1.20$. At any rate, both the real and the imaginary parts have the correct sign and order of magnitude, which, incidentally, is quite different from the one for $m = \infty$. This is a satisfactory conclusion of the exploratory computations reported in this section. A more careful analysis of the entire problem may reveal further essential features which have been overlooked above.

For practical purposes we may estimate from the graphs that the terms in $S(0)/x$ that are not accounted for, including the ripple, usually remain <0.5 for $x > 100$. This means that those terms introduce in

$$Q_{ext} = \frac{4}{x^2} S(0)$$

an uncertainty $< 2/x$. So the semiempirical formula

$$Q_{ext} = 2 - \frac{8m^2}{(m + 1)(m^2 - 1)} \cdot \frac{\sin\{2x(m - 1)\}}{x} + 1.8x^{-2/3}$$

probably gives the correct value within 1 per cent for all values $x > 100$, $m < 2$.

The paper of Boll, Gumprecht, and Sliepcevich (1954), in which numerical results for spheres with $m = 0.8$, 0.9, and 0.93 are given, permits an independent determination of the coefficient. The curves lie systematically lower than the curve for $m = 1 \pm \varepsilon$, as is seen in Fig. 32 (sec. 11.22). This reversal in sign is explained by the fact that $r_1 = r_2 = +1$ for grazing reflection against a medium with $m < 1$. The formula that seems to represent the numerical data reasonably well is

$$Q = 2 + \text{refraction term} - (2.5 \pm 0.6)x^{-2/3}.$$

The situation at the edge is not strictly comparable with that for $m > 1$ because of the occurrence of total reflection. This may have something to do with the fact that the coefficient 2.5 ± 0.6 turns out a little higher than the coefficient 1.8 found above.

17.3. Surface Waves

17.31. Surface Waves in the Literature

The literature on electromagnetic waves contains a lot about surface waves. In this book the subject is not approached from the outlook of the mathematician who requires rigor and mathematical clarity. The point of view is rather that of the practical physicist who wishes to form a workable picture of any effect which comes to his attention and is numerically important. Eventually, these points of view will not lead to different results, but the approach is different. Numerically important effects in the scattering by spheres and cylinders that involve surface waves are discussed in secs. 17.4 and 17.5. If this exploratory investigation suggests new rigorous attacks[6] to the applied mathematician, this chapter will have served its purpose.

The problem that has had the greatest attention is the propagation of waves along the surface between a conductor with finite conductivity and a dielectric (or vacuum). Already at the beginning of this century attempts were made to explain the propagation of radio waves beyond the horizon by means of such surface waves.

One type of wave that is a clear and correct solution of Maxwell's equations is a wave with a forward tilt, following a plane earth and attenuated in the horizontal as well as in the vertical direction. It is polarized with the **H**-vector horizontal. These waves are called Zenneck waves, as Zenneck suggested in 1907 that such waves are set up by the radiation from a vertical dipole antenna. This suggestion was supported by Sommerfeld's computations (1909), but Weyl in 1916 suggested a different transformation by which this component vanishes. Present views are that Weyl was correct and that the Zenneck waves are not present in the radiation from an antenna and, consequently, of no importance for long-range radio propagation.

There are stringent reasons why the Zenneck waves are not important for our discussion. A strong effect of a surface wave is observed in the backscatter by perfectly conducting spheres and cylinders (sec. 17.41); it is observed in both directions of polarization. The Zenneck wave does not vanish if $m \to \infty$, but it loses its character of a surface wave. The tilt vanishes, the attenuation vanishes, and the field in the conductor vanishes. What is left is an ordinary plane-polarized wave in vacuum alongside a perfectly conducting surface that contains the **H**-vector. Evidently such a wave is of no use to us; it does not even exist with the other direction of polarization.

---

[6] The unpublished work of Franz, which came to the author's attention after this chapter had been written, completes an important step in this direction.

Weyl's approximation also contained a component that can be interpreted as a surface wave. This component vanishes if $m \to \infty$, so it is not of any use either in connection with the backscatter problem just mentioned. It bears some resemblance, however, to the surface waves at the interface between two dielectrics that will be discussed below.

The existence of a surface wave at the separation between two non-dissipative media was first discussed by von Schmidt (1938) for pressure waves (sound, earthquake waves). The analogous light waves were discussed with great care by Ott (1942, 1949). In these studies the surface wave is generated by a point source (dipole) located in the optically dense medium. For convenience we shall talk about a glass-air surface. A source in the glass at A (Fig. 83) emits a spherical wave, i.e., rays

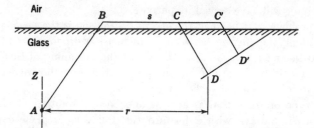

Fig. 83. Generation of a surface wave and a head wave by a source A in the medium with larger refractive index (according to Ott).

in all directions. Among them are rays that hit the surface steeply and are partially refracted, partially reflected. Among them are also rays that hit the surface not steeply enough and are totally reflected. A ray AB at the limiting angle, which forms the transition, gives by Snell's law a refracted ray which grazes the surface. Its intensity is 0 by Fresnel's formula, but Ott's solution shows it to have a sizable strength. At any point (C or C') the latter ray may send energy back into the glass under the same limiting angle in the direction CD or $C'D'$. The observable effect finally is a wave with front DD' traveling into the glass. As there is rotational symmetry around AZ, this is a conical wave. Because of its resemblance to the head wave of a ship or projectile it has been called the Schmidt head wave.

Let $BC = s$ and the projected distance of A and D be r (Fig. 83). The phase of the wave at D agrees with expectation: it is delayed with respect to that in A by the optical path $m \cdot (AB) + 1 \cdot (BC) + m \cdot (CD)$. The intensity is proportional to $k^{-2} r^{-1} s^{-3}$. If A and D are close to the surface, this means a proportionality to r^{-4}, i.e., a much faster attenuation than in a free spherical wave. The polarization with $\mathbf{E}$ in the vertical plane, i.e., $\mathbf{H} \parallel$ surface, gives a stronger head wave than the polarization with $\mathbf{E} \parallel$ the interface. The intensity ratio is m^4 if the source strength is taken equal.

Maecker (1949) experimentally demonstrated these surface waves with light and also summarized the theory in a more popular form. He emphasizes that there must be a good many other methods by which such surface waves may be set up. In particular, it is quite easy to see that if a glass sphere with a large radius is placed in a parallel beam of light we start at the edges with the situation of a grazing ray like BC in Fig. 83, and this situation will recur whenever a refracted edge ray reaches the surface again.

Another way to demonstrate these waves is to send a plane wave, which is sharply cut off at one side, from the glass side to the plane glass-air interface. If the angle of incidence is the limiting angle, the surface wave transports some energy sideways and causes a measureable displacement of the reflected plane wave. The effect diminishes at either side of the limiting angle. The theory is discussed by Artmann (especially 1951a), and experiments have been performed by Goos and Hänchen (1947, 1949).

17.32. Waves at the Surface of a Perfect Conductor

The brief review in the preceding section shows that the older literature contains no hints of the existence of waves that can follow the surface of a perfectly conducting body. Yet such waves exist and have been brought forward quite clearly by Franz and Deppermann (1952, 1954). They were also implicitly contained in the work of Fock (1946).

The reason why these "creeping waves" may have been missed before is that they occur only if the surface is curved. The work of Sommerfeld and Weyl had dealt mostly with plane surfaces, and the extension to a spherical surface which had been worked out with great ingenuity by van der Pol and Bremmer (sec. 12.35) had seemed to offer nothing essentially new.

The starting point in the work of Fock and of Franz and Deppermann is an integral equation for the field at the surface. The method has been explained in sec. 17.23 for one direction of polarization. For convenience we may restrict the discussion again to the situation depicted in Fig. 79 (cylindrical bodies). The surface wave is injected at and near the edge point E and creeps further into the shadow region along the curve ES. When it is advanced far enough into the shadow region it has become a damped periodic wave whose properties are independent of the particular injection mechanism. Fock gave the correct solution of surface wave *and* injection mechanism (for one polarization) but did not interpret the asymptotic form of his function $G(\xi)$ as a surface wave. Franz and Deppermann (1952) gave the correct solution for the edge point[7] and fully discussed the solution of the homogeneous integral equation, which describes the surface waves.   They realized that the least damped surface wave should be the only one of practical value but did not assume the right amplitude.   This error was corrected later (Deppermann and Franz, 1954), and the corrected formula is in complete agreement with Fock's result.

The strongest surface waves occur for linear polarization with **H** tangent to the surface (polarization 2).   The quantitative formulae for those waves will be briefly summarized.

Let the radius of curvature R be a constant; $x = kR$. Let Q be a point at the surface far in the shadow region at an angle φ beyond E. Its distance from E along the surface is $R\varphi$, its straight distance from the wave front through E is $R \sin \varphi = z$ (this z is measured opposite to the z in sec. 17.23).   If the incident field has the amplitude 1 at E, Fock finds the amplitude at Q (with the sign of i changed)

$$v(Q) = e^{-ikz}\, 1.83 \times 10^{0.383\xi}\, e^{i(0.51\xi + \xi^3/3)},$$

where $\xi = -z(k/2R^2)^{1/3}$. Franz and Deppermann (also with the sign of i changed) find

$$v(Q) = 1.83\, e^{-ix\varphi - 2.332(\frac{1}{2}\sqrt{3} + \frac{1}{2}i)x^{1/3}\varphi/24^{1/3}}.$$

As $R\varphi = z + z^3/6R^2 + $ negligible terms, the exponents of e in both expressions reduce to

$$-ix\varphi - (0.699 + 0.403i)x^{1/3}\, \varphi.$$

This simple result may be interpreted as follows.   The *phase* is as if the wave travels along the curved surface with a velocity $c/(1 + 0.403x^{-2/3})$. A more useful interpretation is, however, that the wave travels with the velocity c effectively at the distance $R(1 + 0.403x^{-2/3})$ from the center of curvature, i.e., at the distance $0.403x^{-2/3} R = 0.403k^{-2/3} R^{1/3}$ from the surface.   This agrees very well with the functional dependence of the distance ET (Fig. 76) on k and R as estimated in sec. 17.21.

The *damping* is such that the amplitude reduces to $e^{-2.19x^{1/3}}$ of its initial value in completing a half circle (π radians) around the cylinder. For $x = 8$ this means a reduction to 0.018 of the initial amplitude.   So the surface wave is strongly damped and has little numerical significance for $x > 8$.

For $R \to \infty$ this surface wave (still with polarization 2) becomes an undamped plane Maxwell wave of infinite extent in the radial direction, i.e., the same type of wave into which the Zenneck wave degenerates if the conductivity is made infinite.

---

[7] Except for a small numerical error: the value of $G(0)$ should be 1.399, in agreement with Fock's result (Franz, private communication).

A word about the *damping of surface waves generally* may be added. The damping of the Zenneck waves or of the Weyl-Norton waves (sec. 17.31) is due to Joule losses in the imperfect conductor. No exponential damping factor is found in the equivalent waves at the surface of a dielectric (Ott). The surface waves discussed in the present sections are damped for a quite different reason. In bending around the surface they continually spray energy forward tangentially away from the surface. This "spray" can be computed by Huygens' principle in the same manner as explained in sec. 17.23. Detailed formulae have been derived by Franz and Depperman.

The equations for the surface wave for polarization 1 (**E** tangent to the surface) were also solved by Franz and Deppermann. This wave is numerically less important. It vanishes if $R \to \infty$, as is expected from the fact that the boundary conditions of a total conductor (sec. 9.14) do not permit a plane wave to travel along its surface with **E** tangent to the surface.

The surface waves traveling around partially conducting bodies have not yet been studied in detail. We may expect that they are damped by both factors. A good many further investigations might be suggested here, but it is not necessary to anticipate further developments.

The tentative conclusions are summarized in Table 37.

Table 37. Properties of Surface Waves on Conducting Bodies*

	Flat Surface	Curved Surface
Total conductor	Polarization 1: no wave	Polarization 1: weak surface wave (Franz and Deppermann)
	Polarization 2: plane wave, undamped	Polarization 2: strong surface wave damped by spray (Franz and Deppermann)
Partial conductor	Polarization 1: no wave	Polarization 1: weak surface wave
	Polarization 2: Zenneck wave damped by Joule losses	Polarization 2: strong surface wave damped by spray and by Joule losses

* Polarization 1 means **E** $\parallel$ surface; polarization 2 means **H** $\parallel$ surface

17.4. Backscatter

17.41. Backscatter by Cylinders and Spheres with $m = \infty$

To compute the "spray" of a surface wave in a given direction is a relatively simple problem. It just requires the use of Huygens' principle in one of its rigorous forms.   The field is given at all points of the surface. The field is sought in a point P at a very large distance in a given direction. The major contribution of the integration over the surface comes from the region near the point (or points) Q where the tangent is directed to P, for it is near these points that stationary phase occurs in the integral. For large particles we may assume that the contributions from other parts of the surface are negligible and that the intensity sprayed into the direction of P depends only on the intensity of the surface wave and the local radius of curvature at Q.

The procedure just described is, in fact, identical to the one applied in sec. 17.23, where we computed the "edge function" for forward scattering.   The only difference is that in that section we computed the spray right from the region where the injection occurred, whereas in the backscatter problem the surface wave has already covered an angle π and thus has its asymptotic form, independent of the injection mechanism.

The spray in any direction is also determined by the surface wave that has completed 1, 2, or more, turns around the body.   This holds true for forward or back scatter or for any other direction.   These contributions may be added as a geometrical series, but they are numerically insignificant for $x > 4$ and will not further be considered.

The application of the surface wave to the problem of the radar cross section of totally reflecting cylinders and spheres has been made by Franz and Deppermann.   We shall not follow the full derivation but just give the results in a form consistent with our notation and with omission of terms of a smaller order in x, some of which were included in Franz's and Deppermann's calculations.

Cylinders, case II ($\mathbf{H} \,\|$ axis): The full amplitude function at $\theta = \pi$ is $T_{refl}(\pi) + T_{surf}(\pi)$, with

$$T_{,efl}(\pi) = \left(\frac{\pi x}{4}\right)^{1/2} e^{-\pi i/4 + 2ix},$$

$$T_{surf}(\pi) = 2.92 \left(\frac{\pi}{4}\right)^{1/2} x^{1/3} e^{-2.20x^{1/3} - i(\pi/3 + \pi x + 1.27x^{1/3})}.$$

The addition of the second term to the amplitude thus increases the radar cross section found on the basis of geometrical reflection by the factor

$$\frac{\sigma}{G} = \left| 1 + 2.92 x^{-1/6} e^{-2.20x^{1/3} - i(\pi/12 + 5.14x + 1.27x^{1/3})} \right|^2.$$

Spheres: The radar cross section found from geometrical reflection ($= \pi a^2$) is increased by

$$\frac{\sigma}{G} = \left| 1 + 2.72 x^{1/3} \, e^{-2.20 x^{1/3} - i(-5\pi/12 + 5.14x + 1.27x^{1/3})} \right|^2 .$$

The contribution of the more strongly damped surface wave corresponding to cylinders, case I, has been omitted in this equation.

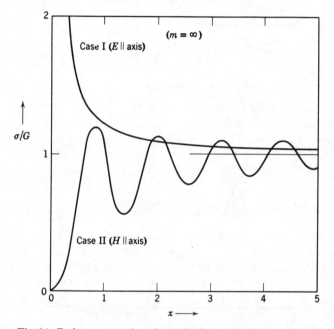

Fig. 84. Radar cross section of a perfectly conducting circular cylinder of infinite length divided by the value following from geometrical optics. The direction of propagation of the incident wave is supposed to be perpendicular to the axis of the cylinder; $x = 2\pi a/\lambda$, where a is the radius.

It is of some interest to compare these expressions with the values obtained by direct computation by the usual method. The radar cross section of totally reflecting spheres was shown for the range $x = 0$ to 3 in Fig. 61 and the corresponding phase in Fig. 62 (sec. 14.32). The radar cross sections of a perfectly conducting cylinder were computed from the formulae in sec. 15.33 and are shown in Fig. 84. The maxima and minima in both figures are spaced by $\Delta x = 1.18$, approximately. The computations for cylinders show quite clearly that the surface wave is important only for polarization 2. A slight wavering in the curve for polarization 1 suggests a weak surface wave also in the other polarization. This is in good agreement with the discussion in sec. 17.32.

The positions of the maxima and minima are determined by the phase of the extra factor in the formulae above. Let us write

$$2\pi f = \pi/12 + 5.14x + 1.27x^{1/3}$$
$$f = 0.04 + 0.82x + 0.20x^{1/3};$$

then the positions are (n integer)

Cylinders, maximum at $f = n$

Spheres,   maximum at $f = n + \frac{1}{4}$

Cylinders, minimum at $f = n + \frac{1}{2}$

Spheres,   minimum at $f = n + \frac{3}{4}$.

The positions computed from these equations check very well with those shown in Figs. 61 and 84, which means that the asymptotic formulae are useful from $x = 2$ or 3 on. For a precise agreement higher terms in $1/x$ have to be added. The spacing between the maxima should be

$$\frac{1}{df/dx} = \frac{1}{0.82 + 0.07x^{-2/3}},$$

which is 1.18 near $x = 3$ and climbs to 1.20 near $x = 10$.

The physical meaning of f is (apart from the small delay 0.04 at ejection) *the difference between the two light paths* shown in Fig. 85, *expressed in wavelengths*. As (sec. 17.32) $q \approx (1 + 0.40x^{-2/3})a$, the path difference is

$$2a + \pi q = \lambda x \left(\frac{1}{2} + \frac{1}{\pi} \right) + 0.20 \, \lambda x^{1/3},$$

which upon division by λ gives $f - 0.04$. Surface waves traveling around the sphere have a path that is effectively shorter by $1/4 \, \lambda$ as they have to pass the focus at F. This phase gain is of precisely the same nature as the gain of an ordinary wave upon passing a focal line (sec. 3.21).

The amplitude of the fluctuation also requires some comment. The formula given above for cylinders gives good agreement with the amplitude shown by Fig. 84. The formula for spheres gives an amplitude that seems to be too small by a factor 2. This is still unexplained.

The results for spheres differ in two important respects from those for cylinders. The first one is that the fluctuations for spheres have a far greater amplitude than those for cylinders. The first minimum even comes close to zero. The fluctuations are far stronger than we could obtain, e.g., by averaging the results of the two polarizations for cylinders. The physical reason is the same as the one which led in sec. 12.21 to infinite backscatter from spheres according to geometrical optics, namely, the cooperation of the radiation coming in equal phase from a full circle. In the case of a cylinder the radiation comes from two opposite sides, and the direction of backscatter has no unique position among all directions Consequently, the scattered intensities at an arbitrary angle for cylinders should have a character quite similar to that shown in Fig. 84. In the case of spheres, however, the direction of backscatter is an exceptional

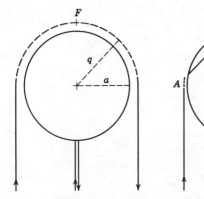

Fig. 85. Contribution of the surface wave to backscatter from a perfectly conducting cylinder or sphere.

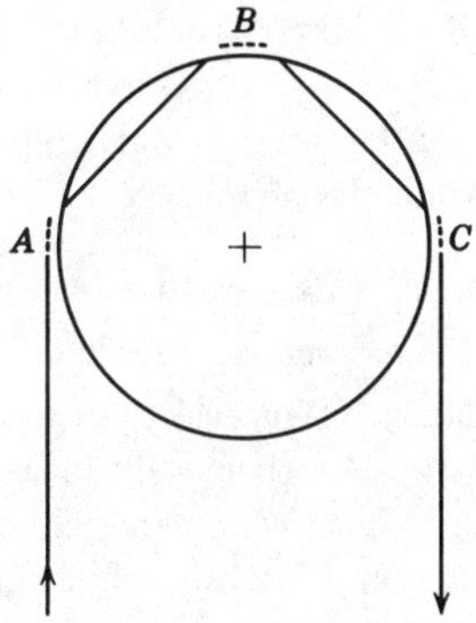

Fig. 86. Contribution of the surface wave to backscatter from a dielectric cylinder or sphere.

direction. Here the analysis of sec. 13.31 is relevant. The present example is, in fact, a perfect example of a "glory" with $C_1/C_2 = 0$, superposed on and interfering with the centrally reflected light.

This semiquantitative essay on the surface waves should not replace the exact treatment found in the papers cited. It may, however, serve to make the latter results more easily accessible.

17.42.   The Glory for $m = 1.33$

The glory observed in the scattering pattern of water drops has been discussed in some detail in secs. 13.31 to 13.33. A question remains whether the existence of surface waves can shed any new light on the matter.

One problem was that a ray as depicted in Fig. 48 did not exist at the refractive index $m = 1.33$. If at the injection A, or at the total reflection B, or at the ejection C, the wave could skip $14°$ of the circumference, the existence of such a ray would be a fact (Fig. 86). This is indeed possible. The situation at B and C resembles very closely the situation in Ott's theory for generating surface waves (sec. 17.31), viz., illumination from the dense medium under the limiting angle of total reflection. The situation at A is different, but the analysis of the preceding sections shows that surface waves are also generated there. So there are three opportunities to skip the 14° or parts of it. The intensity must be considerable. We can also attempt to explain the polarization in this manner. If the surface wave occurs only at B, Ott's theory gives a factor n^2 as ratio of the amplitudes in the two polarizations, polarization 2 (with $\mathbf{H} \perp$ plane of the drawing) being the stronger one. In the terminology of sec. 13.3: $C_2/C_1 = n^2 = 1.7$. Apart from the fact that C_2 has indeed the larger absolute value, this does not agree with the empirical result, $C_2/C_1 = -4$, obtained in sec. 13.33.

The present section is only exploratory; a detailed theory along these general principles should explain the glory phenomenon entirely.

17.5. Ripples in the Extinction Curve

17.51. Empirical Data

The qualitative success of explaining some phenomena in the numerical results of the Mie theory by means of surface waves makes us wonder if there are further phenomena which would admit of a similar explanation.

One intriguing feature, which has not been explained in the preceding pages, is the ripple on the extinction curve for dielectric spheres (in the range $m = 1.2$ to 2.0). The main features of these curves are understood: the asymptotic value $Q = 2$ for $x \to \infty$ (sec. 8.22); the big fluctuations whose maxima and minima occur at fixed values of $\rho = 2x(m - 1)$ (sec. 11.22); and, finally, the fact that the mean curve about which these fluctuations occur lies above $Q = 2$ (sec. 17.26). The ripple is a phenomenon of rapid fluctuations with an amplitude ± 0.15 in Q, numerically insignificant for practical applications.

The ripple has been shown to exist with certainty in the work of the Cambridge Air Force Research Center (CAFRC).[8] Most other computations had been made with steps in x too big to reveal the ripple. The CAFRC tables were made on the IBM 701 Electronic Data Processing Machine. They cover the range $x = 0$ to 30, with narrow intervals of x, and were made for $m = 1.33$, 1.40, 1.44, 1.486, and 1.50. This material is sufficient to reveal some of the general properties of the ripple. These are:

1. The ripple is not confined to any particular value of m. It appears with approximately equal strength in all five curves.

[8] Thanks are due Dr. R. Penndorf for kindly making this material available to the author before publication.

2. The ripple has a definite periodicity: peaks or valleys recur at intervals about 0.80 in x for all values of m concerned.

3. The ripple is not a pure sine wave. It resembles somewhat the beat pattern of two sine waves of slightly different periods, but this is too coarse a description. In some regions it shows double peaks in the interval 0.8, and in some it shows single ones, as indicated below.

	Double Peaks of Equal Amplitude near	Single Peaks near
$m = 1.33$	$x = 20\frac{1}{2}$	$x = 30$
$m = 1.40$	$x = 17\frac{1}{2}$	$x = 27\frac{1}{2}$
$m = 1.44$	$x = 14$	$x = 23\frac{1}{2}$
$m = 1.486$	$x = 12$	$x = 20$
$m = 1.50$	$x = 12$	$x = 19\frac{1}{2}$

4. Since these values of x are nearly proportional to $1/(m-1)$, the character of the ripple (double hump or single hump) has a tendency to recur at the same value of $\rho = 2x(m-1)$, i.e., at a fixed position in the major fluctuations. However, this is not precise, e.g., the double peaks of equal magnitude occur for $m = 1.33$ in the second minimum of the extinction curve, for $m = 1.44$ on the slope from second maximum to second minimum, and for $m = 1.50$ on the second maximum.

5. The ripple also has an imaginary component, as shown by the graph for $m = 1.33$ in the complex domain (Fig. 53, sec. 13.42).

17.52. Tentative Explanation by Means of Surface Waves

The ripple has to be explained as an interference effect between the forward diffracted wave (chaps. 12 and 13) with a wave not hitherto noticed. There are no reflected or refracted rays which could have escaped our attention. The central ray with two internal reflections, which covers the diameter three times, has a phase lag of $2x(3m-1)$ with respect to a ray that passes the sphere. Thus the period between peaks should be $\pi/(3m-1)$, which ranges from 1.05 for $m = 1.33$ to 0.90 for $m = 1.50$, so this is not the explanation. Moreover, its computed amplitude is too low.

Let us see what surface waves can do. Figure 87 shows a ray QA incident at the edge of a sphere or cylinder, e.g., a water drop or water jet. This ray will initially generate a wave AA' traveling along the surface. At any point along the surface the wave will lose energy by two effects: (1) the refraction into the droplet along AB or $A'B'$, as indicated in Ott's theory of the surface waves for dielectrics, and (2) the spray in the forward direction along AP or $A'P'$, etc. It is very unlikely that the original surface wave will be able to travel all the way around the sphere or cylinder and still produce an observable effect.

Suppose that the first effect (refraction into the drop) produces the the stronger attenuation. Then, for the mere geometry of the problem, this energy comes back to the surface. It comes back at B if it was refracted at A, and it comes back at B' if it was refracted at A'. This means that at B a new surface wave is started, which is reinforced by further contributions at any further point B' and travels in the direction $BB'B''$. All these contributions are in phase. Some distance beyond B (the exact amount cannot be estimated) it must reach its maximum amplitude.

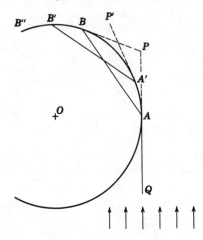

Fig. 87. Reinforcement of the surface wave on a dielectric body by waves traveling through the body.

The short cut AB, or $A'B'$, through the body means a longer optical path. If a is the radius, m the refractive index, and $\cos \alpha = 1/m$, we find from Snell's law that AB subtends an angle 2α. The optical length is $m \cdot AB = 2a \tan \alpha$ and equals the vacuum length of $AP + PB$, where P is the intersection of the tangents at A and B. The increase in optical length over a wave that travels the shortest way around the body thus is

$$2a \tan \alpha - 2a\alpha = ha.$$

We may now easily guess what happens. In traveling around the body the wave will take N short cuts and each time lose phase corresponding to an optical path ha. The full optical path around the body will then be

$$2\pi a + Nha.$$

Probably N will be close to the maximum number that is possible: e.g., for $m = \sqrt{2}$ we might have four short cuts, each subtending a right angle, but it is more probable that only three short cuts occur. Table 38 gives a survey of the numerical values of h and the probable values of N. The less probable values are given in parentheses.

Table 38. Expected Period and Beat Period of the Ripple

m	2α (radians)	$2 \tan \alpha$	h	N	$\left(1 + \dfrac{Nh}{2\pi}\right)^{-1}$	$\dfrac{2\pi}{h}$
$1 + \varepsilon$	$2\sqrt{2\varepsilon}$	$2\sqrt{2\varepsilon}$	$\frac{2}{3}(2\varepsilon)^{3/2}$			
1.1	0.860	0.917	0.057	7	0.94	110
				(6)	(0.95)	
1.2	1.171	1.328	0.157	5	0.89	40
				(4)	(0.91)	
1.333	1.446	1.764	0.318	4	0.83	19.8
				(3)	(0.87)	
1.414	1.571	2.000	0.429	(4)	(0.78)	14.6
				3	0.83	
1.5	1.682	2.236	0.554	3	0.79	11.3
				(2)	(0.85)	
2	2.094	3.464	1.370	3	0.60	4.6
				(2)	(0.70)	

Constructive or destructive interference with the main component (the diffracted light) will occur if this optical path is increased by a full wavelength. Therefore, the spacing of the peaks in the ripple should be given by

$$\Delta(2\pi a + Nha) = \lambda,$$

and so

$$\Delta x = \Delta \frac{2\pi a}{\lambda} = \frac{1}{1 + Nh/2\pi}$$

The table indicates that in the range $m = 1.33$ to 1.50, for which the CAFRC tables were computed, the probable values of Δx range from 0.83 to 0.79. This is in fair agreement with the empirical value 0.80 and lends support to the suggested explanation.

The very uncertainty of the problem, viz., the uncertainty in N, also explains the existence of a complicated beat pattern. For instance, it is possible that the surface wave travels along the surface from A to B without being too strongly attenuated. The character of the interference of the new surface wave generated at B and B' by the emergent radiation and the wave that has stayed at the surface is then determined by the question whether the difference of optical path ha is or is not an integer number of wavelengths. Similarities in the interference pattern will recur if

$$\Delta(ha) = \lambda,$$

so

$$\Delta x = \Delta \frac{2\pi a}{\lambda} = \frac{2\pi}{h}.$$

The values given in the last column of Table 38 agree quite well with the values of x at which a double peak occurs. Just why at these values the beat pattern should have *this* character cannot be judged from the crude explanation given here. The rough agreement of the periodicities suggests that an exact theory of the ripple should be based on the type of surface waves discussed in this section.

Also this section is merely an exploration. By the new investigations of Beckmann and Franz (compare sec. 12.35), it has become possible to treat the combined effect of the surface waves and the short cuts through the sphere rigorously by means of a residue series based on Watson's transformation. No numerical results are yet available.

References

The problem of the optical edge has been presented most clearly by

K. Artmann, *Z. Physik*, **127,** 468 (1950),

and a summary of his results and an experimental application are found in

K. Artmann, *Ann. Physik*, **7,** 209 (1950).

The formulation of the light scattered by a cylinder with one or two edges (17.11 and 17.12) in terms of an edge function is also given by Artmann. The diffraction by one edge has been discussed as well by

V. A. Fock, *Abhandl. Sowj. Physik*, **2,** 7 (1951),

where references to his earlier work are found.

The formulation for a sphere (17.13) is new. The heuristic argument (sec. 17.21) differs from Artmann's but leads to the same result.

Attempts to derive from the complete solution asymptotic formulae that would include other terms than the geometrical-optics and diffraction terms have been made many times since Debye first presented his asymptotic series for the Bessel functions; such efforts are by

G. Jöbst, *Ann. Physik*, **76,** 863 (1925),

for metallic and totally reflecting spheres, and for cylinders by

H. Spohn, *Physik. Z.*, **21,** 444, 469, 501, 518 (1920).
H. Pfenninger, *Ann. Physik*, **83,** 753 (1927).

Further attempts were made in Artmann's first paper for cylinders, $m = \infty$ (sec. 17.22), and by

T. Ljunggrén, *Arkiv Fysik*, **1,** 1 (1949).

for cylinders, arbitrary m (sec. 12.35).

The integral equations for the currents on the surface of a perfectly conducting body in diffraction theory have been formulated and dis cussed by

A. W. Maue, *Z. Physik*, **126**, 601 (1949).

F. E. Borgnis and C. H. Papas, *Randwertprobleme der Mikrowellenphysik*, Berlin, Springer Verlag, 1955.

They have been solved for the fields near the edge of a cylinder (sec. 17.23) by

V. A. Fock, *J. Phys. U.S.S.R.*, **10**, 130 (1946),

a summary of whose results is contained in

V. A. Fock, *Phil. Mag.*, **39**, 149 (1948).

A later paper,

V. A. Fock, *J. Phys. U.S.S.R.*, **10**, 399 (1946),

which contains results for absorbing cylinders has not been available to the author.   The numerical results of Fock's solution have been applied to the computation of the edge term in sec. 17.23.   This result was first presented by

H. C. van de Hulst, *Trans. I.R.E.*, **AP-4**, 195 (1956).

Closely related studies were made by

R. D. Kodis, *Trans. I.R.E.*, **AP-4**, 580 (1956),

and by several other speakers at the Ann Arbor Symposium on Electromagnetic Wave Theory, reported in these transactions.

Some references on the radiation from antennas on or slots in the surface of a circular cylinder are:

P. S. Carter, *Proc. I.R.E.*, **31**, 671 (1943).

S. Silver and W. K. Saunders, *J. Appl. Phys.*, **21**, 153 and 745 (1950).

S. Sensiper, "Cylindrical Radio Waves," Hughes Aircraft Co., *Technical Memorandum* 310, June 15, 1953.

J. R. Wait and S. Kahana, *Can. J. Technol.*, **33**, 77 (1955).

The discussion of numerical results for perfect conductors, metallic bodies, and dielectric bodies (secs. 17.24–17.26) is entirely new.   Results for $m < 1$ are quoted from

R. H. Boll, R. O. Gumprecht, and C. M. Sliepcevich, *J. Opt. Soc. Amer.*, **44**, 18 (1954).

A discussion on surface waves (sec. 17.31) is found, e.g., in

P. Frank and R. van Mises, *Die Differential- und Integralgleichungen der Mechanik und Physik* (2nd ed.), p. 918, Braunschweig, Vieweg, 1935.

J. A. Stratton, *Electromagnetic Theory*, p. 584, New York, McGraw-Hill Book Co. 1941, where also the classical papers by Zenneck, Sommerfeld, and Weyl are discussed. Work on waves at a surface between two dielectrics is contained in

O. von Schmidt, *Physik, Z.*, **39**, 869 (1938).
H. Ott, *Ann. Physik*, **41**, 443 (1942).
H. Ott, *Ann. Physik*, **4**, 432 (1948).
H. Maecker, *Ann. Physik*, **4**, 409 (1948).
E. Gerjuoy, *Commun. Pure and Applied Math.*, **6**, 73 (1952).
K. O. Friedrichs and J. B. Keller, *J. Appl. Phys.*, **26**, 961 (1955).

The related problem of the displacement of the reflected wave was worked out in experiment and theory by

F. Goos and H. Lindberg-Hänchen, *Ann. Physik*, **1**, 333 (1941); **5**, 251 (1949).
K. Artmann, *Ann. Physik*, **8**, 270 and 285 (1951).

The surface wave ("creeping wave") on perfectly conducting cylinders and spheres (sec. 17.32) were recognized and worked out by

W. Franz and K. Deppermann, *Ann. Physik*, **10**, 361 (1952),
K. Deppermann and W. Franz, *Ann. Physik*, **14**, 253 (1954),
W. Franz and R. Galle, *Z. Naturforsch.*, **10a**, 374 (1955),

from the integral equations by Maue (loc. cit). This work is partly parallel to Fock's earlier work. Also given is the correct explanation for the backscatter by totally reflecting cylinders and spheres (sec. 17.41).

The discussion of the glory (sec. 17.42) and the tentative explanation of the ripple (sec. 17.52) are new. The numerical data from which we have quoted in sec. 17.51 have only partly been published:

B. Goldberg, *J. Opt. Soc. Amer.*, **43**, 1221 (1953).

PART III

Applications

18. SCATTERING AND EXTINCTION EXPERIMENTS AS A TOOL

As the foregoing chapters contain more complicated formulae than an average experimenter or research worker likes to handle, the present chapter starts with a fresh introduction of the scattering problems to those scientists whose interests are primarily in the applications. The following chapters (19 to 21) form separate guides for various fields so that, e.g., a chemist or astronomer may find which parts of the preceding theory are of interest to him.

18.1. Introductory Remarks

It may first be repeated from chap. 1 that the scattering problems treated in this book are limited as follows.

1. The book does not consider (as a rule) the media in which the scatterers are so densely packed that their mutual interactions complicate the problem.

2. The book does not treat the scattering of light by media or clouds in which the light is scattered several times in succession before it emerges from the medium (multiple scattering). Even the very simplest law of scattering for the individual particles (isotropic scattering) leads to complex mathematics in the multiple-scattering problem.

3. The book *does* give the dispersion and attenuation of a beam of light traversing a medium, if the individual particles are far apart and if their scattering properties are known (chap. 4).

Composite media consisting of small scattering particles embedded in a homogeneous material are present in nature in great variety (e.g., fog, interstellar dust) and are the desired or undesired product of man's technique in an even greater variety (industrial smoke, colloidal solutions). In many of these examples the scattering of light is the easiest means of detection and an obvious means for further investigation of the particles. The observations or experiments can be interpreted most directly if the particles are not too densely packed and the investigated sample is small enough to avoid multiple scattering.

Most of the computations reported in the preceding chapters were initiated with a view to a specific application. Sometimes the same mathematical problem was posed by research workers in two or three different fields. For example, the formulae in sec. 7.2 were independently developed for X-ray analysis of materials and for the diffusing properties of opal glass. Other problems, which did not yet have a direct application, were discussed for the sake of completeness.

In the following chapters the attempt has been made to restore the original order of research in these fields:

a. A problem belonging to a specialized domain of science is posed.

b. The phenomena of scattering and/or extinction involved in this problem are described.

c. This leads to an estimate of the orders of magnitude of size, wavelength, refractive index, etc.

d. Reference is then made to one of the earlier chapters, in which a detailed discussion of this particular type of scattering is found and a few examples are given.

This completes the task of this book for each of the applications that will be mentioned. It does not, however, complete the entire research in each field. Certainly the following steps will ordinarily follow:

e. Comparison of the detailed scattering computations with the detailed experiments or observations.

f. Discussion of deviations and changes to be made in the working hypothesis, both on the basis of this comparison and on the basis of the physics and chemistry of the scattering particles.

The study of the scattering phenomena thus has the character of a tool rather than the aim of research. It can be a good aid, but it cannot replace a thorough knowledge of the field involved. For that reason the applications in the following chapters necessarily serve as *illustrations* of a method, not as full *discussions* of a subject. For the same reason the author has avoided any attempt at a complete review of all applications or a tabulation of empirical results. This must be left to the experts in the particular field.

There is one type of application, which is rapidly growing in importance, in which the order is reversed. In the theory of the *artificial (composite) media* the problem is not to determine the properties of the scatterers by experiment but to manufacture scatterers with such properties that the medium has prescribed constants. An example is found in sec. 19.42.

The most common situation in which a research worker takes recourse to the literature on scattering is when he contemplates the use of scattering experiments in order to further his knowledge on certain types of small particles. He then will put these three questions:

When are scattering measurements most useful?

What kind of measurements should be made?

What properties of the scatterers can be inferred from these measurements? The answers to these questions are reviewed in the following sections.

18.2. When Are Scattering Measurements Most Useful?

The following circumstances, separately or combined, may make an investigator decide on a scattering method instead of other means of investigation.

A. *The particle size is of the order of the wavelength of (visible) light.* For particles of a much smaller or a much larger size scattering measurements are less effective. Very small particles exhibit Rayleigh scattering, and none of the easily measured characteristics depends on their size. Only the comparison of a quantity that is a linear function of the particle volume, e.g., the refractive index of the composite medium, with a quantity that is a quadratic function, e.g., the intensity of scattered light, makes a determination of the size possible (secs. 6.53 and 19.12).

Very large particles again show small differences in their scattering characteristics. Moreover, they can be examined in a microscope. This much more powerful and direct method will be preferred to scattering studies, whenever possible. Here other circumstances (*B* and *C*) may lead the research worker to consider scattering.

B. *The particles are not readily accessible.* This condition holds for all of astronomy and most of meteorology. All observational data about the dust in interstellar space or in a planetary atmosphere are data on the scattering and extinction by those dust clouds. This is the reason why the need for accurate computations of the scattering properties has been most acutely felt in astronomy. The majority of the computations reported earlier in this book has been made for astronomy. Even before 1900 such computations had been made with a view to the light pressure exerted on comet tails. Later the interest shifted to interstellar matter, where first metallic spheres, later dielectric spheres, and still later ellipsoids and cylinders were in fashion (sec. 21.4).

The haze and clouds in the sky are at any time accessible for balloon and airplane measurements with laboratory instrumentation. It is much easier, however, to make optical measurements or radar observations from a ground station. A different situation, where similar problems arise, occurred with a research group studying the processes in a blast furnace. It was necessary to determine the size and number of the soot particles escaping into the chimney. When collected on a metal probe they coagulate and become indistinguishable. A microscope cannot be brought into the hot gas stream, but it is easy to send in light or ultraviolet radiation and to measure the intensity and directional pattern of the scattered light.

Another typical example of inaccessible particles occurred in the Langley 11-inch hypersonic windtunnel. Because of the rapid expansion of the air in the nozzle of this tunnel, the temperature of the air may drop below that at which it liquefies. A fog of liquid air droplets is then observed in the test section. The drops have radii of the order of 5×10^{-6} cm, i.e., $x = 0.6$ with green mercury light. This has been determined from the transmission curve and by means of dissymmetry and polarization measurements (Durbin, 1951).

C. *A rapid method for routine measures is wanted.* If this consideration enters, the choice is often the scattering method. For instance, the color of a colloidal suspension in the chemical industry may be a sensitive check on the correct size. Other checks, involving polarized light, or the intensity ratio of light scattered at two different angles, may be even more sensitive and are just as easy with modern techniques. Computations as described in this book may help to decide the best method.

18.3. What Kind of Measurements Should be Made?

Whereas astronomers have to collect and interpret all possible data, a laboratory worker is in a distinctly different position. If the object, say a colloidal solution, is given, he has a choice among a variety of measurements. Included within the topic of this book are scattering, extinction, and refraction measurements. In all these measurements the wavelength distribution and the state of polarization of the incident light may be chosen at will. The theoretical interpretation is simplest if a definite wavelength and a definite state of polarization are used. The laziest way to set up an experiment is with white, natural light (white = distributed over a wide range of wavelengths in the visible spectrum; natural = unpolarized; see sec. 5.13). There may be practical reasons for preferring white or natural light, but it usually pays to insert a color filter, or Polaroid filter, or both in the beam. They may be placed in the illuminating, or in the receiving system, or in both. Measuring relative intensities with an accuracy of 1 per cent should be no problem with modern photoelectric cells.

18.31. Scattering

In the following we shall assume that the scattered light of a great many (incoherent) particles is measured simultaneously. Fig. 88 shows the basic parts of the instrument. Practical apparatus for light scattering has been described by Stein and Doty (1946), Debye (1946). Speiser and Brice (1946), Zimm (1948). Most of the remarks made below hold also if the scattered light of one particle (Tyndall light) is viewed in an ultramicroscope or in a photoelectric particle counter (sec. 19.3). It is conventional in a good part of the literature to call the plane of Fig. 88, which contains the directions of propagation of the incident and scattered light, the horizontal plane, and to describe linear polarization by the position, vertical or horizontal, of the electric vector. The vertical position of the electric vector corresponds to what we have called the perpendicular position (index r) in sec. 4.4 and subsequently.

If the particles are spherical and if the incident beam is made vertically polarized by a polarizing filter, then the scattered light is also vertically polarized (unless multiple scattering occurs) and is proportional to i_1. So the addition of a vertically polarizing filter in the receiving system should make no difference (except for absorption in the filter), and the addition of a horizontally polarized filter should reduce the intensity to zero.

The symmetry properties for non-spherical particles are discussed in chap. 5. Simple practical consequences of these symmetry properties have been indicated by Krishnan (1934–1936) and have been checked by many later authors. In the case of spherical particles, where we have separate scattering diagrams

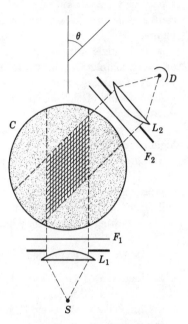

Fig. 88. Schematic arrangement for measuring scattered light. S = light source, L_1 and L_2 lenses with diaphragms, F_1 and F_2 color and polarization filters, D detector, e.g., photocell, C cell for solution with cross-hatched area = scattering volume.

for vertical electric vector and for horizontal electric vector, the first diagram, of $i_1(\theta)$, usually shows sharper details than the second, of $i_2(\theta)$, and both show more details than the diagram for unpolarized light.

The simplest measurement from the theoretical point of view is the measurement of the intensity scattered at a certain angle (per unit solid angle) at a certain wavelength and in a certain polarization. The scattered intensity is given for a single particle in secs. 2.1 and 4.1, for an assembly of particles, e.g., the cross-hatched area in Fig. 88, in secs. 2.6 and 4.22. Such a "simple measurement" is hard to attain, for besides a range of wavelengths there may be a range of angles involved, owing to the apertures or slit widths in the illuminating and receiving device. In certain experiments convergent light was used, which by itself gives a range of angles.

Since an absolute measurement of the scattering cross section of a particle for a certain angle θ involves a difficult calibration of the light source, reflection losses, etc., it is more usual to make relative measurements:

a. One angle with respect to another: dissymmetry measurements or, more completely, the scattering diagram.

b. Horizontal versus vertical polarization at a given angle, i.e., measurements of the degree of polarization of the scattered light if the incident light is natural.

c. One wavelength against another: color measurements, or also higher-order Tyndall spectra (sec. 19.22).

The great variety of particles necessitates a great variety in the actual methods and instruments. The most convenient choice may be left to the ingenuity and taste of the reader. Table 39 (sec. 18.4), the collection of scattering diagrams in Fig. 25 (sec. 10.4), and the examples in the following chapters may aid in making the choice.

18.32. The Extinction

Extinction can be measured by observing the intensity J of a light source as seen through a container with scattering particles. If J_0 is the intensity of the same light source seen through the same container without scattering particles, the ratio is

$$J/J_0 = e^{-\gamma l}$$

where l is the path length through the particulate medium and γ is variously called the extinction coefficient (secs. 4.22, 4.3, etc.), the turbidity, or the attenuation coefficient; its dimension is length^{-1}.

Figure 89 shows a diagram of an instrument for measuring the extinction. A lens is needed at least at one end of the path in order to make sure that only the light from the source is measured and not part of the scattered light. This ideal cannot be fully realized: although it is mathematically possible to define the extinction of light from a point source (sec. 4.2), any practical instrument measures the dimmed light of the source *plus* a certain amount of light scattered at small angles. In the instrument in Fig. 89 these angles are defined by the apertures in front of the lamp and of the photocell.

Some authors have found it convenient to introduce an effective, or apparent, absorption coefficient, which is found by putting for J, in the formula above, the measured energy of source *and* scattered light in the small reception cone. This apparent absorption coefficient depends on the reception cone of the instrument. For small particles and particles comparable to the wavelength the difference is minute. For large particles, the diffracted light is so strongly concentrated near the forward direction that a good part, or even all of it, falls in the reception cone of the instrument. The apparent extinction cross section thus falls from twice the geometrical cross section to this cross section itself, if the instrument is fixed and the particles grow in size, while the factor 2

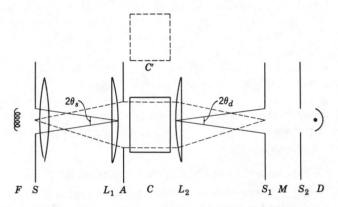

Fig. 89. Schematic arrangement for measuring attenuation (after Lewis and Lothian). F = filament of lamp, S and A diaphragms, L_1 and L_2 lenses, C and C' cells for specimen and blank solvent, S_1 and S_2 slits of monochromator M, D detector, e.g., photocell.

remains true for the total extinction (sec. 8.22). A detailed computation of the apparent extinction for reception cone half-angles from 0° to 1.4° has been made by Gumprecht and Sliepcevich (1953). A simple rule may be stated if the reception cone of the instrument is itself due to diffraction, as is nearly realized in many astronomical telescopes. The ratio of apparent cross section to geometrical cross section then is 2 if the objective of the telescope is much larger than the (interstellar) particles; it would approach 1 only if the particles become larger than the objective.

The extinction measurement may be made into a relative one by comparing the extinction of two wavelengths (color of the setting sun!) or in two directions of polarization. If in the latter case a difference is found the particulate medium is dichroic (secs. 19.4 and 21.4).

It is much more of a problem to compare the extinction with the scattered intensity (at one angle or integrated over all angles). Yet this is the only direct method to determine the absorption inside the particles, or their albedo (sec. 2.2). As quite different quantities have to be compared, the method requires great caution.

18.33. Refractive Index of Composite Medium

One further quantity not usually mentioned in connection with scattering problems but sometimes susceptible of accurate measurement is the refractive index of the material made up by the medium and the embedded particles. Theoretically, this refractive index is intimately related to the extinction: both can be computed with equal ease and form together the complex refractive index of the composite medium (secs. 4.3 and 4.4). The refractive index may be measured with a refractometer or interferometer in the optical domain, or in a wave guide in the microwave domain.

If the refractive index is different for different states of polarization the composite medium is birefringent and/or rotates the plane of polarization. Applications in the domain of molecular scatterers are plentiful. In microwave technique, artificial dielectrics are becoming more and more important (Süsskind, 1952). The full formulae for particles of arbitrary size are in sec. 4.41.

18.4. What Can Be Inferred from These Measurements?

Scattering or extinction measurements allow us to determine (at least in principle) size and distribution of sizes, shape and orientation, and chemical composition of the particles. However, not every property can be inferred from the measurements with similar accuracy. Some of them will be mentioned in order of practicability. Particles of molecular size have been discussed in sec. 6.21; see also sec. 19.12.

1. *The concentration*, or number of particles per unit volume. If the scattering or extinction of a single particle are known, one simple measurement suffices to give the concentration. The most common way is to employ the measured extinction coefficient, which is

$$\gamma = NC_{ext},$$

and to find N by means of an assumed value of C_{ext}. The scattered intensity may be used in a similar manner (sec. 4.22). Usually this problem is combined with the problem of determining size and/or composition, which are needed first in order to know the value of the cross section C_{ext}.

2. *The size* is the most important parameter and the one which is easiest to determine; for spherical particles with diameter $2a$, the parameter $x = 2\pi a/\lambda$ is the most important one in the theory. The most promising methods have been compiled in Table 39. The scattering

diagrams (Fig. 25) may serve for a closer illustration. Table 39 serves for dielectric spheres with a moderate refractive index (e.g., water drops in air). The limits of the ranges are somewhat arbitrary and depend on the precision of the instrument.

Table 39. Survey of Size Criteria for Water Drops

	Range of x	Range of Diameters for Red Light	Quantities Sensitive to Changes in Size
Small drops (Rayleigh scattering)	0–0.5	0–0.1 μ	The ratio of any quantity quadratic in V (scattered intensity, extinction) to any quantity linear in V (refractive index of composite medium, excess density of composite medium over solvent). See secs. 6.21 and 19.12 for details; V = volume of drop.
3-term region	0.5–1	0.1 μ–0.2μ	Angle of maximum polarization, or degree of polarization at 90°, or color of scattered light at any angle, or ratio of scattered intensities at 45° and 135° (dissymmetry). See secs. 10.3 and 19.12.
Region of complicated patterns	1–10	0.2 μ–2μ	Dependence of extinction on wavelength; width of main scattering lobe between half-intensity angles; angles of any distinguishable maxima or minima in the scattering diagram; higher-order Tyndall spectra. See secs. 19.22, 20.3, and 21.2 for examples.
Anomalous diffraction	10–50	2μ–10μ	Angular positions of minima and maxima in the diffraction pattern. See sec. 13.41.
Big drops (ordinary diffraction)	50–∞	10 μ–∞	Diameters of diffraction rings, or angles between them; position of rainbow maximum and angles between the adjoining maxima; diameters of glory rings; colors of all these phenomena if the light source is white. See secs. 8.31, 13.24, and 13.33.

The quantities given in Table 39 may be completed by many more; the ingenuity of the research worker and the exact aim he has in mind will be decisive for the exact method chosen. The methods may be standardized, if desired, and have to some extent been standardized in chemistry.

The method will be different for metallic particles, especially if they are small. In the Rayleigh region a sensitive criterion of size is the ratio of scattering to absorption or total extinction. The transition region is narrower than for dielectric spheres, and anomalous diffraction does not occur; again the width of the main lobe or the dependence of extinction on λ are the best criteria. For large particles the diffraction rings form the best size criterion.

If the particles are not in vacuum or air, the wavelength $\lambda = \lambda_0/m_0$ should be used, where λ_0 is the wavelength in air and m_0 is the refractive

Table 40. Values of x for Aerosols and Hydrosols*

	Aerosols			Hydrosols		
Radius a	$\lambda_0 = 0.40$ $\lambda_0^{-1} = 2.5$ $2\pi/\lambda = 15.7$	0.50 2.0 12.6	0.67 1.5 9.4	$\lambda_0 = 0.40$ $\lambda_0^{-1} = 2.5$ $2\pi/\lambda = 20.9$	0.50 2.0 16.8	0.67 1.5 12.6
0.01	$x = 0.16$	0.13	0.09	$x = 0.21$	0.17	0.13
0.02	$x = 0.31$	0.25	0.19	$x = 0.42$	0.34	0.25
0.04	$x = 0.63$	0.50	0.38	$x = 0.84$	0.67	0.50
0.05	$x = 0.79$	0.63	0.47	$x = 1.05$	0.84	0.63
0.1	$x = 1.57$	1.26	0.94	$x = 2.1$	1.7	1.3
0.2	$x = 3.1$	2.5	1.9	$x = 4.2$	3.4	2.5
0.4	$x = 6.3$	5.0	3.8	$x = 8.4$	6.7	5.0
0.5	$x = 7.9$	6.3	4.7	$x = 10.5$	8.4	6.3
1	$x = 15.7$	12.6	9.4	$x = 21$	17	13
2	$x = 31$	25	19	$x = 42$	34	25
4	$x = 63$	50	38	$x = 84$	67	50
5	$x = 79$	63	47	$x = 105$	84	63

* a, λ_0, and λ are expressed in microns.

index of the medium in which the particles are embedded. Table 40 permits a quick orientation about the values of x for aerosols and hydrosols.

3. *The distribution of sizes* usually is very hard to determine, especially when other unknown parameters, e.g., shape or composition of the particles, also enter. An attempt to determine the size distribution can be made by performing very accurate measurements of scattering and/or extinction and by comparing the results with computations made for various assumed distribution functions of a.

In practice it is found advisable to do at least some presampling of the sizes in order to remove the particles that are very much smaller or larger than average. Also, a narrow range of wavelengths is strongly recommended, for a range in λ has much the same effect as a range in a on the distribution function of the parameters $x = 2\pi a/\lambda$.

4. *The shape* of the scattering particles in a medium with random orientation of the scatterers usually is hard to determine. As an example we may mention the problem of the diffraction rings by water drops or randomly oriented ice needles. The difference in the intensity distribution and the positions of the maxima and minima for corresponding sizes is noticeable but small (Fig. 16, sec. 8.32).

Nevertheless, sensitive checks may be devised in a variety of ways, especially with the use of polarized light. A standard method for very small particles, or molecules, is the determination of the depolarization factors (sec. 6.52), from which at least one relation between the elements of the polarizability tensor may be obtained. Other methods, also based on checks for the presence of the elements S_3 and S_4 in the scattering matrix (sec. 4.41) have been proposed by Kastler. Similar checks are available in the microwave domain. For instance, a radar unit emitting and receiving circular polarization can discover a target only if it is non-spherical.

Problems of the surface condition of the scattering particle (rough or polished) may be classed also under "shape." This question arises only in connection with particles $\gg \lambda$. The distinction is fairly easy. The difference between a "white" and a "polished" totally reflecting particle was explained in sec. 8.42. Similarly, it will be clear that a smooth, solid glass sphere will give something like rainbow effects, whereas a roughened glass sphere (surface of ground glass) will not.

5. *The orientation* of particles which are elongated, or have an otherwise peculiar form, can be determined fairly easily both by scattering and by extinction methods. It is simply necessary to compute the scattering properties for various orientations and to apply the theory of chaps. 4 and 5 if the orientations are not at random. An example is the determination of the orientation of the interstellar particles from the observed effect of interstellar polarization (sec. 21.4).

References*

E. J. Durbin, *NACA Techn. Note* 2441, August, 1951.

R. S. Stein and P. M. Doty, *J. Am. Chem. Soc.*, **68**, 159 (1946).

P. P. Debye, *J. Appl. Phys.*, **17**, 392 (1946).

R. Speiser and B. A. Brice, *J. Opt. Soc. Amer.*, **36**, 364 (1946).

B. H. Zimm, *J. Chem. Phys.*, **16**, 1099 (1948).

R. S. Krishnan, *Proc. Indian Acad. Sci.*, **A1**, 211 (1934); **A2**, 221 (1935); **A3**, 126 (1936).

R. O. Gumprecht and C. M. Sliepcevich, *J. Phys. Chem.*, **57**, 90 (1953).

C. Süsskind, *J. Brit. Inst. Radio Engrs.*, **12**, 49 (1952).

* Given in the order in which they are cited in this chapter.

19. APPLICATIONS TO CHEMISTRY

19.1. Scattering by Small Particles and Macromolecules

19.11. Tyndall Light

Colloidal solutions may be distinguished from real solutions by a simple experiment based on their relatively strong scattering. If a strong ray of light is passed through a sol in the dark, the path of the light through the sol is clearly visible from all sides. The phenomenon is most clearly seen with very small, non-absorbing particles, like a mastic emulsion or a very fine smoke. The scattered light is blue, and when seen perpendicularly to the ray it is fully polarized with the electric vector perpendicular to the plane through the incident and scattered ray. When the incident light is plane-polarized, the path is clearly seen from aside by an observer located in the H-plane; it is not seen by an observer located in the E-plane.

These particulars correspond with the law of Rayleigh scattering, discussed in detail in chap. 6. The scattered light was first studied by Brücke (1852) and Faraday (1857) and in great detail by Tyndall (from 1869 on). Lord Rayleigh explained it as the radiation by the induced electric dipoles of the particles, in a series of papers in 1871. Most of this work was directed towards finding an explanation of the blue sky, and the experiments were made with smoke. It was not until 1899 that Rayleigh suggested that, instead of embedded droplets, the air molecules themselves might be the scattering particles. This interpretation was supported by a determination of the number of particles from the refractive index and the extinction coefficient (sec. 6.53). It is in agreement with the known number of Avogadro. Present determinations of the number of Avogadro from scattering, both in the laboratory and in the entire atmosphere, give the correct value within 1 per cent.

The first laboratory experiments on Rayleigh scattering by pure gases date from 1913, the refinement of anisotropic particles and depolarization (sec. 6.52) from 1918 (Rayleigh, 1918). An account of theory and experiments is given by Cabannes (1929). The extra factor f in the extinction formula is not of extreme importance but should be used in accurate work. The application of these formulae has been adequately reviewed by Oster (1948). They hold for small aspherical colloidal particles just as well as for anisotropic molecules.

The deep blue color of the Tyndall light is lost if the volume of the sol is so large that multiple scattering becomes important, or if the individual particles are too large. In the latter case it makes an important difference if the particle is a randomly coiled molecule packed in the form of a loose ball (sec. 19.12) or if it is a liquid or solid sphere with a refractive index different from the medium (sec. 19.2). Suppose that the relative refractive index is 1.25 or 1.5. The first symptoms of the size increasing beyond the Rayleigh domain then are

a. The ratio of the intensities in a forward angle ($\theta_1 < 90°$) and the symmetrical backward angle ($\theta_2 = 180° - \theta_1$) becomes > 1; there is more light in the forward direction.

b. The polarization maximum shifts to an angle $\theta > 90°$.

c. Consequently, the polarization at $\theta = 90°$ falls below 100 per cent.

They were described by Rayleigh in 1881. The mathematics behind these phenomena is given in the form of the three-term formulae (electric-dipole, magnetic-dipole, and electric-quadrapole radiation) in sec. 10.3. The second symptom is particularly suitable for size-estimates but becomes, like the other ones, useless when still higher terms enter (see Table 39, sec. 18.4).

The blue color is also modified if the refractive index of the particles changes selectively with λ in the visual domain. This is true with some metal sols or smokes (sec. 19.21). The full polarization at $\theta = 90°$ is then retained, provided that the particle size is sufficiently small.

19.12. Molecules and Macromolecules

There is an important body of theoretical and experimental research on light scattering by chain molecules and high polymers. This subject comprises a substantial part of the review paper by Oster (1948), and a separate monograph on it by Zimm and Doty is in preparation; therefore, we shall not discuss the entire subject. Nevertheless, a short section must be included to show that, despite strong differences in terminology and in notation, the special cases discussed in chaps. 6 and 7 of this book are also relevant to *dilute solutions of macromolecules.*

If the overall dimensions of the molecules are very much smaller than the wavelength, say $<1/20 \lambda$, the theory of chap. 6 is applicable. The principle of the method for determining molecular weight is explained in secs. 6.21 and 6.53, and only the notation has to be adapted. The turbidity τ measured in chemistry is identical to our extinction coefficient γ.

Let c be the concentration (gram/cm³), $N_A =$ Avogadro's number, and M the molecular weight; then the number of solved molecules per cm³ is $N = N_A c/M$. Further, let λ_0 be the wavelength in vacuum, n_0 be the refractive index of the solvent and n of the solution; then our $\tilde{n}$ is the relative refractive index n/n_0, and our λ is λ_0/n_0. The turbidity equation for very dilute solutions is then, according to sec. 6.53

$$\tau = HcM,$$

where

$$H = \frac{32\pi^3 n_0^2 (n - n_0)^2}{3 N_A \lambda_0^4 c^2} f.$$

H contains only known quantities (N_A) or measurable ones ($n - n_0$). Thus by measuring τ and c the molecular weight can be determined.

Instead of the turbidity the scattering at 90°, which is simply related to it, may be measured. Also white light may be used, if monochromatic light is not sufficiently intense (Kerker, 1952).

If the molecules are larger than $1/20\ \lambda$ and are of the type of straight rods (virus) or coiled chains (polystyrene) they still are exposed to essentially undisturbed radiation. Hence the scattering follows from Rayleigh scattering by the principle that in the chemical literature is called internal interference. It is identical to what we have called Rayleigh-Gans scattering (sec. 7.11). The factor to be added to the formula for the scattered intensity, i.e., the square of our $R(\theta,\varphi)$, might be called the Debye factor (compare sec. 7.5). The factor to be added to the small-particle law for the turbidity or extinction is sometimes called the dissipation factor; its computation for spheres is found in sec. 7.22.

Representative sizes for polystyrene particles, as mentioned by Zimm (1948), are: molecular weight of entire chain 1.000.000; length of one lirk of the order of 10 A; all links coiled into a ball with diameter 1000 A and higher.

Williams and Backus (1949) describe a sample of very uniform polystyrene particles with diameter $2a = 2590$ A $\pm$ 40 A. If $\lambda_0 = 5000$ A, i.e., in the solvent about 3500 A, we have $x = 2\pi a/\lambda = 2.3$. This is small enough to make the Rayleigh-Gans approximation permissible. It is just about the value for which the first zero of intensity appears in the backscattering (sec. 7.21). Hence most experiments in this field deal with a relatively simple scattering diagram in which the dissymmetry factor $I(45°)/I(135°)$ characterizes the size sufficiently. Many authors have found it useful to extrapolate the measured intensities to $\theta = 0$ in order to apply the unmodified Rayleigh formula. For details of technique we refer to the papers quoted. The much more difficult subject of solutions that are not very dilute is beyond the scope of this book.

Another illustrative example is the scattering by tobacco mosaic virus. The particles are thin rigid rods having a length comparable with the wavelength of visible light. In the study by Oster, Doty, and Zimm (1947) the sizes of the rods as determined by the electron microscope were: average length $= 0.270\ \mu$, diameter $= 0.015\ \mu$. The diameters are sufficiently thin to make the Rayleigh-Gans theory applicable. The relevant formula for the intensity scattered by randomly oriented rods is in sec. 7.34. The length may be determined from optical dissymmetry measurements. At $\lambda_0 = 0.546\ \mu$ in air, i.e., $\lambda = 0.409\ \mu$ in the solvent water, the intensity ratio at $\theta = 42.5°$ and $137.5°$ was found to be 1.94. From a graph based on the formula of sec. 7.34 it then follows that $l/\lambda = 0.66$, so that the length is $0.270\ \mu$ in excellent agreement with the electron-microscope studies. The diameter cannot be determined directly, but the estimate made from the molecular weight $M = 40 \times 10^6$, which follows from the measured turbidity, agrees again with the electron-microscope data.

19.2. Hydrophobic Solutions

19.21. Colors of Gold Solutions

A useful distinction in the types of colloidal solutions is the distinction between lyophobic solutions and lyophilic solutions (hydrophobic and hydrophilic, if the solvent is water). In hydrophobic solutions the colloidal particles do not join with the water molecules; in hydrophilic solutions they are hydrolized as, for example, $Fe(OH)_3$ particles, or swell and form a hydrogel. All hydrophobic solutions and some hydrophilic solutions are good examples of the idealized problem of separate particles in a homogeneous medium, a subject dealt with throughout this book.

Among the most strictly lyophobic solutions are the *colloidal gold solutions in water*. Their colors have attracted the attention of the experimenters from the time they were first made. Also, the very first time that a full solution of the scattering of light by a homogeneous sphere was made (Mie, 1908), the author's aim was to explain the colors of gold solutions. The general solution is now often called the Mie solution, and it is historically correct to mention gold solutions as its first application.

Gold solutions can be made in a variety of ways. The reader may refer to Freundlich (1932) for the early literature. Gustav Mie in his long paper in 1908 derives the rigorous scattering formulae for homogeneous spheres including only the first three terms, the coefficients of which we have denoted by a_1, b_1, and a_2. He computes for a number of wavelengths and sizes the cross sections for extinction, scattering, and absorption.   He observes also that the scattering (but not the extinction) is determined almost completely by the term with a_1, which represents dipole scattering; this means, incidentally, that the wavelength dependence of total scattering is the same as that of scattering under 90°.   Some polarization values and scattering diagrams are also computed in Mie's paper.   The numerical values are based on the refractive indices determined by Hagens and Rubens, multiplied by 3/4 in order to obtain the refractive index relative to water (see Table 26, sec. 14.22).   His computations are compared with the qualitative experiments by Steubing, and the results are so successful that they are still worth being reproduced

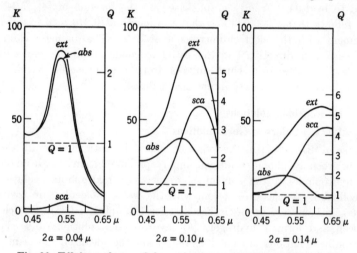

Fig. 90. Efficiency factors Q for extinction, scattering, and absorption by spherical gold particles in water, according to Mie's computations. Abscissa = wavelength in air; $2a$ = diameter.

in some detail.   The terms and notations of Mie are translated below into our terms and notations.   The coefficients tabulated by Mie have the same definition as those adopted later by Schalén (sec. 14.21).

Mie's terms and notations	Notations and terms in this book
K' (total scattering)	Q_{sca} $(3/4a)$ (scattering)
K'' (pure absorption)	Q_{abs} $(3/4a)$ (absorption)
K (absorption)	Q_{ext} $(3/4a)$ (extinction)
$K = K' + K''$	$Q_{ext} = Q_{sca} + Q_{abs}$

Numerical results for three sizes of spheres are shown in Fig. 90. Both K and Q are given as the ordinates.   Particles with $2a = 0.04$ μ form a *red gold sol*, because the extinction has a maximum at 0.53 μ.   The scattered light has a yellow-green color but is very weak. These are also the properties of very small particles, which reflect the peculiarities of the refractive index, not of a size effect. The second figure shows that a white source shining through a solution with $2a = 0.10 \; \mu$ is seen with a purple color, because the extinction maximum has shifted to $0.58 \; \mu$; also the color of the scattered light has changed to orange-brown, and its intensity has strongly increased. The third figure shows the curves for $2a = 0.14 \; \mu$, i.e., a typical *blue gold sol:* the extinction maximum has shifted to the red, at $0.61 \; \mu$. The scattering has become stronger than the absorption, and the scattering diagrams already have a strong forward peak (Mie effect).

Seraphin (1952) quotes more recently determined values for the refractive index of gold. His photoelectric measurements, made from 2900 A to 5000 A, show good agreement with the computations, except for very small particles, diameter 30 A, which show an additional absorption in the ultraviolet. This effect almost vanishes for diameters about 100 A and must be due to a real modification of the optical properties of gold, as also found in Wolter's experiments (1939, 1940) with thin layers. Further data on the refractive indices of the metals may be found, e.g., in Mott-Jones (1936) and Weisz (1948).

The ruby red color of small gold particles (when viewed in transmission) has of old been employed in the manufacture of various types of ruby glass.

Other metals which form lyophobic solutions are: mercury, silver, platinum. The refractive index of these metals has no peculiar variations in the visible region, so that the scattered light is blue and the transmitted light yellow or red when the particles are small. Extensive computations for silver and mercury by means of the Mie formulae have been made by Feick (1925). We refer again to Table 26, sec. 14.22, for the refractive indices and the range of sizes for which computations were made. A range of varying colors is observed when the particles grow larger, but the correspondence with the Mie theory is not too good. It seems probable that the discrepancy is to some extent caused by a non-spherical shape of the particles. Gans worked out the theory for ellipsoids that are small compared to the wavelength (sec. 6.32), and he and others have interpreted the measurements on metal sols in terms of this theory (see Freundlich loc. cit.). However, the extension of the Mie theory (including higher terms than dipole scattering) for ellipsoidal particles has not yet been brought to numerically useful results (sec. 16.11). In a set of papers Wiegel (1929, 1930a,b) has studied the size distribution in silver sols by various methods, including microphotographs and Debye-Scherrer exposures. Another study by the same author (1953) confirms the deviations from the Mie theory for silver sols made by the peroxide method and shows by means of electron-microscope photographs that the particles are disk-shaped.

We may conclude that the application of the Mie theory to metal sols meets two serious complications:

A. The form of the particles may not be spherical.

B. The refractive index of the particles may not be the same as it is for metal in bulk.

Perhaps too much attention has been paid to the first complication and too little to the second one. In modern research no doubt should be left about the deviating *forms*, for it is always possible to observe representative samples of a colloidal solution in an electron microscope. Such sample measurements, together with a good scattering theory, may provide the basis of routine methods for optical checks on the sizes and form of the particles.

19.22. Sulfur Hydrosols

Sulfur hydrosols are easy to make, for example, by mixing dilute solutions of hydrochloric acid and sodium thiosulfate. The molecularly dispersed sulfur thus formed condenses into droplets of super cooled sulfur, and these droplets grow as the sol ages. The refractive index relative to water is 1.44. If no foreign nuclei are present, the droplets are formed only after a certain supersaturation is reached. Their initial radii then are of the order of 0.01μ. For visual light (e.g., $\lambda_0/n_0 = 0.40$ A) this means $x = 2\pi a/\lambda$ of the order of 0.17, so that these droplets are in the domain of Rayleigh scattering.

Keen and Porter (1913) noted the changes of color of the transmitted light as the sols age. These differences correspond to the different slopes of the extinction curves (Fig. 32, sec. 11.22) for different values of *x*. At the same time the scattered light begins to show more and more maxima in its angular pattern, i.e., more and more colored bands if white light is used for irradiation. These additional bands are termed higher-order Tyndall spectra. The first traces of their appearance were found by Ray (1921, 1923), and by Raman and Ray (1922). Theoretical scattering diagrams have been computed by Lowan (see references, chap. 10).

La Mer and his associates have made an extensive study of this field and also of sulfur aerosols (sec. 19.3). For a review, see La Mer (1948) and Sinclair and La Mer (1949). In carefully controlled conditions monodisperse sulfur sols are prepared, which allow many chemical problems to be studied, in connection with the growth, the vapor pressure, etc. We shall review only the optical phenomena. In accordance with secs. 18.31 and 18.32 these may be separated into two classes.

Transmission. The correctness of Beer's law for these sols, i.e., the strict proportionality between $\log (I/I_0)$ and the concentration, has been tested by Dinegar and Smellie (1952). The size of the particles may be determined by comparing two plots: a theoretical one of Q_{ext} against $2\pi a/\lambda$ and an experimental one of $\log(I/I_0)$ against $1/\lambda$. If both are plotted on logarithmic scales for abscissae and ordinates the plots may be shifted to make them coincide. The shift of abscissae gives the radius a, the shift of ordinates gives, after some reduction, the concentration. The method cannot be applied for the very smallest particles (sec. 18.4) but is useful in the range $a = 0.3\ \mu$ to $0.8\ \mu$ (Barnes and La Mer, 1946). A variant, which may be useful, is to multiply the ordinates by x^{-p}, where p is suitably chosen so as to bring a maximum at a convenient place (La Mer, Inn, Wilson, 1950).

If data for only a small range of wavelengths are available it is also possible to use the local slope of the extinction curve as expressed by the exponent α (compare sec. 20.21).

A study of the scattering and transmission by sulfur hydrosols in the ultraviolet down to 2050 A, where the particles are not purely dielectric, has been made by Kenyon and La Mer (1949). The relevant theory is discussed in chap. 14 of the present book.

We may illustrate the method of fitting the observed transmission to the theoretical extinction curve by an investigation in which an even wider range of wavelengths was used. Lewis and Lothian (1954) have measured the transmission by aerosols and hydrosols of barium sulphate from $0.4\ \mu$ to $2\ \mu$. Figure 91 shows their results on $BaSO_4$ particles with radius $a = 4.6\ \mu$ in water. The observed and theoretical curves are drawn in the same figure and not in a logarithmic scale. In this investigation the size of the particles was measured by microscope and their number counted. Thus the efficiency factor Q was measured directly, and no arbitrary scale factor in the vertical direction was applied. The scale factor in the horizontal direction was chosen so as to give the best correspondence. The empirical relation was

$$\rho = 11\lambda_0^{-1}.$$

As the theoretical expression for ρ is

$$\rho = \frac{4\pi a(m-1)}{\lambda} = \frac{4\pi a(n/n_0-1)}{\lambda_0/n_0} = \frac{4\pi a}{\lambda_0}(n - n_0),$$

we obtain

$$4\pi a(n - n_0) = 11\ \mu\ .$$

With $a = 4.6\ \mu$ this gives $n - n_0 = 0.19$, so that the refractive index of the particles (relative to air) is

$$n = 1.33 + 0.19 = 1.52\ .$$

The accuracy of this method might even warrant the use of a further decimal. The small relative refractive index $(m = 1.52/1.33 = 1.14)$ in this example makes the transmission method useful for particles that are about four times larger than the limit for which it would be useful in sulfur hydrosols.

Absolute turbidity measurements on monodisperse polystyrene and polyvinyl toluene lattices with diameters from 0.1 μ to 7 μ were made by Heller, Epel, and Tabibian (1954). The resulting diameter matches the

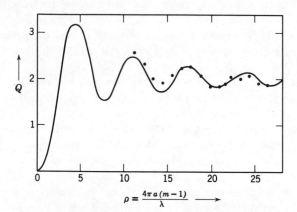

$$\rho = \frac{4\pi a (m-1)}{\lambda} \longrightarrow$$

Fig. 91. Extinction by a hydrosol of barium sulfate particles with radii 4.6 μ, as measured by Lewis and Lothian for a wide range of wavelengths (dots). The curve represents the theory for m close to 1.

correct one, which was determined by electron microscopy, with errors of the order of 10 per cent.

Scattering. The experiments based on the properties of the scattered light may be put in two classes.

A. Small particles. The patterns for sizes $0.5 < x < 2$ differ sufficiently from Rayleigh scattering, without showing the complicated maxima and minima that appear for larger x. Table 39 (sec. 18.4) shows that we might use the dissymmetry or the polarization as a size criterion. The method of the polarization ratio has been worked out for practical application and proves effective for radii $0.05 \mu < a < 0.2 \mu$ (Kerker and La Mer, 1950). It may also be used with white light, as shown by the computations of Kerker (1950) and by later experiments on aerosols (Kerker and Hampton, 1953). The intensity ratio between the horizontal and vertical component of the scattered light is measured at one angle. The graph of this ratio as a function of size is shown in Fig. 92 for a relative refractive index $m = 1.44$ and the angle $\theta = 110°$. This angle proved somewhat more practical than $\theta = 90°$. The ratios for white light agree approximately with those for monochromatic light with $\lambda = 0.54\ \mu$.

B. *Larger particles.* As soon as several maxima turn up in the angular distribution of the scattered light, any one parameter does not give a unique dependence on size. La Mer and his coworkers have shown that monodisperse sulfur hydrosols may show up to nine maxima. Only

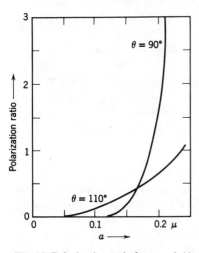

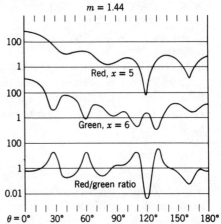

Fig. 92. Polarization ratio for $m = 1.44$ versus radius at two scattering angles. The curves have been computed for white light (after Kerker).

Fig. 93. Logarithmic plots of the scattered intensity at two different wavelengths and of their intensity ratio, all as a function of the scattering angle. The particle radius is $0.38\ \mu$.

vertical polarization (polarization 1) is used. As the human eye is more sensitive to color differences than to intensity contrasts, the bands show up even better with white light, which gives colored bands in the angular spectrum. These are called higher-order Tyndall spectra, and La Mer defines the number of these orders seen as the number of red bands observed (La Mer and Barnes, 1946).

The interpretation, of course, is less clear in this case than for monochromatic light, as a red band may be due to a maximum in the red pattern or a minimum in the green pattern. The latter are sharper and more numerous (because of the shorter wavelength), so that, as a rule, La Mer's orders may be interpreted as green minima. This rule is not foolproof. A sample computation for $m = 1.44$ (van de Hulst, 1948) is shown in Fig. 93. Similar graphs may be constructed at will from Fig. 25, sec. 10.4.

The experiments (Barnes, Kenyon, Zaiser, and La Mer, 1947; Johnson and La Mer, 1947) indicate that sulfur hydrosols show the following number or orders:

Radius (micron):	0.2	0.3	0.4	0.5	0.6
Number of orders:	2	4	6	8	10

By means of a crude instrument, in which the sol is visually inspected through a small telescope, the positions of the orders may be determined with a precision of 1°, from which the radii follow with a precision of 0.003 μ or about 1 per cent.

We may finally mention an investigation of vanadium pentoxide hydrosols, with $m = 1.20 - 0.17i$ relative to water at $\lambda_0 = 0.546 \mu$, in which absorption and light-scattering measurements were combined with electron-microscope studies (Kerker, Jones, Reed, Yang, and Schoenberg, 1954). It furnishes a good illustration of the insufficiency of Gans' theory as the rod-shaped particles grow. The length in a well-aged sol is about 2 μ.

19.3. Aerosols

The science of aerosols, i.e., airborne particles of microscopic and submicroscopic size, has become increasingly important in recent years. Aerosols of primarily *meteorological* importance are discussed in chap. 20, those of biological origin (spores and bacteria) are simply mentioned. The problem of *air pollution* exists in industrial areas all over the world. A variety of liquid and solid aerosols are involved in this problem. The solid aerosols, e.g., industrial dust, consist of very irregular particles and are strongly polydisperse. Both properties make the application of any of the refined methods of light scattering impossible, or at any rate unattractive. Nevertheless, the simple rules that the scattering is proportional to concentration and that the coarser particles have the more strongly forward-directed pattern still hold true. On these principles powerful instruments for daily checks may be devised. [See Drinker and Hatch (1936) for a general review of industrial dust.]

The optical methods are most useful if applied after a preliminary size selection by other means. Ellison (1954) has made a study of light scattering by polydisperse dust. An instrument for measuring polydispersity by settling combined with light scattering is described by Gumprecht and Sliepcevich (1953). The combined scattering and settling method has been tested by electron-microscope studies on sulfur and mercury aerosols by Kerker, Cox, and Schoenberg (1955).

Aerosols manufactured under controlled conditions for a specific use have particles much more uniform in size, usually in the form of liquid droplets. They range from the very dense aerosols that are used as smoke screens to the sols of toxic fluid droplets of very minute concentration that have been used in biological warfare. Also insecticidal liquids have to be sprayed in the form of aerosols of controlled particle size in order to be most effective.

The development of an aerosol generator that can produce monodisperse aerosols of a great variety of liquids is due to Sinclair and La Mer in 1941. The principle is to have a stream of hot air collect the vapor of a liquid and then let the air cool slowly while it streams through a chimney (La Mer, Inn, and Wilson, 1950). Convenient materials for experiments on light scattering are, 99 per cent H_2SO_4, dibutylphthalate (DBP) and dioctyl-phthalate (DOP). A standard DOP smoke with radii 0.3μ has been used in tests of smoke filters for gas masks. The refractive index is $m = 1.49$. Durbin (1951) has used ammonium chloride smokes with radii about 0.4μ in order to test various methods of size determination.

The methods for size determination of such aerosols by optical means are strictly similar to those for hydrosols, explained in sec. 19.22. They have been tested in detail by La Mer, Inn, and Wilson (1950), Inn (1951), Kerker and Hampton (1953). Large particles offer no problem, for both the transmission and the higher-order Tyndall spectra can be applied to full advantage. The polarization ratio is useful to the lower limit of radius $a = 0.08 \mu$. This limit is higher than for hydrosols as the wavelength in the "solvent" air is higher. In both cases this lower limit corresponds to $x = 2\pi a/\lambda =$ about 1. It would seem that the dissymetry method could be used somewhat further down, but La Mer and associates state that no optical method of measuring particle size is practical for $a < 0.08 \mu$. They found the elegant solution to make such particles grow in a controlled manner. This is done by exposing the H_2SO_4 particles to water vapor or the DBP particles to the vapors of a volatile solvent, as toluene or xylene. The radius may be increased by a factor 10, which by Rayleigh's law would increase the scattering area of each particle by a factor 10^6. Actually, a factor 5 of radius growth is recommended, and the purpose is to bring the particle size outside the Rayleigh domain. So the actual gain in scattering efficiency is smaller. The size of the grown particles is easily determined by means of the polarization ratio, and the original size is found by division by 5.

Another problem is to measure concentration and size distribution of the particles in aerosols of very low concentration. This problem, of obvious importance in the testing of gas masks, was studied by Gucker and by O'Konski. A Tyndall meter is insufficient for the concentrations of 10^{-9} g/liter and lower that have to be investigated. The better method is to count the pulses of light scattered by individual particles as they flow through the beam of light. These pulses may be recorded on an electronic counter (Gucker, O'Konski, Pickard, and Pitts, 1947; Gucker and O'Konski, 1949a,b).

One version of the photoelectric counter (Gucker and Rose, 1954) utilizes the fact that most particles scatter light more strongly in the

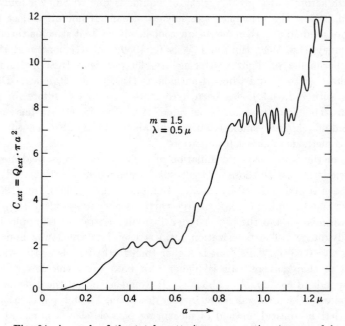

Fig. 94. A graph of the total scattering cross section (expressed in square microns) versus radius (expressed in microns).

forward direction; light scattered between 1° and 20° is collected and measured on a photomultiplier cell. One to a thousand counts per minute may be made. A pulse height selector can be set to receive only the light scattered by particles in a narrow range of sizes. The calibration curve of pulse height versus particle size may be obtained by test samples of known size or may be computed directly by Mie's theory. The authors have not made an integration of Mie intensities between 1° and 20° but refer to the plots of total cross section versus a as a useful guide. Such curves may be constructed from the extinction curves (e.g., Fig. 24) by multiplying Q by πa^2; Fig. 94 gives an example. Another version (O'Konski and Doyle, 1955) is a general-purpose instrument that may serve either as a counter or as an integrating photometer. The lower limit of sensitivity for single particles is a diameter of

0.3 μ. Uniform test aerosols, available in diameters from 0.13 μ to 1.17 μ, are used for calibration. Diameters may thus be determined with a standard deviation of 8 per cent.

A well-known method of producing an aerosol of small liquid drops is by adiabatic expansion of the saturated vapor. It is the method used in a Wilson camera. Paranjpe and his associates have studied this method for many years and have obtained remarkable results. The drop size is adjusted by bringing a controlled amount of fine smoke into the chamber. Paranjpe and Shingre (1951) give scattering diagrams of drops with $a = 1.88$ μ measured with monochromatic light, $\lambda = 0.5893$ μ, so that $x = 20.0$. The measurements were made photoelectrically with intervals of 5° from $\theta = 10°$ to $\theta = 160°$, and are for chloroform ($m = 1.449$), benzene ($m = 1.504$), aniline ($m = 1.590$), and carbon bisulphide ($m = 1.632$). Theoretical values for the same x, θ, and m are also presented. The result is an almost perfect agreement between the positions of the maxima (orders) that are measured and computed. Chloroform shows 14 maxima in the measured range, all predicted by the computations.

An important paper for the measuring technique is Lyot's study (1929) of the polarization of scattered, or diffusely reflected, light. The polarization values are accurate to 0.1 per cent, if the polarization is small. His experimental results include data for water drops of sizes $2a = 1$ mm, 300 μ, 35 μ, 5 μ, and 2.5 μ, probably with a considerable dispersion of sizes in each case. The very large drops show precisely the dependence of polarization on angle that may be expected from the geometrical-optics theory (sec. 13.11). The agreement is illustrated in Fig. 99, sec. 20.3, where further discussion is presented.

19.4. Anisotropic Media

19.41. Dichroism and Birefringence

The optical anisotropy of a colloidal solution may be studied by the same methods by which the anisotropy of an uniaxial crystal is studied. The possible effects are birefringence, dichroism, rotation of the plane of polarization (= circular birefringence), and circular dichroism. All these effects may be formally combined into one, namely, the effect that the complex refractive index m of a plane wave traveling in the same direction is different for different states of polarization of the wave. The mathematical formulation was given in sec. 4.41. The occurrence of linear and circular birefringence and dichroism was discussed in secs. 5.41 and 5.42 in its dependence on the symmetry properties of the scattering particles.

The present discussion will be confined to *linear* birefringence and dichroism. Provided that the solvent is not itself anisotropic, such anisotropy exists only if *both* the following conditions are fulfilled.

a. The scattering particles are anisotropic. This may be due to their form, or they may consist of an anisotropic material.

b. The particles should show preferred orientations in space, for a random orientation cannot give birefringence (sec. 5.41).

The preferred orientation may have various causes: streaming of the solution, ultrasonic waves traversing the solution, or an external electric or magnetic field. Each of these causes defines one preferential direction which takes the place of the axis in an uniaxial crystal. Let the direction of propagation of the plane-polarized light wave be taken perpendicular to the axis; then the complex refractive index is

$$m_{ord} = n_{ord} - in'_{ord} \text{ for the ordinary ray } (E \perp \text{ axis}),$$

$$m_{ext} = n_{ext} - in'_{ext} \text{ for the extraordinary ray } (E \parallel \text{ axis}).$$

Birefringence exists if the real parts differ; it is called positive if $n_{ext} - n_{ord} > 0$. Dichroism exists if the imaginary parts differ. For either ray the value of m is related to the scattering properties of the individual bodies by means of the fundamental relation (sec. 4.3):

$$m = n_0 \left(1 - i \sum S_{1,2}(0) \cdot 2\pi k^{-3}\right).$$

Here we have supposed that the solvent has the refractive index n_0, that the wave number is $k = 2\pi n_0/\lambda_0$, and that the summation is made over all particles (number N) that are present in a unit volume (1 cm³). We must take $S_1(0)$ for the ordinary ray and $S_2(0)$ for the extraordinary ray, if the plane of reference is taken through the "axis" of the solution. We assume that $S_3(0)$ and $S_4(0)$, or their sums over all particles, vanish. The functions $S_1(0)$ and $S_2(0)$ have to be computed the hard way from the scattering theory. For details, the reader may wish to refer to the following sections:

Ellipsoids $\ll \lambda$ or with small phase shift: secs. 6.32 and 7.31.

Small ellipsoids of anisotropic material: sec. 6.33.

Small perfectly conducting ellipsoids: sec. 6.4.

Long cylinders much thicker than λ: no birefringence in the approximation of sec. 8.32, but some in the more refined theory of sec. 17.2.

Long circular cylinders of isotropic material and arbitrary diameter compared to λ: chap. 15.

Ellipsoids, arbitrary size (incomplete theory): sec. 16.11.

The formula given above holds for colloidal solutions in which a relatively small total volume is occupied by the embedded particles. This means that the optical theory is based on the set of conditions (II) in sec. 4.5.

It may be noted that somewhat different formulae apply under the conditions (I) of sec. 4.5, namely, if the particles are very small and the volume concentration relatively high. For this case and the presence of oriented ellipsoids Wiener (1909) has suggested approximate formulae which have proved fairly accurate in practice. A more exact set of formulae was derived for a medium containing spheres by Bruggemann (1935) and for a medium containing ellipsoids by Niesel (1952).

For a general review of artificial birefringence we refer to Freundlich (1932), and to Peterlin and Stuart (1943). The theory of flow birefringence has been treated by Scheraga and coworkers (1951, 1952). The theory that governs the orientation of colloidal particles in the case of ultrasonic waves was given by Oka (1939, 1940). The acoustic birefringence may be used as a practical method to measure the intensity of ultrasonic waves. The preceding chapters contain the material from which it is possible to estimate which particles would be most effective for this purpose.

The properties of a colloidal graphite solution with oriented particles have been studied by Cayrel and Schatzman (1955).

Liquid droplets in a homogeneous electric field become elongated in the direction of the field. The eccentricity increases with increasing field strength and with increasing size. O'Konski and his associates have shown by theory (1953) and experiment (1955) that the eccentricity becomes as high as 0.6 (axial ratio 1.25) for water drops with radius 0.9 mm in a field of 11 kV/cm.

19.42. Sheet Polarizers

An important application of the preceding theory is found in sheet polarizers. Natural dichroism of a material is, like natural birefringence, due to its anisotropic molecular structure. The ratio of the two absorption coefficients for two linear polarizations is called the dichroic ratio. It may reach the value 10 in natural dichroic crystals (tourmaline). The commercial Polaroid sheet polarizers consist of small needles of a dichroic material embedded in a medium; dichroic ratios > 100 are attained. The theory for the absorption by such needles, small compared to λ, is given by Clark Jones (1945) and in sec. 6.33. In these polarizers the dichroism results mainly from the dichroism present in each crystal.

Another method of producing a dichroic material is to embed oriented needles of a metal that is *not* dichroic by itself in a medium. The composite material then is said to have *form-dichroism*, just as a composite medium containing oriented dielectric needles exhibits form-birefringence.

Such metal polarizers have been made (a) by reducing metallic salts to metals in oriented, linear high polymers, and (b) by strongly deforming a gelatine sheet containing initially metallic particles with random orientation. Still other methods are mentioned by Berkman, Boehm, and Zocher (1926). These authors also present qualitative measurements of birefringence and dichroism, based on the transmitted colors.

Quantitative measurements on samples of silver, gold, and mercury polarizers are reported by Land and West (1946). The most common effect, illustrated by silver, is that red light suffers the greater absorption when vibrating parallel to the long axes of the needles (positive dichroism: $R > 1$). In violet light the situation is reversed (negative dichroism: $R < 1$). At an intermediate wavelength $R = 1$, and the material is not dichroic.

The theory of such composite media is described in sec. 4.41. If the particles were $\ll \lambda$, the formulae from sec. 6.32 would have to be used, with the result that

$$R = \gamma_1/\gamma_2 = \text{Im } \tfrac{1}{2}(m^2 - 1)/\text{Im } \left(\frac{m^2 - 1}{m^2 + 1}\right).$$

Wiener has found that by this formula silver needles in air give an inversion of the dichroism near 0.380μ. When imbedded in a medium with $n = 1.33$ the computed inversion point is at 0.480μ. Neither the values of R nor the position of this inversion point should change with size as long as the needles are sufficiently thin, for these results are due merely to the variation of the refractive index with wavelength.

However, the fact that a shift of the inversion point with increasing size is observed shows that many of the needles were not sufficiently thin. The curves in Fig. 69 (sec. 15.51) show that another effect may become important, namely, the inversion due to size that is presented by metal wires even if the change of refractive index with λ is neglected. The actual samples probably show both effects mixed. Only a numerical computation of Q_1 and Q_2 as a function of size for each of the values of the refractive index involved would enable us to make a detailed comparison of observations and theory.

19.5. Miscellaneous Applications

Applications to the domain of physics and chemistry and to related branches of industry are by no means limited to those discussed in the preceding sections. A few further possibilities are mentioned in Table 41, and more might be added. It is obviously impossible to summarize all these topics even briefly. However, reference to the most nearly relevant chapter of the present book and to one or two papers in the literature may be helpful to the reader.

Table 41. Some Miscellaneous Scattering Problems

Object of Study	Shape of Obstacle	Typical Values of			Relevant Chapter	Reference
		m	x	$2x(m-1)$		
Small-angle X-ray scattering	Varied	Very nearly 1	Large	Very small	7	a
Scattering by imperfections in a solid body	Varied	Near 1	Small	Small	7	b
Opal glass	Sphere	0.9	Medium	<1	7, 11	c
Optical study of soot particles in a flame	Varied	Complex	Small	Complex	14	d
Cellulose fibers	Circular cylinders	1.3	Large	Large	13, 15	e
Viruses	Varied	$1.1 - 1.2$	Medium	<1	7, 19 (sec. 19.12)	f
Blood corpuscles	Varied	$1.1 - 1.2$	Large	Varied	11	g
Optical transmission through wire gratings	Circular cylinders, etc.	$1.4 - 1.4i$	Large	Large	15, 17	h
Radar reflection from wire gratings	Circular cylinders, etc.	Near ∞, partly absorbing	Small	Near ∞	15, 16	i
Radar reflection from meteor trails	Cylinder, finite length	0 to 1	Medium	Medium	15	k

a. A. Guinier and G. Fournet, *Small-Angle Scattering of X-Rays*, New York, John Wiley & Sons, 1955.

b. A. Peterlin, *Kolloid-Z.*, **120**, 75 (1951).

c. J. W. Ryde and B. S. Cooper, *Proc. Intern. Illumination Cong.*, pp. 387 and 410, Cambridge, Cambridge Univ. Press, 1931.

d. H. Senftleben and E. Benedict, *Ann. Physik*, **60**, 297 (1919); G. Naeser and W. Pepperhoff, *Kolloid-Z.*, **125**, 33 (1952).

e. P. H. Hermans, *Contribution to the Physics of Cellulose Fibres*, Amsterdam, Elsevier Publ. Co., 1946.

f. G. Oster, *Science*, **103**, 306 (1946); F. T. Gucker, *Science*, **110**, 372 (1949).

g. D. Verveen, Diameter-, dikte-, volume- en oppervlaktebepaling van Erythrocyten met behulp van lichtbuigingseffect, "Thesis Utrecht," Nijkerk, G. F. Callenbach N.V., 1949.

h. See chap. 15 references.

i. See chap. 16 references.

k. T. R. Kaiser and R. L. Closs, *Phil. Mag.*, **43**, 1 (1952); T. R. Kaiser, *Meteors* (Special Suppl. No. 2, *J. Atm. and Terrest. Phys.*), 55, London, Pergamon Press, 1955.

References*

Some references relevant to very small particles and high polymers (secs. 19.11 and 19.12) are:

Lord Rayleigh, *Phil. Mag.*, **41**, 107, 274, and 447 (1171) (*Sci. Papers* 8 and 9).

Lord Rayleigh, *Phil. Mag.*, **47**, 375 (1899) (*Sci. Papers* 247).

Lord Rayleigh, *Phil. Mag.*, **12**, 81 (1881) (*Sci. Papers* 74).

J. Cabannes, *La diffusion moléculaire de lumière*, Paris, Les Presses universitaires de France, 1929.

Lord Rayleigh, *Phil. Mag.*, **35**, 373 (1918) (*Sci. Papers* 430).

G. Oster, *Chem. Revs.*, **43**, 319 (1948).

B. H. Zimm and P. M. Doty, *Theory and Application of Light Scattering*, New York, John Wiley & Sons, in preparation.

M. L. Kerker, *J. Chem. Phys.*, **20**, 1653 (1952).

B. H. Zimm, *J. Chem. Phys.*, **16**, 1099 (1948).

R. C. Williams and R. C. Backus, *J. Appl. Phys.*, **20**, 224 (1949).

G. Oster, P. M. Doty, and B. H. Zimm, *J. Am. Chem. Soc.*, **69**, 1193 (1947).

References to metal sols (sec. 19.21) are:

G. Mie, *Ann. Physik*, **25**, 377 (1908).

H. Freundlich, *Kapillarchemie*, Band II, Leipzig, Akad. Verlagsgesellschaft M.B.H., 1932.

B. Seraphin, *Ann. Physik*, **10**, 1 (1952).

N. F. Mott and J. Jones, *The Theory and Properties of Metals and Alloys*, chap. 3. London, Oxford Univ. Press, 1936.

K. Weisz, *Z. Naturforsch.*, **3a**, 143 (1948).

R. Feick, *Ann. Physik*, **77**, 573 (1925). (This page occurs twice in this volume; the reference appears on the second p. 573.)

H. Wolter, *Z. Physik*, **113**, 547 (1939).

H. Wolter, *Z. Physik*, **115**, 696 (1940).

Sulfur sols and similar sols without appreciable absorption (sec. 19.22) have been investigated by:

B. A. Keen and A. W. Porter, *Proc. Roy. Soc. London*, **A89**, 370 (1913).

B. Ray, *Proc. Indian Assoc. Cultivation Sci.*, **7**, 10 (1921); **8**, 23 (1923).

C. V. Raman and B. Ray, *Proc. Roy. Soc. London*, **A102**, 151 (1922).

V. K. La Mer, *J. Phys. & Colloid Chem.*, **52**, 65 (1948).

D. Sinclair and V. K. La Mer, *Chem. Revs.*, **44**, 245 (1949).

R. H. Dinegar and R. H. Smellie, *J. Colloid Sci.*, **7**, 270 (1952).

M. D. Barnes and V. K. La Mer, *J. Colloid Sci.*, **1**, 79 (1946).

V. K. La Mer, E. C. Y. Inn, and I. B. Wilson, *J. Colloid Sci.*, **5**, 471 (1950).

A. S. Kenyon and V. K. La Mer, *J. Colloid Sci.*, **4**, 163 (1949).

P. C. Lewis and G. F. Lothian, *Brit. J. Appl. Phys.*, Nottingham Conf. Suppl. (1954).

W. Heller, J. N. Epel, and R. M. Tabibian, *J. Chem. Phys.*, **22**, 1777 (1954).

M. L. Kerker and V. K. La Mer, *J. Am. Chem. Soc.*, **72**, 3516 (1950).

M. L. Kerker, *J. Colloid Sci.*, **5**, 165 (1950).

M. L. Kerker and M. I. Hampton, *J. Opt. Soc. Amer.*, **43**, 370 (1953).

V. K. La Mer and M. D. Barnes, *J. Colloid Sci.*, **1**, 71 (1946).

* Given in the order in which they are cited in this chapter.

H. C. Van de Hulst, *J. Colloid Sci.*, **4**, 79 (1949).

M. D. Barnes, A. S. Kenyon, E. M. Zaiser, and V. K. La Mer, *J. Colloid Sci.*, **2**, 349 (1947).

I. Johnson and V. K. La Mer, *J. Am. Chem. Soc.*, **69**, 1148 (1947).

M. L. Kerker, G. L. Jones, J. B. Reed, C. N. P. Yang, and M. D. Schoenberg, *J. Phys. Chem.*, **58**, 1147 (1954).

Some references to aerosols (sec. 19.3) aré:

P. Drinker and T. F. Hatch, *Industrial Dust*, New York, McGraw-Hill Book Co., 1936.

J. M. K. Ellison, *Brit. J. Appl. Phys.*, Nottingham Conf. Suppl., 1954.

R. O. Gumprecht and C. M. Sliepcevich, *J. Phys. Chem.*, **57**, 95 (1953).

M. L. Kerker, A. L. Cox, and M. D. Schoenberg, *J. Colloid Sci.*, **10**, 413 (1955).

V. K. La Mer, E. C. Y. Inn, and I. B. Wilson, *J. Colloid Sci.*, **5**, 471 (1950).

E. J. Durbin, *NACA Techn. Note* 2441, August, 1951.

E. C. Y. Inn, *J. Colloid Sci.*, **5**, 368 (1951).

M. L. Kerker and M. I. Hampton, *J. Opt. Soc. Amer.*, **43**, 370 (1953).

F. T. Gucker, C. T. O'Konski, H. B. Pickard, and J. N. Pitts, *J. Am. Chem. Soc.*, **69**, 2422 (1947).

F. T. Gucker and C. T. O'Konski, *Chem. Revs.*, **44**, 373 (1949); *J. Colloid Sci.*, **4**, 541 (1949).

F. T. Gucker and D. G. Rose, *Brit. J. Appl. Phys.*, Nottingham Conf. Suppl. (1954).

C. T. O'Konski and G. J. Doyle, *Anal. Chem.*, **27**, 694 (1955).

M. M. Paranjpe and M. V. Shingre, *J. Univ. Bombay*, **19**, part 5, No. 29, 1951.

B. Lyot, *Ann. observ. Paris-Meudon*, **8**, No. 1, see pp. 125–134, 1929.

From the extensive literature on anisotropic systems (secs. 19.41 and 19.42) we cite:

D. A. G. Bruggemann, *Ann. Physik*, **24**, 636 (1935).

W. Niesel, *Ann. Physik*, **10**, 336 (1952).

H. Freundlich, *Kapillarchemie*, Band II, Leipzig, Akad. Verlagsgesellschaft M.B.H., 1932.

A. Peterlin and H. A. Stuart, *Doppelbrechung insbesondere Künstliche Doppelbrechung*, Leipzig, Becker und Erler, 1943.

H. A. Scheraga, J. T. Edsall, and J. O. Gadd, *J. Chem. Phys.*, **19**, 1101 (1951).

R. Cerf and H. A. Scheraga, *Chem. Revs.*, **51**, 185 (1952).

S. Oka, *Kolloid-Z.*, **87**, 37 (1939).

C. T. O'Konski and H. C. Thacher, *J. Phys. Chem.*, **57**, 955 (1953).

C. T. O'Konski and R. L. Gunther, *J. Colloid Sci.*, **10**, 563 (1955).

R. Cayrel and E. Schatzman, *Ann. astrophys.*, **17**, 555 (1954).

R. Clark Jones, *Phys. Rev.*, **68**, 93 and 213 (1945).

S. Berkman, J. Boehm, and H. Zocher, *Z. physik. Chem.*, **124**, 83 (1926).

E. H. Land and C. D. West, *Colloid Chemistry*, vol. **6**, p. 160, New York, Reinhold Publ. Corp., 1946.

20. APPLICATIONS TO METEOROLOGY

The light of the sun and the stars is weakened as it penetrates the earth's atmosphere. The study of this effect, the astronomical extinction, as a function of the wavelength, is one of the means of investigating the scattering properties of the atmosphere. More detailed information is contained in the distribution of light in the daylight sky, again as a function of wavelength. Both types of measurements, extinction and scattering, may be repeated with artificial light sources, e.g., searchlight beams, and are then useful for investigating dense media such as fog or rain. More recently, radar has been added as a powerful method.

20.1. The Extinction Components

The extinction in the clear sky has three components caused by:

a. Rayleigh scattering by the air molecules.
b. Scattering by the aerosol (haze and dust).
c. Selective absorption by the air molecules.

The scattered light observed from the clear sky is due to components a and b only.

Component a follows the familiar λ^{-4} law. Small variations in it are due to seasonal changes of the total extent of the atmosphere (air mass) and to changes in the water-vapor content and possibly in the number of very small water drops with $a < 0.02\ \mu$.

Component b is due to scattering by a large variety of particles, usually with radii $< 1\ \mu$. It is subject to very strong variations in space and time, and its presence contributes greatly to the splendor of the colors of the sky.

Component c includes, besides molecular absorption bands of O_2, H_2O, CO_2, etc., the continuous Chappuis bands of ozone that cover a good part of the visual spectrum and cause an apparent bulge in the λ^{-4} law if the aerosol is absent.

In sec. 20.2 a few examples of investigations of the aerosol are reviewed. For discussion of the other components we may refer to reviews by Pernter and Exner (1910), Dorno (1919), Minnaert (1940), Middleton (1941), van de Hulst (1949), Sekara (1951), Neuberger (1951). Even though the pattern of single Rayleigh scattering is quite simple, the problem of handling the secondary and multiple scattering correctly with the polarization becomes extremely complicated and was first successfully solved by Chandrasekhar (1950), and by Chandrasekhar and Elbert (1954). See also Deirmendjian and Sekara (1954, 1955).

Multiple scattering, including diffuse reflection at the earth, was treated in a more elementary fashion for the case of isotropic single scattering by van de Hulst (1948, 1949). Further theoretical approaches and some observational data are given in sec. 19 of Volz's thesis (1954).

The aerosol mentioned above consists of a light haze and is hardly noticed in daily life. The visual (and sometimes nasty) mist, fog, clouds, and falling rain have a much larger liquid water content and at the same time a much coarser particle size, about $5\,\mu$ to $20\,\mu$ for fog, $200\,\mu$ to $2000\,\mu$ (2 mm) for rain. The optical phenomena caused by such large drops are quite different and will be separately discussed in sec. 20.3. One striking consequence of the larger size is that the penetration power of infrared radiation through fog is hardly better than that of visual light. Only radio and radar waves give a great improvement (sec. 20.41).

In our discussion of the extinction we assume that the true extinction coefficient is meant and can be measured. This requires a point source and a good optical instrument by which the light of the source can be measured separately from the scattered light around it. Middleton (1949) has shown that this requirement is not met in the usual working conditions with telephotometers in fog and has computed the corrections.

20.2. The Atmospheric Haze

20.21. Extinction Law

The extinction by haze and light mist can be studied most readily by means of observations made with a horizontal beam. If, for the moment, we assume n particles of *one* radius a per cm³, then the extinction coefficient per unit length is (sec. 2.6)

$$\gamma = \pi a^2 n Q,$$

where Q is the efficiency factor for extinction, defined in sec. 2.4 and discussed at great length in chaps. 6 to 17. So, a beam traveling a path of l cm is reduced to a fraction

$$e^{-\gamma l} = 10^{-0.434\gamma l}$$

of the intensity it would have had in the absence of any extinction. After subtracting the molecular extinction and the selective absorption from the observed extinction a residual is left, owing to haze. Such a residual is present even on days and sites selected as "very clear."

Similar observations may be made on the extinction of sunlight. Then l has to be replaced by h sec z where h is the effective thickness of the haze layer and z the zenith distance of the sun.

The graph of Q for water (refractive index $m = 1.33$) in Fig. 32 shows that the first maximum occurs at $x = 2\pi a/\lambda = 6$. So the wavelength λ_{max} at which this first maximum occurs is very nearly the same as the radius.

Before quoting some investigations of the extinction curve of haze we should emphasize first that the dispersion of particle sizes is often so wide as to make the interpretation virtually impossible and second that the refractive index may in dry air at times be more nearly 1.50 than 1.33, owing to the content of semicrystalline salts (Junge, 1952, and Volz, 1954).

Horizontal beam. Vassy (1939) and Dessens (1946) distinguish from photometry four types, which may be interpreted by a slightly different effective particle size:

1. Extinction roughly neutral; $a = \lambda_{max} > 0.8\ \mu$.
2. Extinction maximum at 5000 A; $a = \lambda_{max} = 0.5\ \mu$.
3. Extinction decreases gradually from 6000 A to 4000 A; $a = \lambda_{max} = 0.6\ \mu$.
4. Extinction maximum at 4000 A; $a = \lambda_{max} = 0.4\ \mu$.

The extinction data were the first evidence that droplets with a diameter about $1\ \mu$ occur in a clear atmosphere, even in very dry climates and up to altitudes of 3000 meters. Dessens (1947) has also collected such drops on the web of a small spider and has observed them microscopically. The usual distribution function of the radii shows a maximum near $a = 0.5\ \mu$.

Vertical extinction. Götz (1944) concludes from extinction measurements that the most common radii are $a = 0.3\ \mu$ for light haze and $a = 0.4\ \mu$ for heavy haze.

As most meteorological studies are based on extended observational material, it is obviously desirable to express the extinction characteristics by only a few parameters. Many authors therefore use the empirical formula of A. Ångström (1929):

$$\gamma \sim \lambda^{-\alpha}.$$

Evidently, $\alpha = 4$ for Rayleigh scattering, and $\alpha = 0$ for neutral extinction. The empirical values of α for the atmospheric haze, as discussed, e.g., in large numbers by Schüepp (1949), usually range from 1 to 2. If $\alpha < 1$ (relatively large particles) there usually is a bright white region (aureole) around the sun. The sky may be fairly bright blue in spite of strong turbidity, provided that $\alpha > 2$.

The same representation of the extinction law in a limited wavelength region by a simple power of λ has also been applied in different fields. (Heller, Klevens, and Oppenheimer, 1946; Heller and Vassy, 1946; La Mer, 1948). It is, of course, quite easy to derive the values of α from the theoretical extinction curves. Figure 95 shows these values as a

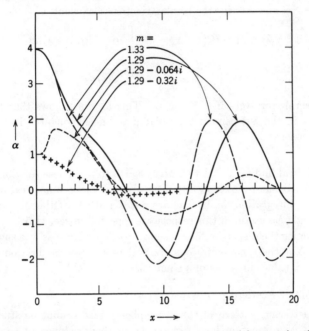

Fig. 95. Values of the exponent α, in the exponential approximation to the extinction law, computed for monodisperse particles of four different refractive indices as a function of the size parameter $x = 2\pi a/\lambda$.

function of x, and of a for $\lambda = 5000$ A, for the refractive indices 1.33, 1.29, $1.29 - 0.064i$, and $1.29 - 0.32i$. The ripple has been disregarded. Such curves may be used as calibration curves, i.e., to derive the radius a from the observed exponent α, only if it is certain that the aerosol is fairly monodisperse.

It is somewhat of a problem why the formula of Ångström is at all applicable to water drops with radii of the order of 0.5 μ, as the theoretical extinction curve does not resemble a pure power law. As a solution Götz (1944) suggests that the simultaneous presence of absorbing dust, for which he takes without any good reason the $m = \infty$ curve (Fig. 28, sec. 10.62), obliterates the large fluctuations of the extinction curve of water drops. It is more likely that this obliteration is caused by the superposition of a very wide range of sizes.

It is easy to show that a power law, in which

$$dN = C \cdot a^{-v-1} da$$

drops are present per cm^3 in the size range da, will give the same wavelength dependence, no matter what form the extinction curve $Q(x)$ has. For we have (sec. 2.6) for the total extinction coefficient

$$\gamma(\lambda) = \int \pi a^2 Q(a)\, dN = \pi C \int_0^\infty a^{-v+1} Q(2\pi a/\lambda)\, da$$

$$= \pi C \left(\frac{\lambda}{2\pi}\right)^{-v+2} \int_0^\infty x^{-v+1} Q(x)\, dx.$$

The integral converges if $2 < v < 6$. This formula shows that the exponent α in the λ-dependence is related to the exponent v in the size distribution by

$$\alpha = v - 2.$$

The empirical finding $\alpha = 1$ to 2 gives $v = 3$ to 4. These relations have been pointed out by Volz (1954), who also has made numerical integrations with integration limits that are not 0 and ∞. Other integrations which might be useful in this respect are reported in sec. 11.5.

In practice the dispersion in sizes may not be so large as was suggested above. Yet these computations show that a direct calibration of the observed α-values in terms of a single size is unreliable.

20.22. Scattering Pattern

The scattering pattern of the atmospheric haze cannot be studied as easily as its extinction law, for, owing to the existence of secondary scattering, the skylight cannot simply be separated into a component due to molecular scattering and a component due to haze. Nevertheless, photometry of the distribution of light on the daylight sky can give a fairly precise impression of the scattering pattern of the aerosol, especially since this is usually limited to the lower 3000 meters of the atmosphere. Measurements of its polarization would be much more difficult and have, to the author's knowledge, not been attempted.

Another method is to study the light scattered from a searchlight beam. The arrangement is then strictly as indicated in Fig. 88 (sec. 18.31); only the container is absent.

Measurements of one type or the other have been published by Hulburt (1941), Reeger and Siedentopf (1946), Bullrich (1947), Volz (1954), Strzalkowski (1955), and others. The problem in their interpretation is again the large dispersion in drop sizes. Table 42 (van de Hulst, 1949) gives the theoretical values of $(1/2x^2)(i_1 + i_2)$ for a number of values of x, based on computations explained in secs. 9.3 and 10.4. As shown in detail in sec. 13.12, the fluctuations will be smoothed out by the dispersion in sizes. The relative intensities measured by Reeger and Siedentopf in the range from 30° to 170° are in fair agreement with any value of x from 2.5 to ∞. For example, from 90° to 30° the intensity increases by a factor 15. Those from 10° to 30° indicate $x = 5$, i.e., $a = 0.4\ \mu$ for the wavelength 5000 A. This is in good agreement with the inference from the extinction curve (sec. 20.21). Volz (1954) discusses the searchlight measurements in some detail and also notes that the intensity ratio $I(90°)/I(20°)$ ranges from 5 for very clear sky (visual range 100 km) to 70 for light fog (visual range < 1 km).

The rapid increase of brightness if we approach the sun to within a few degrees is caused mainly by the coarser particles in the atmosphere.

Table 42. Theoretical Values of $(1/2x^2)\ \{i_1(\theta) + i_2(\theta)\}$ for $m = 1.33$

x	a for $\lambda = 0.5\ \mu$	θ				
		0°	10°	30°	60°	Av. 90°–180°
0.6	0.048	0.0060	0.0059	0.0051	0.0035	0.0034
1.0	0.08	0.053	0.052	0.044	0.027	0.018
1.5	0.12	0.29	0.28	0.22	0.11	0.035
2.0	0.16	0.99	0.95	0.70	0.26	0.015
2.5	0.20	2.37	2.24	1.39	0.32	0.04
3.0	0.24	4.62	4.30	2.33	0.13	0.03
3.6	0.29	8.5	7.7	3.2	0.13	0.06
4.0	0.32	12.3	10.8	3.6	0.13	0.05
5.0	0.40	23.4	19.1	3.1	0.41	0.07
6.0	0.48	34.8	25.7	1.57	0.26	0.08
8.0	0.64	46.0	24.4	2.4	0.40	0.10
10.0	0.80	36	8.6	2.2	0.37	0.01
12.0	0.96	23	4.2	1.1	0.14	—
15.0	1.2	104	16.9	0.7	0.24	0.02
20.0	1.6	120	3.0	0.5	0.15	0.32
25.0	2.0	244	0.2	0.9	0.13	0.06
30.0	2.4	231	14.9	1.2	0.12	0.06
35.0	2.8	445	2.3	1.5	0.18	0.09
40.0	3.2	401	1.2	1.97	0.20	0.07
Large x (diffraction) (geometrical optics).		$x^2/4$	0	0	0	0
		4.09	3.37	1.15	0.11	0.05

If these particles (drops or ice needles) are fairly monodisperse, we observe diffraction coronae around the sun. Otherwise, it is a bright white "aereole." Some photometric data about this aereole were given by van de Hulst (1949), many others by Volz (1954).

Volz (1954) has attempted to obtain useful meteorological results by collecting data during many different days and circumstances. The

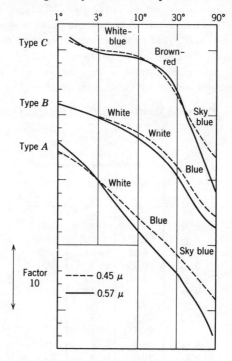

Fig. 96. Typical intensity distributions and colors in the clear sky as a function of the angular distance from the sun (after Volz). The logarithmic intensity scale has an arbitrary zero for each of the three types.

sky brightness at 0.2° to 120° from the sun is measured in different colors. The color- and angle-dependence combined permit (more clearly at Arosa than at Frankfurt) the distinction of the following types that appear to be correlated with the physical history of the air masses (see Fig. 96):

Type A. Polar cold air. The sky becomes gradually brighter and whiter towards the sun; this occurs without steps.

Type B. Occurs after the dissolution of extensive stratus clouds. The scattering often is very strong. There is a bright white disk about the sun with radius about 35°.

Type C. Occurs rarely in air masses after heavy rains. Blue-white area around the sun, followed by a reddish rim at $\theta = 30°$.

The correct interpretation of the differences is by no means obvious. Volz computes a number of theoretical curves on the basis of the Mie formulae and on the distribution law of the sizes as mentioned in sec. 20.21. The relative overall pattern is then the same for all λ, or also the spectral distribution is the same at all θ. The same principle is the basis for the discussion by Möller, de Bary, and Krog (1953). The actually observed patterns are explained by Volz by assuming certain deviations from the standard distributions with $v = 3$ and $v = 4$. This seems indeed a better method than to choose simply the one drop size that fits the observations best. Although the particle radii from 0.1 μ to 30 μ are considered, Volz concludes that only the distribution in the range from 0.15 μ to 0.6 μ determines the distinction between types *A*, *B*, and *C*, mentioned above.

It might be interesting, therefore, to make further numerical integrations in order to find the patterns for a distribution in which the sizes range in a factor 1 to 2 or 1 to 3 only. One interesting point may be brought up at once. For each refractive index specific angles have a tendency to show a maximum and others to show a minimum. For instance, Fig. 52 (sec. 13.41) shows at once that in a range of sizes ($x = 9$ to 16, so that $a = 0.7$ μ to 1.3 μ for $\lambda = 5000$ A) waterdrops have a minimum near $\theta = 22°$. In another range ($x = 19$ to 26, so that $a = 1.5$ μ to 2.1 μ) the first minimum occurs near $\theta = 10°$.

Further, by reference to Fig. 34 (sec. 11.33), we find that a strong maximum occurs at $\rho = 2x(m - 1) = 6$, $z = x \sin \theta = 4.8$. For water drops ($m = 1.33$) this occurs if $x = 9$, $a = 0.7$ μ and is then situated at $\theta = 32°$. For refractive index $m = 1.5$ it occurs if $x = 6$, $a = 0.5$ μ and is then situated at $\theta = 42°$. Neighboring sizes will also give a maximum at or near those angles, and this maximum will persist in the integrated scattering pattern of the haze. It is thus possible that a sudden drop beyond $\theta = 35°$ as noted by Volz (type *B*) simply reflects a peculiarity of the scattering pattern of any mixture of water drops with sizes of the order of 0.5 to 1 μ.

These semiquantitative considerations cannot replace a full integration of the Mie intensities. Further research in this direction may give interesting results.

20.23. Aerosols from Volcanic Eruptions and Forest Fires

Two special contributions to the world-wide aerosol may be mentioned now. One is volcanic dust, the other oil droplets arising from distillation of the wood during big forest fires.

The most famous optical phenomena due to volcanic dust were seen after the eruption of the Krakatoa in 1883. Unusual light distributions in the sky and beautiful sunsets were seen in most parts of the world and were still noticeable three years later. A century earlier the large eruption at Iceland had caused the sun and moon in June, and July, 1783, to be seen "as red as the juice of cherries" (Melmore, 1953).

A more recent eruption, for which some photometric material is available, is the Katmai eruption in June, 1912. Götz (1944) mentions that extinction data from Mount Wilson, Algiers, and Sweden give surprisingly similar results. He interprets them as a superposition of haze ($m = 1.33$, radius not given but presumably 0.5 or 0.6 μ), and dust ($m = \infty$, radius 0.17 to 0.22 μ).

The scattering pattern of volcanic dust causes the so-called Bishop's ring, a reddish ring around the sun with radius $\approx 30°$. Volz applies the same name to any reddish ring with radius 10° or larger. The limits are 20° and 30° for an aerosol of type C, and 10° and 30° for the aerosol causing "the blue sun" (see below). The most obvious interpretation would be with a narrow range of sizes, such that the first minimum in the diffraction pattern for green light gives the hue of its complementary color, red. This is in analogy with an old theory for the colors of the diffraction corona, in which it has been customary to interpret the outer edge of the first red zone as the first minimum for "white light" with $\lambda = 0.56 \mu$. This gives $a = 0.56 \mu$ for a red zone reaching to 35°. A quite different interpretation based on a very wide range of sizes was proposed by Volz. The idea is that a low exponent v in the size distribution gives a low exponent α in the wavelength distribution (sec. 20.21) and thus a reddish color, so that the observation of a reddish color at a certain θ points to a low exponent v in the range of sizes of the particles that are the most effective scatterers at this angle. Thus the variations in color with θ are translated into variations of v with a. A numerical integration (e.g., for two wavelengths) may be made to check whether the proposed size distribution really fits the observations. We must refer to Volz's work for detailed results.

On September 23, 1950, great forest fires raged in the Canadian province of Alberta. The blue color of the sun that was a consequence of it aroused much attention. Helen S. Hogg (1950) reports that on the following days the sun in Ontario was seen to have a bluish color. The smoke particles were blown to Europe and reached Great Britain on September 26, where they were 9000 to 13000 meters high (Bull, 1951). Observations are reported by Guthrie (1950) and Wilson (1951). Photometry of the sun was performed by Wilson at a moment when the clouds cleared, and the sun had a deep indigo color. The sun was weaker than normal at that altitude (about 20° above the horizon) by 10.9 magnitudes at 6300 A, 10.0 magnitudes at 4400 A, and 10.2 magnitudes at 4000 A; 10 magnitudes is a factor 10000 in intensity. The minimum extinction at 4400 A may be explained by a predominant particle size a given by

$$a = \frac{0.26\ \mu}{m - 1},$$

where m is the real refractive index. This follows at once from the relation (sec. 11.22):

$$\rho_{min} = \frac{4\pi a}{\lambda}\ (m - 1) = 7.4.$$

As m must be in the vicinity of 1.5 we find $a = 0.5\ \mu$.

Wilson found it necessary, however, to ascribe 8 magnitudes to a "neutral" extinction by very coarse particles. Perhaps it is more likely that a very wide range of particle sizes of the kind discussed in sec. 11.5 was present.

Volz (1954) mentions that a Bishop's ring was also seen here and that its extension was 15°, which is interpreted with a distribution function of the radii with upper limit 0.6 μ. This seems in reasonable agreement with the preceding estimate, but more precise numerical data would be desirable.

20.3. Optical Phenomena in Clouds, Fog, and Rain

The drops of clouds, fog, and rain are very much larger than those in the haze described in the preceding section. Evidently, a cloud may also contain a good proportion of smaller droplets, but the radii of the drops that dominate the extinction and scattering characteristics are in the range from 5 μ to 20 μ.

This means that the *extinction is virtually constant* throughout the ultraviolet, visual, and near-infrared region. Just where the maximum of the extinction curve will fall is difficult to predict, as the rule $\lambda_{max} = a$ holds only for refractive index 1.33, and the infrared absorption bands modify the refractive index and give it an imaginary part. This difference will at any rate not improve the transparency of a cloud in the far infrared, for the extinction for absorbing spheres (Fig. 54, sec. 14.22) reaches its maximum for a smaller value of x (larger λ) than that for dielectric spheres (Fig. 24, sec. 10.4). The entire question has little interest as the wavelength region from 5 to 20 μ is unattractive to work in anyway, and the selective absorption by water vapor and other air molecules is large.

Good measurements in the wavelength range 0.4 μ to 5.5 μ have been made by Arnulf, Bricard, and Veret (1950). The extinction remains constant within ± 10 per cent in this range.

Measurements are feasible again in the millimeter and centimeter waves. Cloud drops and even raindrops are then $<\lambda$, and the Rayleigh formula gives a fair approximation (sec. 20.41). In the present section

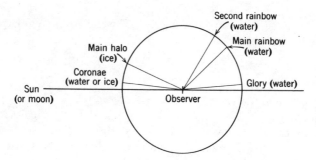

Fig. 97. Survey of the most prominent optical phenomena due to water drops and ice crystals in the atmosphere.

we shall consider only light in the visual and near-infrared region ($\lambda < 2\,\mu$).

As the extinction curve is not a good criterion for drop size, we may consider the scattering characteristics. Of course, direct measurements, for instance, by high-speed photography of individual drops, are also quite good, but they miss the advantage (compare sec. 18.2) of giving a rapid survey of the dominant drop size in a fairly large volume of air. The overall scattering characteristics are the same for any drop with $a \gg \lambda$, except for the following phenomena that are size-dependent:

    A. Diffraction coronae (secs. 8.31 and 13.41),

    B. Rainbows (sec. 13.2),

    C. Glory (sec. 13.3).

Fig. 97 gives a rapid orientation of these phenomena. More details may be read from Fig. 42, sec. 13.11.

The angle between successive diffraction rings (A) and the angle between glory rings (C) are proportional to x^{-1}, or to λ/a. The angles between successive maxima ("supernumerary bows,") at the bright side of the rainbow (B) and also the shift of the rainbow from the position according to geometrical optics are proportional to $x^{-2/3}$ or to $(\lambda/a)^{2/3}$. Fig. 47 gives a survey of the positions of those maxima.

All three phenomena belong to the strikingly beautiful phenomena of nature. The incident light is then sunlight, consisting of radiation from a wide wavelength region and, moreover, coming from a source that subtends an angle of about half a degree. The resulting colors, as they appear to the normally color-sensitive eye, have been computed for diffraction coronae and rainbows by various authors (Prins and Reesinck, 1944; Buchwald, 1943). The results of these computations may be used to estimate the drop size from the coronae and rainbow observed in nature. The usual but cruder method is to assume that the red or

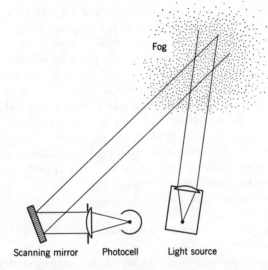

Fog

Scanning mirror   Photocell   Light source

Fig. 98. Instrument for measuring drop sizes by scanning the intensity in an artificial rainbow (after Malkus, Bishop, and Briggs).

reddish bands agree approximately with the minima in the pattern for green light, to which the eye is most sensitive. Fraunhofer gave as the "effective wavelength of white light" corresponding to this rule, 0.57 μ (Pernter-Exner, 1910, p. 460).

If really quantitative results are desired it is better to take an artificial point source and to select a narrow band of wavelengths by means of a filter or spectrograph. All three phenomena have been used in this way to measure drop sizes in clouds and fogs. A possible arrangement for the instruments devised by Malkus, Bishop, and Briggs (1948) is sketched in Fig. 98.

In quantitative interpretations of the light distribution in the *diffraction coronae* we have to be aware of the phenomena of anomalous diffraction (sec. 11.33). The height of the second maximum (the first bright ring) and the position of the first minimum are subject to strong fluctuations for any $x < 50$, i.e., $a < 4 \mu$. Details may be read from Figs. 51 and 52. The height of the next ring and the positions of the next minima may not attain their normal values before $x = 100$, i.e., $a = 8 \mu$. Beyond the domain of Fig. 52, the simple interference theory of sec. 11.31 is about correct.

These considerations show that, as a rule, the diffraction coronae cannot be interpreted with the simple textbook theory at angles $> 5°$. This is also illustrated by the spectral photometry of diffraction coronae of water drops made by Teucher (1939). His data are for two angles, $\theta = 9°10'$ and $14°20'$, and cover the range $a = 4.4 \mu$ to 11μ, $\lambda = 0.43 \mu$ to 0.66μ, i.e., $x = 40$ to 160. This is precisely the range in which the anomalous behavior of the second and third maxima occurs.

A few quantitative observations on the *glory* have been discussed in sec. 13.33 as they were needed in support of the theoretical approach, for this is the only phenomenon for which a complete theory is lacking. For further observations on natural glories, see Peppler (1939), Diem (1942), and Ahrenberg (1948). There seems no reasonable doubt that the theory proposed in secs. 13.31 and 17.42, namely, the radiation from a toroidal wave front originating from grazing incidence, one internal reflection, and a piece of surface wave, is essentially correct. The differences with the diffraction coronae are:

a. Different ratios of the radii of the rings.

b. Strong polarization.

c. More rings seen, as intensity decreases as $(180° - \theta)^{-1}$ against the law $\sim \theta^{-3}$ for diffraction coronae.

Unfortunately, no measurements have come to the author's attention in which more than one of the size criteria (coronae, rainbow, glory) were applied to the same fog or cloud. Such measurements are obviously desirable as a further check on the theory of the glory.

We may finally mention some investigations in which especially the polarization of the scattered light was investigated. Lyot (1929) used a very sensitive visual polarimeter for the investigation of light scattered by various solid surfaces and aerosols (compare sec. 19.3). His results for droplets with estimated radii of 1 mm and 0.25 μ are shown in Fig. 99. The dashed curve in this figure represents the data of Table 21 (sec. 13.11). Striking features are the negative polarization of the twice refracted component, the positive polarization of the reflected light, and superposed on it the two positive peaks due to the first and second rainbow. The curve gradually changes with decreasing size, as illustrated by the experimental curve for $2a = 5 \mu$ ($x = 30$). The rainbows are washed out and shifted as predicted in sec. 13.24 (Fig. 47). Although numerical data for this range of x have been computed, a detailed comparison of experiment with theory cannot be made owing to interpolation troubles, as explained in sec. 13.12.

A similar but less complete investigation was made by Pokrowski (1927). In the first paper he measures the polarization of light scattered by very fine droplets produced by condensation of vapor streaming from a nozzle. In the second paper a similar investigation is made for coarse droplets (diameter $> 100 \mu$) produced by an atomizer. The results of the latter work are shown also in Fig. 99; they probably are less accurate than the measurements of Lyot for intermediate sizes, which are not shown.

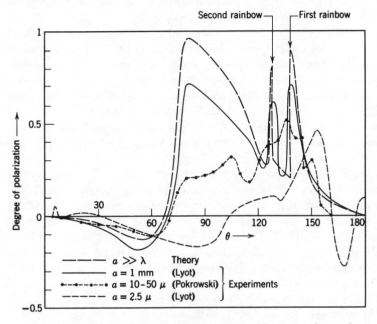

Fig. 99. Degree of polarization of light scattered by water drops of various sizes, according to theory and experiments.

Bricard (1940) published a study of natural fogs. The predominant radii found by him are

Stratus:	$a = 4.2 \mu$
Cumulus:	$a = 5.4 \mu$
Stratocumulus:	$a = 7.6 \mu$
Nimbostratus:	$a = 9.8 \mu$.

In the optical part of this work he reviews the various classical theories; only his explanation of the glory seems inacceptable. Two shorter papers (Bricard, 1941a,b) report on intensity data for the angle ranges $\theta = 10°$ to $50°$ and $\theta = 140°$ to $172°$. Bricard finds the polarization of the rainbows in agreement with theory and with the laboratory measurements shown in Fig. 99, viz., positive, i.e., with the electric vector tangent to the arcs of the rainbow. Further, he reports in this paper that the maxima have shifted to the place of the minima if the rainbow is observed through a sheet of Polaroid that transmits only the weaker component ($E \perp$ arcs). An explanation for this phenomenon was suggested in sec. 13.24. Somewhat disconcerting is Bricard's report that the rings of the glory have the same sign of polarization as the rainbows. This is opposite to the laboratory and open-air data reported in sec. 13.33.

So far the clouds (cirrus, cirrostratus) consisting of small *ice particles* have not been mentioned. For these clouds the same fundamental remark holds, namely, that the sizes are $\gg \lambda$ so that the laws of geometrical optics may be applied.

The hexagonal form of the ice needles, combined with their tendency to assume preferential orientations in a quiet atmosphere, causes the beautiful halo phenomena. For a thorough description and theory we may refer to the books of Pernter-Exner (1910), Minnaert (1940), Visser (1943).

The diffraction coronae caused by ice needles are similar to those caused by water drops but not quite the same. A glance at Table 6 (sec. 7.4) shows that the minima in the diffraction pattern of cylinders [function $E(u)$] are equally spaced whereas in the pattern for spheres [function $F(u)$] the angle from the centre to the first minimum is considerably larger than from the first to the second minimum. This difference must be noticeable also in the observed coronae around the sun and moon, although a direct comparison is made difficult because of the white light and the extension of the source.

It has been thought for theoretical reasons that the intensity in the outer rims of an ice-cloud corona would fall off less steeply than in a water drop corona. Meyer (1950) showed that this conclusion is incorrect; his argument is repeated in sec. 8.32. Meyer also gives observational evidence against the existence of any glory phenomenon in ice clouds.

20.4. Radar Meteorology and the Attenuation of Microwaves

The use of ever shorter radar waves (10 cm, 3 cm, 1.25 cm, 6 mm) during and after World War II made the influence of rain and fog more and more noticeable. Two effects are of practical importance.

1. The attenuation of the radiation by rain and fog reduces the power in one-way transmission, and even more so of an echo as the radiation travels the path twice. The same effect necessitates corrections in radio astronomy at very short wavelengths.

2. The backscatter from clouds or rain storms is received as a radar echo. The storms seen on a radar screen may at times blot out the echoes from ships, etc. The storms themselves may be classified, and their development studied by this method. This is the new specialized field of radar meteorology.

20.41. Attenuation by Fog and Rain

Detailed reviews with many tables and figures have been published by Ryde (1947) and Goldstein (1951). These tables give estimates of all relevant quantities for ready use. In order to show which important factors determine these results, we shall in the following discussion give a few examples and comments.

For very small drop sizes the formulae of chap. 6 may be applied, viz., as water is a partly absorbing material, the formulae of sec. 6.13. They show that the extinction is determined mainly by absorption and that it is proportional to volume. For such particles the attenuation by a cloud is proportional to its liquid water content and thus may be expressed in db/km per g/m³. Table 43 shows Ryde's values for this quantity. They are very strongly dependent on temperature as the imaginary part of m^2 changes strongly with temperature. The attenuation is roughly twice the tabulated value for 0° C and half this value for 40° C. The author has estimated from Lowan's tables at what value of $x = 2\pi a/\lambda$ the deviation from this volume proportionality becomes 10 per cent. These values (see also Table 25, sec. 14.21) and the corresponding drop radii are also given in Table 43.

Table 43. Microwave Attenuation by Very Small Water Drops

λ (cm)	Attenuation by Small Drops of 18° C (db/km per g/m³)	Deviation Becomes 10% for $x = 2\pi a/\lambda$	a (cm)
0.2	7.14	0.18	0.006
0.5	1.65	0.12	0.010
0.7	0.876	0.10	0.011
1.0	0.438	0.09	0.014
1.25	0.280	0.08	0.016
2.0	0.112	0.06	0.019
3.0	0.050	0.05	0.024
5.0	0.0178	0.03	0.024
10.0	0.0045	0.03	0.032

This means that for most *clouds* and *fogs* (drop radii < 0.01 cm $= 100$ μ) the approximation by the volume law is satisfactory but that all *rains* require a computation by means of further terms of the Mie series. Even a fine drizzle, in which drops of $a = 0.05$ cm are common, is above the limit for $\lambda = 10$ cm; here the small-particle approximation cannot be applied in spite of the fact that the wavelength is 200 times the radius!

For wavelengths > 10 cm it might be a better approximation to neglect the imaginary part and to use the scattering cross sections for $m = 9$ invariably. The practical need for a formula is small, as the attenuation at these wavelengths is negligible anyway.

Increasing drop size gives initially a larger extinction cross section for a given weight. The ratio reaches a maximum if a is 0.01λ to 0.02λ, ($|mx| \approx 1$) and then drops slowly down. This follows from the values of $Q(x)/x$, where $Q(x)$ is the efficiency factor for extinction (Fig. 60, sec. 14.31). As the extinction cross section of a drop is $\pi a^2 Q(x)$ cm^2 and the weight is $(4/3)\pi a^3$ gram, the cross section is $(3\pi/2\lambda) Q(x)/x$ cm^2/gram. By adding a factor 0.434, this becomes the extinction coefficient in db/km per g/m^3, the limiting values of which for small x are given in Table 43.

The wide range of drop sizes makes it a little hard to obtain a quick survey of what various rains will do. We shall show by one example what integrations are involved.

Let a be the radius and let $n(a)$ da be the number of drops per m^3 in the range of radii da. In Table 44 the interval is $da = 0.025$ cm. All data in this table are based on the paper by Goldstein. The integration for the total rate of precipitation is shown in column 2. The contributions of each size interval are proportional to the product of the relative volume 10^{-6} $(4/3)\pi a^3 \cdot n(a)$ da and the velocity of fall v. The proportionality constant is 3.6, and the total rate of precipitation is 100 mm/hr (very heavy rain). Division by the values of v given in column 3 gives the relative volume itself, and upon integration the mass of liquid water $M = 3.95$ g/m^3. The density of water is taken 1 g/cm^3. Dividing the integral in column 2 by the integral in 4 we obtain the effective velocity of fall 7.0 m/sec. Column 5 gives the individual values of

$$\frac{C_{ext}}{4/3 \, \pi a^3} = \frac{3\pi}{2\lambda} \frac{Q(x)}{x}$$

for $\lambda = 1$ cm and for the refractive index appropriate to $\lambda = 1$ cm at 18° C. Multiplication of the individual values in columns 4 and 5 gives those in 6. Upon integration they give the total cross section in cm^2 per m^3. Division of the integrals in 4 and 6 gives the "effective" value in column 5. The extinction coefficient as used throughout this book

Table 44. Attenuation of Waves of $\lambda = 1.0$ cm by a Very Heavy Rain at 18° C

a (cm)	$(4/3)\pi a^3 v n(a)\,da$ $(10^{-6}$ m/sec)	v (m/sec)	$(4/3)\pi a^3 n(a)\,da$ (g/m³)	$\dfrac{C_{ext}}{(4/3)\pi a^3}$ (cm²/g)	$C_{ext}n(a)\,da$ (cm²/m³)
0.025	0.3	2.1	0.14	1.9	0.3
0.05	1.1	3.9	0.28	4.4	1.2
0.075	2.5	5.3	0.47	8.9	4.2
0.10	3.9	6.4	0.61	15.3	9.3
0.125	4.7	7.3	0.64	15.4	9.8
0.15	5.0	7.9	0.63	13.8	8.7
0.175	4.2	8.4	0.50	12.4	6.2
0.20	2.5	8.7	0.29	10.8	3.1
0.225	1.6	9.0	0.18	9.5	1.7
0.25	0.8	9.2	0.09	8.5	0.8
0.275	0.6	9.4	0.06	7.7	0.5
0.30	0.4	9.5	0.04	7.2	0.3
0.325	0.2	9.6	0.02	6.7	0.1
Integral	27.8		3.95		46.2
"Effective value"		7.0		11.7	
Multiplier	3.60		1		0.434
Result	$p = 100$ mm/hr		$M = 3.95$ g/m³		$\gamma' = 20.1$ db/km

thus is $\gamma = 46.2 \times 10^{-6}$ cm^{-1}, but upon multiplying by 0.434×10^6 the more familiar units are obtained: $\gamma' = 20.1$ db/km.

This completes the discussion of the example.

The only parameter that can be readily measured in practice is the precipitation rate p, and even this is subject to large local and temporal variations. It is therefore convenient to compute the ratio of γ' to p. The results reported by Goldstein and completed by interpolation are given in Table 45.

It is seen from this table that the attenuation is not strictly proportional with the precipitation rate in the sample rains on which the computations were based. For $\lambda < 1$ cm the ratio γ'/p goes down with increasing precipitation, and for $\lambda > 1$ cm it goes up. The explanation lies in the fact that the drops in a drizzle are smaller on the average than in a moderate rain and larger again in a very heavy rain. This has two effects: (a) a change in velocity of fall, by which the ratio of M to p decreases with increasing precipitation, and (b) a change in the factor $C_{ext}/(4/3)\pi a^3$.

Only the latter effect is important. At 1 cm the most abundant drops are also most effective, so that a shift to either side has

Table 45. Ratio of Attenuation to Precipitation Rate at 18° C

	Drizzle 0.25 mm/hr	Moderate Rain 4 mm/hr	Very Heavy Rain 100 mm/hr
	γ'/p	(db/km per mm/hr)	
$\lambda = 0.3$ cm	1.20	0.65	0.33
$\lambda = 0.5$	0.64	0.50	0.29
$\lambda = 0.6$	0.42	0.40	0.27
$\lambda = 1.0$	0.15	0.19	0.20
$\lambda = 1.25$	0.086	0.12	0.15
$\lambda = 2.0$	0.024	0.042	0.06
$\lambda = 3.0$	0.0089	0.017	0.032
$\lambda = 5.0$		0.0028	0.005
$\lambda = 10.0$		0.0003	0.0003

not much effect. At 3 cm the most abundant drops have a smaller size than those which are most effective; consequently, the increasing size in a heavier rain will increase the effectiveness. At $\lambda = 0.3$ cm the most abundant drops are already larger than those which are most effective; consequently, the increasing size in a heavier rain will decrease the effectiveness.

20.42. Radar Observations of Rain and Clouds

The backscatter of microwaves from raindrops and ice crystals is of great meteorological importance since it makes radar meteorology possible.

A striking feature of the computed radar cross sections of water drops (sec. 14.32) is the existence of large fluctuations (Fig. 61). These fluctuations are incorrectly called resonances in Goldstein's review; the correct explanation by means of the surface wave creeping around the drop is discussed in sec. 17.41. The practical importance of these fluctuations for present-day radar meteorology is small, as the first maximum occurs at $x = 2\pi a/\lambda = 1.0$, which means drops of radius 1 mm for 6 mm radar. Hence, in most circumstances the common drops just reach (or do not quite reach) the first maximum, and only very large drops reach the first minimum.

For very small drop sizes we may use the Rayleigh approximation in which

$$\frac{\sigma}{G} = \frac{3}{2} Q_{sca} = 4 \left| \frac{m^2 - 1}{m^2 + 2} \right|^2 x^4.$$

Here σ is the radar cross section, and G the geometrical cross section πa^2. This approximation gives results of the correct order of magnitude up to $x = 1$. For instance, comparing the curve of Fig. 61 ($\lambda = 3$ mm, $m = 3.41 - 1.94i$) with the Rayleigh approximation we find that the deviation stays below 20 per cent for $x < 0.93$. This limit is much higher than the values of x at which the *attenuation* formula for small drops breaks down (sec. 20.41). A graph of the ratios of the actual value of σ to the Rayleigh approximation for three values of m is shown in Fig. 100.

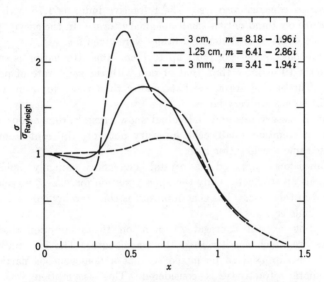

Fig. 100. Ratio of the radar cross section of a water drop to that computed by the Rayleigh approximation for three wavelengths. Abscissa is $x = 2\pi a/\lambda$, $a =$ radius of drop.

The fact that the radar cross section may become more than twice the value following from the Rayleigh approximation is due to magnetic dipole resonance, as has been explained in sec. 14.31. The resonance effect is striking only if the imaginary part of the refractive index n' is small. As n' decreases strongly with increasing temperature (for $\lambda \geqslant 3$ cm) the resonance effects must increase with increasing temperature. If it is impractical to repeat the detailed computations for many values of m, the theory in sec. 14.31 may give an impression of the expected magnitude of the effect.

Since a factor 2 is unimportant in most practical computations, the Rayleigh approximation may be used up to $x = 1$. The large drops in a rain are relatively more important in the echo than in the attenuation (sec. 20.41).

Numerical example: We may quote the following values for a typical summer thunderstorm from Goldstein's report. Precipitation rate: 30 mm/hr; liquid water content: $M = 1 \, \text{g/m}^3$. Effective drop radius for echo computations: $a = 0.18$ cm, which gives $n = 41$ drops per m^3. At $\lambda = 9.2$ cm we obtain $x = 0.123$, and on Rayleigh's approximation: $\sigma/G = 4 \times 0.93 \times (0.123)^4 = 8.5 \times 10^{-4}$, and $\sigma = 8.6 \times 10^{-9} \, \text{m}^2$. The cross section per unit volume thus is $n\sigma = 3.5 \times 10^{-7} \, \text{m}^{-1}$, i.e., 65 db below the level of 1 meter^{-1}.

The problems of radar meteorology are not limited to rain drops. We have to consider also *ice*. The refractive index is 1.75 with little dependence on λ and only a small imaginary term. In the survey paper that Ryde wrote in 1947 detailed tables were given for:

a. Hailstones, spherical, radii up to 1 cm; the attenuation is smaller by a factor 10 or more than that of rain with the same rate of precipitation. The echoes are somewhat smaller than those for rain, except when the stones are very large.

b. Small ice crystals and individual snow crystals, dimensions up to 2 mm; attenuation small, echo intensity not very different from rain with the same precipitation rate.

c. Snowflakes composed of crystal aggregates, roughly spherical, dimensions up to 2 cm. Only the approximation for size $\ll$ wavelength was used in this case. The echo intensity is considerably enhanced over that of single crystals.

d. Melting crystal aggregates, again on the assumption that the particles are much smaller than the wavelength. The ice and water and air are supposed to be mixed so that a homogeneous particle of intermediate refractivity is considered. This assumption was later criticized (sec. 20.43). An extension to particles of larger size might be made with the ordinary Mie theory (chaps. 10 and 14).

20.43. The Bright "Melting Band" in Radar Echoes

Among the problems that have attracted the attention of later writers we may mention especially the *bright bands* seen in precipitating clouds if they are observed by ground radar. A good summary of the meteorological conditions under which these bands occur is given by Bowen (1951). The most common one, often called the "melting band," occurs at the bottom of the cloud where ice particles melt before they fall in the form of rain. A less conspicuous band sometimes occurs at a higher level and may coincide with the levels where supercooled water particles freeze. A quantitative study of the melting band was made by Austin and Bemis (1950).

The intense echoes from the bright band are due to a sharp peak in the echo area per unit volume at the melting level. At least four effects have been suggested as important in causing this peak, and it is generally believed that a quantitative theory may be derived from the combined effect of these phenomena. They are:

a. The increase of refractive index upon melting.

b. The increase in fall velocity upon melting.

c. Coalescence.

d. A decrease of radar cross section upon attaining spherical shape.

For a quantitative survey of the manner in which these effects cooperate we may refer to Labrum (1952b). It appears from this study that an inconspicuous bright band would already be caused by effects a and b. This is due to the fact that, comparing the echoes from above and below the melting level, the larger radar cross section of water drops over ice particles is offset by their larger velocity of fall, but the final velocity of fall is only gradually reached after the particles have already attained their large cross section. Coalescence (c) gives an additional factor of increase over the echoes from higher levels. Effects of non-spherical shape (d) give an additional factor of increase over the echoes from lower levels, as has been argued in particular by Atlas, Kerker, and Hitschfeld (1953). A critical discussion would lead too deeply into meteorological problems.

It has proved necessary in this work to compute the radar cross sections of melting snow flakes. As a model an ice sphere covered by a spherical water mantle was considered. The theory for arbitrary ratio of the radii to λ was worked out by Aden and Kerker (1951) and by Güttler (1952). It has been applied to one practical example ($\lambda = 10$ cm, outer radius $= 0.2$ cm, radius of ice core varying) by Kerker, Langleben, and Gunn (1951). The question arises whether the complicated formulae might be replaced by simpler ones in view of the fairly small ratio of radius to wavelength. This is indeed true. In the usual notation we have $x = 2\pi a/\lambda = 0.126$, and by reference to Fig. 100 we find that this may just be small enough to make the Rayleigh approximation (electric dipole scattering) adequate within 10 per cent. The relevant formula for the polarizability α of the body with the spherical shell was given in sec. 6.34. The radar cross section is proportional to the square of α. The results reported by Kerker, Langleben, and Gunn have a character very similar to that shown for hollow water shells in Table 4 (sec. 6.34). In particular, it was found in both cases that a water film one-fifth of the radius thick suffices to make the body scatter almost as effectively as a full water sphere.

The effect contributing to the bright band phenomenon thus is that the melting particles probably go through a stage of maximum effectiveness in which they have two properties causing a high scattering cross section: (1) a non-spherical shape due to the ice core, and (2) a high refractive index due to the water mantle. This effect has been demonstrated by Labrum (1952a) by experiments with melting specimens in a wave guide.

The work on the effect of particle shape by Atlas, Kerker, and Hitschfeld is based entirely on the approximation for small sizes. For that reason their theoretical discussion of the effect of random orientation must be identical to that of Rayleigh (1918), which has been reported in sec. 6.52 (for scattered intensity) and 6.53 (for attenuation). For $\theta = 180°$ we find at once from the formulae of sec. 6.52 the intensity

$$I = k^4 r^{-2}(4A + B)I_0,$$

independently of the polarization of the incident light, and the ratio

$$\frac{\text{Intensity of cross-polarized component}}{\text{Intensity of parallel component}} = \frac{A - B}{3A + 2B},$$

if the incident light is linearly polarized. The latter ratio is called the "depolarization" by Atlas, Kerker, and Hitschfeld. The substitution of the polarizability values for oblate or prolate spheroids from sec. 6.32 into the definitions of A and B (sec. 6.52) is straightforward. From these formulae the author found in a sample case that, for $m = 1.75$ (ice) and axial ratios 1 : 5, the relative values of $4A + B$, which indicate the gain in radar cross section over a sphere of the same volume, are the following.

Spheres	, 1.00,	depolarization = 0.000
Prolate spheroids (1:5),	1.20,	depolarization = 0.024
Oblate spheroids (1:5),	1.31,	depolarization = 0.025.

The effects are far more pronounced if the refractive index of water is used, as might be effectively correct if the ice rods or disks are covered with a thin film of water. In the latter case a factor 7 is gained over equivolume spheres by prolate, and a factor 4 by oblate spheroids of random orientation, if the axial ratio is 1 : 5. The depolarizations are 0.20 (20 per cent) for prolates and 0.08 for oblates. Of course, even greater gains may be realized if the particles have a preferred orientation suitable to present a large polarizability to the electric vector of the incident beam. Numerical values are found in the paper cited.

An earlier report by Kerker (1950) contains the same theory and considers also the cross-polarized component due to multiple scattering. This is found to be unimportant.

It may finally be noted that the theory of sec. 6.52 also contains the ready formulae for incident light of any arbitrary state of polarization. Interesting results can be obtained with circular polarization. We have already remarked (sec. 5.32) that, for symmetry reasons, spherical particles of arbitrary size and illuminated by circularly polarized radiation can show in their backscatter only radiation with the same state of circular polarization. This does not hold true for non-spherical particles. It has thus been suggested (Kennaugh, 1951; White, 1951) that radar systems with emission and reception in opposite circular polarizations may be used as a discrimination against spherical drops. This would furnish a sensitive means of detecting effects of non-spherical shape or of secondary scattering.

References*

General reviews on meteorological optics and studies devoted to multiple scattering by pure air (sec. 20.1) are found in:

J. M. Pernter and F. M. Exner, *Meteorologische Optik*, Vienna, W. Braumüller, 1910.

C. Dorno, *Physik der Sonnen und Himmelsstrahlung*, Braunschweig, Vieweg, 1919.

M. Minnaert, *Light and Colour in the Open Air*, London, G. Bell, 1940.

W. E. K. Middleton, *Visibility in Meteorology* (2nd ed.), Toronto, Univ. of Toronto Press, 1941.

H. C. van de Hulst, "Scattering in the Atmospheres of the Earth and the Planets," chap. 3, *The Atmospheres of the Earth and Planets* (1st ed), G. P. Kuiper, ed., Univ. of Chicago Press, 1949. (Same in 2nd ed., 1952).

Z. Sekara, *Compendium of Meteorology*, T. F. Malone, ed., p. 79, Boston, *Am. Meteor. Soc.*, 1951.

H. Neuberger, *ibid.*, p. 61.

S. Chandrasekhar, *Radiative Transfer*, Oxford, Oxford Univ. Press, 1950.

S. Chandrasekhar and D. D. Elbert, *Trans. Am. Phil. Soc.*, **44**, 643 (1954).

D. Deirmendjian and Z. Sekara, *Tellus*, **6**, 382 (1954); *Nature*, **175**, 459 (1955).

H. C. van de Hulst, *Astrophys. J.*, **108**, 220 (1948).

W. E. K. Middleton, *J. Opt. Soc. Amer.*, **39**, 576 (1949).

Papers on the extinction by atmospheric haze (sec. 20.21) are:

F. Volz, "Thesis Mainz," 1954 (*Ber. deut. Wetterdienstes*, Nr. 13, Band 2, 1954).

C. Junge, *Ann. Meteorol.*, Beiheft (1952).

A. Vassy and E. Vassy, *J. Phys.*, **10**, 75, 403, and 459 (1939).

H. Dessens, *Ann. géophys.*, **2**, 68 (1946); **3**, 68 (1947).

F. W. P. Götz, *Verhandl. schweiz. naturforsch. Ges.*, p. 88, 1944.

A. Angström, *Geograf. Ann.*, **11**, 156 (1929).

W. Schüepp, *Arch. Meteorol. Geophys. u. Bicklimatol.*, Ser. B, **1**, 257 (1949).

W. Heller, H. B. Klevens, and H. Oppenheimer, *J. Chem. Phys.*, **14**, 569 (1946).

W. Heller and E. Vassy, *J. Chem. Phys.*, **14**, 505 (1946).

V. K. La Mer, *J. Phys. & Colloid Chem.*, **52**, 65 (1948).

* Given in the order in which they are cited in this chapter.

The scattering by atmospheric haze (sec. 20.22) is discussed by:

E. O. Hulburt, *J. Opt. Soc. Amer.*, **31**, 467 (1941).

E. Reeger and H. Siedentopf, *Optik*, **1**, 15 (1946).

K. Bullrich and F. Möller, *Optik*, **2**, 301 (1947).

F. Volz, op. cit.

A. Strzalkowski, *Acta Astronomica* (Warszawa) **5**, 95 (1955).

F. Volz, *Photographie u. Wiss.* (Agfa Mitteilungen), **3**, 3 (1954).

F. Möller, E. de Bary, and G. Krog, *Geofisica pura e applicata* (Milano), **26**, 141 (1953).

The aerosols from volcanic eruptions and forest fires (sec. 20.23) were studied by

S. Melmore, *Observatory*, **73**, 105 (1953).

F. W. P. Götz, op. cit.

C. Dorno, op. cit.

H. S. Hogg, *J. Roy. Astron. Soc. Can.* **44**, 241 (1950).

G. A. Bull, *Meteorol. Mag.* **80**, 1 (1951).

W. G. Guthrie, *Irish Astron. J.*, **1**, 122 (1950).

R. Wilson, *Monthly Notices Roy. Astron. Soc.*, **111**, 478 (1951).

F. Volz, op. cit.

For optical phenomena in large drops (sec. 20.3) see chap. 13 and

A. Arnulf, J. Bricard, and C. Veret, *Recherche Aéronaut.* No. **15**, 27 (1950).

J. A. Prins and J. J. M. Reesinck, *Physica*, **11**, 49 (1944).

E. Buchwald, *Ann. Physik*, **43**, 488 (1943).

S. Günther, *Optik*, **5**, 240 (1949).

J. M. Pernter and F. M. Exner, op. cit.

W. V. R. Malkus, R. H. Bishop, and R. O. Briggs, NACA *Techn. Notes* 1622, 1948.

R. Teucher, *Physik. Z.*, **40**, 90 (1939).

W. Peppler, *Wetter*, **56**, 173 (1939).

M. Diem, *Ann. Hydrograph.*, **70**, 142 (1942).

D. L. Ahrenberg, *J. Opt. Soc. Amer.*, **38**, 481 (1948).

H. C. van de Hulst, *J. Opt. Soc. Amer.*, **37**, 16 (1947).

J. Bricard, *Ann. phys.*, **14**, 148 (1940); *Compt. rend.*, **213**, 136 (1941a); 495 (1941b).

R. Meyer, *Ber. deut. Wetterdienstes U.S. Zone*, Nr. 12, 1950.

B. Lyot, *Ann. observ. Paris-Meudon*, **8**, 1, pp. 125-134, 1929.

G. Pokrowski, *Z. Physik*, **43**, 394 and 769 (1927).

S. W. Visser, *Optische verschijnselen aan de hemel*, Gorinchem, Noorduyn, 1943.

General introductions into the field of radio and radar meteorology (secs. 20.41 and 20.42) are:

J. W. Ryde, "Meteorological Factors in Radio Wave Propagation," *Rept. Phys. Meteorol. Soc.*, London, p. 169, 1947.

H. Goldstein, *Propagation of Short Radio Waves*, D. E. Kerr, ed., parts of chaps. 7 and 8. Radiation Laboratory Series, No. 13, New York, McGraw-Hill Book Co., 1951.

J. S. Marshall, W. Hitschfeld, and K. L. S. Gunn, "Advances in Radar Weather," *Advances in Geophysics*, vol. 2, chap. 2, H. E. Landsberg, ed., New York, Academic Press, 1955.

Discussions relevant to special problems, in particular the bright melting band (sec. 20.43), were given by:

E. G. Bowen, *J. Atm. and Terrest. Phys.*, **1**, 125 (1951).

P. M. Austin and E. C. Bemis, *J. Meteorol.*, **7**, 145 (1950).

N. R. Labrum, *J. Appl. Phys.* **23**, 1324 (1952b).

D. Atlas, M. L. Kerker, and W. Hitschfeld, *J. Atm. and Terrest. Phys.*, **3**, 108 (1953).

A. L. Aden and M. L. Kerker, *J. Appl. Phys.*, **22**, 1242 (1951).

A. Güttler, *Ann. Phys.*, **11**, 65 (1952).

M. L. Kerker, P. Langleben, and K. L. S. Gunn, *J. Meteorol.*, **8**, 424 (1951).

N. R. Labrum, *J. Appl. Phys.*, **23**, 1320 (1952a).

M. L. Kerker, "Stormy Weather" Research Group, *McGill Univ. Research Rept.* MW-1, 1950.

E. M. Kennaugh (Ohio State Univ. Research Foundation), Paper at I.R.E. Convention, New York, March, 1951.

W. D. White (Airborne Inst. Lab., Mineola, L.I., N.Y.), Paper at I.R.E. Convention New York, March, 1951.

21. APPLICATIONS TO ASTRONOMY

21.1. Introduction

The development of the light-scattering theories discussed in this book has been closely linked with astronomy from the very beginning. Even before 1900 computations on the radiation pressure exerted on perfectly conducting spheres were made with a view to the theory of comet tails. Since 1930 much attention has been paid to the optical properties of the interstellar grains. Later work on the scattering by cylinders was aimed at explaining the interstellar polarization.

The need for light-scattering theories in astronomy is obvious, as the supplementary methods used in other fields, weighing and counting, electron-microscope studies, velocity of fall in air, etc., are all lacking. It is doubtful whether we would know of the existence of either *interplanetary* or *interstellar* dust, if their scattering and/or extinction had been unobservable. As it is, however, the basic data obtained from interpreting these optical observations are complemented by physical and chemical theories of the properties of such grains in the ambient gas and radiation field. The present review is incomplete, since it summarizes only the optical properties and neglects the physical chemistry.

A subject of practical importance for astronomy is the *scintillation* of the stars. It arises from irregular variations in the refractive index in the atmosphere, i.e., in the troposphere in the optical twinkling and in the ionosphere in the twinkling of the radio stars. The theory in either case is beyond the scope of this book, as it would be artificial to distinguish individual "blobs" as single particles. In this approximation, a scattering theory of the type of chap. 11, combined with a formula for multiple scattering, is called for. From the extensive literature based on more rigorous methods we may cite papers by Pekeris (1947), Booker and Gordon (1950), Hewish (1951), Chandrasekhar (1952), and Fejer (1954).

It seems certain that a similar phenomenon on a much larger scale occurs in the outer corona of the sun. The radio waves emitted by the deeper layers, or by a galactic source that happens to be covered by the corona, are deviated from their original course by a great many local variations in the electron density. This causes a blurring and widening of the image of the radio sun, or of the occulted point source, on the meter wavelengths. The blobs are estimated to have diameters between 1 and 10^5 km. (Hewish, 1955).

21.2. Planetary Atmospheres

Ten planets and satellites in the solar system are known to have an atmosphere. Four planets, Jupiter, Saturn, Neptune, and Uranus, have very thick atmospheres; the first one has been studied most thoroughly. The earth and Venus have moderately thick atmospheres. The four bodies that are certain to have tenuous atmospheres are Pluto, Triton, Mars, and Titan, of which Mars has rightly claimed the greatest attention. The physical meteorology of several of these planets may be as complicated as that on the earth. In all of them, a minute fraction of liquid or solid particles in the gaseous atmosphere may change or even dominate the appearance of the planet. As all direct sources of information, such as spectrographic and photometric studies of the reflected or emitted radiation are affected by the presence of such particles, their study is of the greatest importance. We shall briefly review the present status for the planets Jupiter, Venus, and Mars. Most of the data quoted are from G. P. Kuiper's book (2nd ed.) *Planetary Atmospheres* (1952).

Jupiter. Kuiper concludes from a variety of considerations that the base of the visible atmosphere of Jupiter is formed by the top of a cloud layer consisting of solid ammonia particles. The temperature is about 160° K. The top of this layer may not be flat but may consist of many individual clouds with large interspaces. No estimate of the size of the particles has been made, for the polarization curve for diffuse reflection by the cloud layer can be observed only from $180° - \theta = 0$ to $11°$. It is likely that some of the blue scattering occurring above the cloud layer is not due to the gaseous atmosphere but to ammonia particles with sizes $\ll \lambda$, for ultraviolet photographs show occasional veils and light clouds not seen on infrared photographs. Such particles would evidently obey the laws of Rayleigh scattering (chap. 6).

Venus. Also the visible surface of Venus, which is brilliantly yellow-white (albedo 0.76), is formed by the top of a permanent cloud layer. Here the degree of polarization of the reflected light, which is found to be similar at any time for all illuminated points of the disk, can be measured for phase angles from nearly 0° to nearly 180°. The polarization curve obtained by Lyot (1929) is shown in Fig. 101. It has given rise to much speculation about the chemical composition of the clouds and the range of sizes of their particles.

The optical problem met here is a problem of radiative transfer (or multiple scattering) in a thick layer in which the single particles have complicated scattering diagrams in two polarizations. This problem has not been solved in a numerically useful manner for single diagrams more complicated than Rayleigh scattering. Even in the latter case an enormous amount of analytical and numerical work is required for a precise solution (Chandrasekhar and Elbert, 1954). Useful numerical data for planetary atmospheres based on Chandrasekhar's solutions for simple phase functions have been given by van de Hulst (1952) and Horak (1954).

As a very crude approximation one may assume that any radiation emerging from the atmosphere is unpolarized if it has been scattered two or more times by successive particles. Let this radiation be q

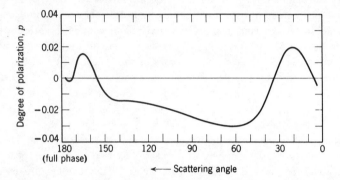

Fig. 101. Degree of polarization of the light reflected by Venus as a function of the scattering angle, according to measurements by Lyot.

times as intense as the radiation emerging after a single scattering. The latter fraction has the degree of polarization of single scattering

$$p = (i_1 - i_2)/(i_1 + i_2),$$

which may be computed from the formulae of this book. The observed polarization then should be $p/(1 + q)$. Clearly, q depends on the directions of incidence and emergence and also on the form of the scattering diagram. Assuming that q is 2.5 to 6, which does not seem unreasonable (van de Hulst, 1952), Lyot interpreted his curve by means of water droplets with diameters about 2.5 μ. The approximate agreement is seen by comparing Fig. 101 with the experimental curve for $2a = 2.5$ μ (not shown in Fig. 99).

Kuiper objects that the yellowish color remains unexplained and that on occasional days the cloud layer of Venus drops so fast that diameters of at least 20 μ are required. His suggestion of fine sand (SiO_2 with iron colorings) has been criticized by Menzel and Whipple (1955).

Another possibility (Kuiper, private communication) may be crystalline NH_4.

Evidently, much remains to be solved. Fuller computations of multiple scattering are needed, and we might learn more from polarization measurements in different wavelengths.

Mars. The studies of the planet Mars have gradually given rise to an enormous collection of observational data and to an equally impressive body of theoretical computations by which the Mars meteorology begins to approach the complexity of the meteorology of the earth. Adequate reviews are available in Kuiper's book (2nd ed., 1952) and in de Vaucouleurs' book (1954). We present here a very brief summary of the data pertinent to the solid particles in the Mars atmosphere.

In visible light Mars shows its solid surface, including the polar caps of a thin layer of H_2O snow and the green and brown areas that may contain life. Blue and ultraviolet photographs, however, show an atmospheric haze, which is definitely not gaseous because of its low albedo in ultraviolet light and because occasional clearings occur. This haze layer is referred to by de Vaucouleurs as "the violet layer."

Three suggestions on the nature of the particles of the violet layer have been made:

a. Ice (solid H_2O), size 0.3–0.4 μ, height 5–10 km (Kuiper).

b. Dry ice (solid CO_2), size 0.3 μ, height 45 km (Hess).

c. Carbon (solid C_3), no suggestions about size or height (Rosen).

One of the observational clues is that the opacity of the violet layer rises fairly sharply if λ decreases from 6000 A to 4500 A. When interpreted on the basis of extinction curves for non-absorbing particles, as given in Fig. 24 (sec. 10.4) or Fig. 32 (sec. 11.22) for spheres and in Fig. 67 (sec. 15.31) for cylinders, this would indeed suggest that $2a(m-1)$ is of the order of 0.2 μ. An estimate of 1.33 for the refractive index m then gives a diameter $2a = 0.6$ μ. The explanation of the maxima and minima in the extinction curve (secs. 11.22 and 13.42) would seem to indicate that particles of a different form, e.g., ice crystals, may show a similar run of the extinction curve as has been computed for smooth cylinders and spheres. The crude data and the small range of wavelengths make it impossible to obtain a more precise interpretation this way. Rosen has suggested that the strong rise of opacity in the blue is not a size effect but is due to the absorption band of carbon particles that has also been observed in N type stars. This suggestion is not supported by physical arguments.

Kuiper has made several size estimates of the haze particles from physical considerations. For instance, particles with a size $2a > 60$ μ

would fall too fast. However, his most accurate estimate is made on purely optical grounds, as follows.

Arguing that at $\lambda = 4700$ A the haze layer is still thin enough to make any contribution of multiple scattering to the reflected radiation negligible, yet thick enough to make the surface virtually invisible, Kuiper explains the data at that wavelength by assuming single scattering by a mixture of ice spheres of different sizes. Both the observed degree of polarization (for which Kuiper adopts $p = +0.05$ at $180° -\theta = 20°$) and the observed reflectivity of the planet (for which Kuiper gives $R = 0.08$) have to be explained.

On the basis of the assumptions made, R may be put equal to the directivity (or phase function) of the particles for $\theta = 160°$. This is defined as the ratio of the energy scattered per unit solid angle in this direction to the average energy scattered per unit solid angle in all directions, i.e., in the notations of the earlier chapters:

$$R = \frac{\frac{1}{2}(i_1 + i_2)I_0 k^{-2}}{Q\pi a^2 I_0/4\pi} = \frac{2(i_1 + i_2)}{Qx^2}.$$

Kuiper gives a table of p and R based on Lowan's tables for $m = 1.33$ and finds by sample computations for different mixtures that a size range centered on $x = 2.0$ to 2.5, i.e., $2a = 0.3$ to 0.4 μ, would fit the observed values of p and R for $\theta = 160°$. A more precise computation might be made by means of an interpolation method, as described in sec. 13.12 (Fig. 43) and used in constructing Fig. 102 (sec. 21.3). In view of the physical uncertainties this seems hardly worth while. It may be noted, however, that data for larger particles than those considered by Kuiper have since been computed by Gumprecht, Sung, Chin, and Sliepcevich (see bibliography, chap. 10). At $x = 10$ these tables give the values for $m = 1.33$, $\theta = 160°$: $p = +0.48$, $R = 0.44$. A relatively small number of particles of this size might dominate the characteristics of the radiation reflected from the planet. The explanation of the strong increase in p and R with respect to the data quoted by Kuiper is that the influence of the main rainbow maximum is felt here. This may be seen from Fig. 47 (sec. 13.24). At $x = 40$ (first rainbow minimum) R has dropped back to 0.16.

Apart from the nearly permanent haze layer occasional clouds occur on Mars. From a careful discussion of all available data de Vaucouleurs distinguishes three types that are indicated by the colors "blue," "white," and "yellow." These terms refer to the methods by which they are observed rather than to their true colors. The main characteristics are:

Blue clouds: highest in the atmosphere, probably denser parts in the haze layer, polarization positive for $180° - \theta < 23°$; nature of the particles uncertain but sizes of the order of 0.4 μ as discussed above.

White clouds: intermediate in height but probably a collective name for various types, one of which may actually be identical to the blue clouds. Those for which the polarization has been measured show negative polarization for $180° - \theta < 20°$ and may consist of coarse ice crystals ($2a = 1$ μ or larger). Compare the curves for water drops shown in Fig. 99, sec. 20.3.

Yellow clouds: low-level clouds, almost certainly consisting of fine desert dust blown up by the winds. No size estimate available.

21.3. Interplanetary Dust and the Zodiacal Light

The general characteristics of the zodiacal light leave no doubt that it consists of sunlight scattered by interplanetary material arranged roughly symmetrically to the planes of the planetary orbits. It is not at once clear, however, what material this is. Initially it seemed obvious to assume coarse dust, or even pieces of rock, as the main scattering agent. On this assumption all scattering would occur by diffuse reflection against the surface of these rocks, which in the simplest case (Lambert's law) may be taken from sec. 8.42; diffraction effects would be unimportant.

The different ideas that are now prevalent have mainly arisen from two facts. (1) The observed degree of polarization of the zodiacal light in the main cone (30° to 60° from the sun) is higher than can be explained by solid particles alone. About half the observed intensity may be due to scattering by free electrons. (2) The assumption that the increase of brightness with decreasing angle from the sun is exclusively due to a spatial increase of density with decreasing distance from the sun cannot be maintained. A substantial part of this increase must be attributed to the Mie effect (predominant forward scattering). At angles $< 5°$, i.e., in the F-corona, most of the light is due to particles with diameters $2a$ of the order of 10 μ and larger; the effect is sufficiently well described by the classical Fraunhofer diffraction formula (secs. 8.31 and 12.32). For angles $> 30°$ the full Mie formulae must be employed. In any event, the scattering diagrams shown in Fig. 25 give enough data for a discussion of the problem. The wide size distribution makes the diffraction rings unnoticeable. It is possible to explain the observed brightness distribution with a fairly wide variety of assumptions about size distribution, composition (albedo), and distribution of the grains in space. For a further review see Minnaert (1955).

An estimate of the intensity of the electron component from the observed polarization of the zodiacal light requires information on the degree of polarization of light scattered by the dust component. For pieces of rock this may by computed by integration from empirical data on the polarization of diffusely reflected light from the surfaces of rocks. For fairly small particles the assumption has often been made that the frequent changes of sign in the degree of polarization following from the Mie formulae (shown clearly in Fig. 44, sec. 13.12) would cancel out upon integration over a reasonable size distribution. This assumption is not justified. It is certainly incorrect in the limit of very large particles (geometrical optics), and a computation is needed to show the result

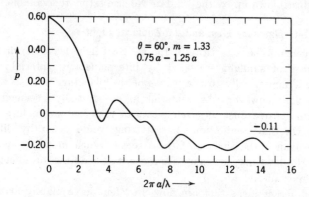

Fig. 102. Degree of polarization of the light scattered in the direction $\theta = 60°$ by a cloud of spherical particles with refractive index 1.33 and sizes ranging from $0.75a$ to $1.25a$;
$$x = 2\pi a/\lambda.$$

for smaller sizes. The results of a sample computation made for $\theta = 60°$ is shown in Fig. 102 (van de Hulst, 1955). It is based on the assumption that the scattered light in each of the directions of polarization (1,2) is proportional to

$$\int_{0.75x_0}^{1.25x_0} i_{1,2}(x)\, x^{-2}\, dx\,,$$

so that the sizes range from $0.75a_0$ to $1.25\,a_0$. It is seen that p oscillates between -0.10 and -0.20 before it approaches the value -0.11 defined by geometrical optics. A further integration (over θ and the space-distribution law) is needed before the astronomical implications can be discussed.

21.4. The Solid Grains in Interstellar Space

About 1930 an old suspicion was verified, namely, that the galactic system contains interstellar material causing a strong extinction of any distant object. It has since become clear that this dense fog renders optical observations of the distant parts of the galaxy (e.g., its nucleus) impossible. All observations show that the distribution of this material is extremely irregular and that much of it is concentrated in dense clouds. On a larger scale, the cloud complexes are definitely associated with spiral arms. This makes it difficult to define an average extinction. In our neighborhood it may be 2.5 mag/kpc for visual light (5500 A). Converting by means of

$$1 \text{ magnitude} = 4 \text{ decibels} = 0.92 \text{ factors } e$$
$$1 \text{ kpc} = 1000 \text{ parsec} = 3.08 \times 10^{21} \text{ cm},$$

this estimate gives $\gamma = 7.5 \times 10^{-22} \text{ cm}^{-1}$.

The most stringent reason to attribute this effect to small solid particles is that other scattering or absorbing agents can be excluded. The amount of them required to explain the interstellar extinction either gives an excessive density (e.g., scattering electrons, sec. 6.23, plus an equal number of protons) or is impossible for other reasons (e.g., absorbing H^- ions). Grains very much smaller or very much larger than the wavelength also give too small a ratio of extinction cross section over mass.[1] This ratio is proportional to a^3 for $a \ll \lambda$ and to a^{-1} for $a \gg \lambda$ ($a =$ radius, $\lambda =$ wavelength).

Data about the size distribution, form, and composition of the grains may be inferred from the observed characteristics, which may be summarized as follows:

Color of transmitted light. The extinction is strong in the ultraviolet and weak in the infrared. The precise law deviates somewhat from a λ^{-1} law and is remarkably similar for all stars investigated. (Fig. 103.)

Linear polarization of transmitted light. This phenomenon, discovered in 1949, is interpreted as due to different values of the extinction coefficients for differently polarized light (dichroism of the interstellar medium). For stars showing the effects most strongly, the difference is about 6 per cent of the full extinction in visual light. The difference (vertical distance between the two curves in Fig. 103) decreases slightly towards the infrared; the ratio polarization/extinction increases towards the infrared.

Color, angular distribution, and polarization of the scattered light. Fully reliable data concerning the scattered light are absent. Reflection nebulae have colors resembling the illuminating star, and high values of the polarization are sometimes observed. The interpretation usually remains uncertain as the position of the star with respect to the nebula

---

[1] J. R. Platt has pointed out that a high ratio of extinction to mass may be reached by particles as small as 10 A if they consist of light elements and have their electronic energy bands unfilled (*Astrophys. J.*, **123**, 486, 1956).

is unknown. On a large scale, there is evidence in the diffuse galactic
light that the interstellar grains have a fairly large albedo ($Q_{sca}/Q_{ext} \geqslant 0.4$)
and a fairly strongly directed phase function ($\overline{\cos \theta} \geqslant 0.4$). Possible
combinations of these two would, e.g., be 0.9 and 0.5, 0.7 and 0.7, or
0.5 and 0.9. The values are uncertain, and the dependence on wavelength is unknown.

A compilation of the quantitative data regarding extinction and
polarization is given in Fig. 103. It is based on the extinction measured
by Whitford (1948), as corrected in the ultraviolet ($\lambda^{-1} > 2.5$) by Miss

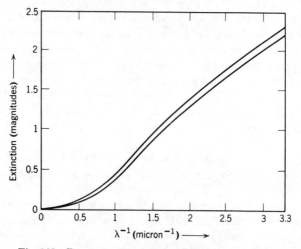

Fig. 103. Representative observational data on interstellar extinction. The extinction values for two directions
of polarization are plotted against the inverse wavelength
for a star of medium obscuration and strong polarization.

Lucienne Divan (1954), and on the polarization measured by Hiltner
(private communication). The ordinate scale is arbitrary and has been
chosen to correspond to a fairly heavily reddened star. The separation
of the two curves, i.e., the polarization effect, is shown at about its
maximum observed strength. The numerical values pertinent to this
compilation are given in Table 46.

The interpretation of these curves has been the subject of many
investigations, see van de Hulst (1955).

Disregarding the polarization effect, we see that the curves have a
good resemblance to many of the extinction curves computed by the
Mie theory, as shown, e.g., in Figs. 24 (sec. 10.4) and 32 (sec. 11.22).
Roughly, one would imagine the maximum to fall between $\lambda^{-1} = 4$ and 5.
Thus, as a rough size estimate we have (e.g., for $m = 1.25$): $2\pi a/\lambda = 8$

Table 46. Typical Dependence of Interstellar Extinction and Polarization
on Wavelength

λ^{-1} (μ^{-1})	λ (μ)	A_1 (mag)	A_2 (mag)	$\frac{1}{2}(A_1 + A_2)$ (mag)	$A_1 - A_2$ (mag)	A_1/A_2
0*		0	0	0	0	?
0.5†	2.000	0.10	0.08	0.09	0.02	1.25
1	1.000	0.44	0.38	0.41	0.06	1.16
1.5	0.667	0.98	0.89	0.94	0.09	1.10
2	0.500	1.41	1.31	1.36	0.10	1.07
2.5	0.400	1.80	1.70	1.75	0.10	1.06
3	0.333	2.17	2.07	2.12	0.10	1.05
3.5*	0.286	2.46	2.36	2.41	0.10	1.04

* Extrapolated values.
† Uncertain values.

for $\lambda^{-1} = 4.5\ \mu^{-1}$ and hence $2a = 0.56\ \mu$. Van de Hulst (1946, 1949) and others have shown that it is not difficult to choose a distribution function of sizes (compare sec. 11.5) that will render the observed curves exactly. So it is not possible to derive a unique size distribution, nor a unique value for the refractive index from the observations. Yet, when a size distribution has been selected, the product of the average size times $m - 1$ may be determined within 5 per cent.

It is even possible to obtain a fair representation of the extinction curve on the assumption of metallic spheres (Güttler, 1952). This would seem to be excluded if we compare the behavior for small λ^{-1} in Fig. 103 with that for small x in Fig. 54 (sec. 14.22). Yet the complex refractive index changes with λ in such a manner as to make the computed curve for iron spheres with $2a = 0.08\ \mu$ resemble at least roughly the observed curve. There is an observational argument against this assumption in the values of albedo and forward scattered component. By entering Fig. 54 at $x = 0.6$, corresponding to $2a = 0.08\ \mu$, $\lambda = 0.42\ \mu$, we find

$$\text{Albedo} = Q_{sca}/Q_{ext} = 0.15,$$

$$\overline{\cos\theta} = \overline{\cos\theta}\ Q_{sca}/Q_{sca} = 0.1.$$

The combination of these values is entirely outside the range that follows from the observations.

Turning to the effect of interstellar polarization, we find again that the observational data about the dependence on wavelength resemble at least roughly the theoretical expectations. This follows from a comparison of the curves in Fig. 103 with those for long dielectric cylinders in Fig. 67 (sec. 15.31). In both sets of curves the separation is about constant over a wide range but diminishes towards the small values of λ^{-1} and $\rho = 4\pi a(m-1)\lambda^{-1}$, respectively. By assuming a suitable distribution function of the radii the irregularities in the theoretical curves will be smoothed out, and a close agreement between theory and observations may be obtained. An admixture of metallic cylinders (Fig. 69, sec. 15.51) would also give the correct wavelength dependence if the observable range would correspond to $x = 0.1$ to 0.6, so $2a = 0.06 \mu$. More elaborate computations would be needed to take the change of m with λ into account.

A difficulty made clear by these comparisons is that approximately the right order of magnitude of the polarization is obtained on the assumptions of very long, fully oriented needles. Any assumption about less perfect orientation, as the spinning around an oriented axis in the Davis-Greenstein theory, and about shapes that deviate less drastically from the spherical shape will appreciably reduce the expected magnitude of the polarization. As no formulae for ellipsoids of arbitrary size are available at present (cf. sec. 16.11), most authors have turned in their interpretations to the formulae for sizes $\ll \lambda$, known as the Gans theory (sec. 6.32). These formulae permit the complete computation for any given distribution function of the orientations and thus give at least a first idea of the reduction in polarization owing to imperfect orientation. Davis (1955a, b) has given the formulae in a form suitable for astronomical application. However, the proper application of these formulae would be in the infrared, where Fig. 103 and Table 46 show that a larger ratio A_1/A_2 has to be explained than in the visual or photographic domain. A way out of this difficulty would be to assume that the polarization is caused by very small grains only, so that the Gans domain would cover the entire observed range of λ. This, however, would give an incorrect dependence of the polarization on λ.

Several experimental studies have been made with a view to explaining the interstellar polarization. Fick (1952, 1953) orients the elongated particles of Fe_2O_3 smokes in a magnetic field of variable strength up to saturation. The approximate sizes are: length $= 4 \mu$, diameter $= 0.5 \mu$. A strong positive dichroism results. Fick's theoretical discussion is confined to the domain of chap. 6, i.e., to sizes $\ll$ wavelength. Cayrel and Schatzman (1955) study the dichroism due to oriented graphite flakes.

The reduction of the amplitudes of the waves emitted by the stars (i.e., the extinction effect) is theoretically accompanied by a modification of the phase of these waves. This modification of the phase must depend on λ and may be different for the two polarizations. Therefore it is of interest to inquire if any observable effects would result.

The pertinent theory is given in secs. 4.41 and 4.5. Disregarding the polarization at first, we see that the formula

$$\tilde{m} = \tilde{n} - i\tilde{n}' = 1 - i2\pi \, Nk^{-3} \, S(0)$$

of sec. 4.3 gives the connection between the imaginary part of the refractive index of interstellar space, n', which is linked to the extinction value in magnitudes A by

$$0.92 \, A = 2k \int n' \, dz,$$

and the real part n which determines the phase retardation φ by

$$\varphi = k \int (n - 1) \, dz.$$

In both formulae the integral is taken over the line of sight between the observer and a particular star. So the ratio is

$$\frac{\varphi}{A} = 0.46 \frac{\text{Im} \, [S(0)]}{\text{Re} \, [S(0)]}.$$

A glance at the curves of $S(0)$ in Figs. 31 (sec. 11.22) and 53 (sec. 13.42) suffices to show that this ratio is of the order of 1. So the phase retardation (or advance) never exceeds more than part of one cycle for a star which can still be observed ($A < 10$). This corresponds to a negligible time; hence the suggestion made in the older literature, that the phase retardation in blue and red light would determine a traveling time which differed by several minutes (Nordmann-Tikhov effect), must be entirely discarded.

The ratio φ/A becomes higher for very small dielectric particles and becomes highest for non-absorbing molecules. The simple relation to the theory of the refractive index of a molecular gas is sketched in sec. 4.5. A comparison with the scattering and phase retardation by the earth's atmosphere shows that the very small particles also fail to give an observable phase retardation in interstellar space.

A differential phase shift between the two components of polarization at a given wavelength could be tested in a more precise manner by the ordinary methods of measuring double refraction. It is quite probable that the dichroism of interstellar space is accompanied by double refraction. The imaginary parts of the amplitude functions $T_1(0)$ and $T_2(0)$, whose real parts have given the data for Figs. 67 (sec. 15.31) and 69 (sec. 15.51), have not been computed. A precise prediction is therefore not possible. Roughly, we may expect a difference of the same order, say, 10 per cent. This would mean that, if we had a source emitting linearly polarized light with strong polarization (something like the

Crab nebula) behind a reasonably heavy obscuring layer in which the orientation of the interstellar polarization was oriented at a 45° angle with the polarization of the source, an observable effect might result. This effect would be an elliptical component of several per cent in the received radiation.

A curious application of the Mie theory, finally, is the computation of the thermal radiation by the interstellar grains, which is the main loss of internal energy of a grain and, therefore, decisive for its temperature. As the emitted waves are in the far infrared, i.e., much larger than the size, we must use the formulae for Q_{abs} following from the Mie theory (chap. 14). Kirchhoff's law states that the radiation is Q_{abs} times the value computed on the basis of black-body emission. Van de Hulst (1946, 1949) estimated on this basis that the temperature of the interstellar grains, metallic or dielectric, would be more likely in the range of 10° to 20° K than at the traditional value of 3° K.

References*

Some references to scattering by irregularities in the refractive index (sec. 21.1) are:

C. L. Pekeris, *Phys. Rev.*, **71**, 268 (1947).

H. G. Booker and W. E. Gordon, *Proc. I.R.E.*, **38**, 401 (1950).

A. Hewish, *Proc. Roy. Soc. London*, **A209**, 81 (1951).

S. Chandrasekhar, *Monthly Notices Roy. Astron. Soc.*, **112**, 475 (1952).

J. A. Fejer, *Proc. Roy. Soc. London*, **A220**, 455 (1954).

A. Hewish, *Proc. Roy. Soc. London*, **A228**, 238 (1955).

Scattering in planetary atmospheres (sec. 21.2) is discussed by:

G. P. Kuiper, *The Atmospheres of the Earth and Planets* (2nd ed.), G. P. Kuiper, ed., chap. 12, Chicago, Univ. of Chicago Press, 1952.

B. Lyot, *Ann. Obs. Paris-Meudon*, **8**, 70 (1929).

S. Chandrasekhar and D. D. Elbert, *Trans. Am. Phil. Soc.*, **44**, 643 (1954).

H. C. van de Hulst, *The Atmospheres of the Earth and Planets* (2nd ed.), G. P. Kuiper, ed., chap. 3, Chicago, Univ. of Chicago Press, 1952.

H. G. Horak, *Astrophys. J.*, **112**, 445 (1950).

G. de Vaucouleurs, *Physics of the Planet Mars*, London, Faber and Faber, 1954.

D. H. Menzel and F. L. Whipple, *Publ. Astron. Soc. Pacific*, **67**, 161 (1955).

Scattering by interplanetary dust (sec. 21.3) was reviewed by:

M. Minnaert, "Les particules solides dans les astres" (Rept. Intern. Astrophys. Symposium), *Mém. soc. roy. sci. Liège*, quatrième série, **15**, 15 (1955).

H. C. van de Hulst, *ibid.*, p. 89, 1955.

and by other authors in the same volume.

* Given in the order in which they are cited in the text.

Some references on the extinction and polarization by interstellar grains (sec. 21.4) are:

A. E. Whitford, *Astrophys. J.*, **107**, 102 (1948).

L. Divan, *Ann. astrophys.*, **17**, 456 (1954).

H. C. van de Hulst, "Les particules solides dans les astres," op. cit., **15**, 393 (1955).

H. C. van de Hulst, *Recherches astron. Obs. d'Utrecht*, **11**, part 1, 1946; part 2, 1949.

A. Güttler, *Z. Astrophysik*, **31**, 1 (1952).

L. Davis, "Les particules solides dans les astres," op. cit., **15**, 584 (1955).

L. Davis, *Vistas in Astronomy*, I, p. 336, London & New York, Pergamon Press 1955.

E. Fick, *Z. Physik*, **138**, 183 (1954); **140**, 318 (1955).

E. Schatzman and R. Cayrel, *Ann. astrophys.*, **17**, 555 (1954).

NAME INDEX

455

SUBJECT INDEX

A CATALOGUE OF
SELECTED DOVER BOOKS
IN ALL FIELDS OF INTEREST

A CATALOGUE OF SELECTED DOVER
BOOKS IN ALL FIELDS OF INTEREST

RACKHAM'S COLOR ILLUSTRATIONS FOR WAGNER'S RING. Rackham's finest mature work—all 64 full-color watercolors in a faithful and lush interpretation of the *Ring*. Full-sized plates on coated stock of the paintings used by opera companies for authentic staging of Wagner. Captions aid in following complete Ring cycle. Introduction. 64 illustrations plus vignettes. 72pp. 8⅝ x 11¼. 23779-6 Pa. $6.00

CONTEMPORARY POLISH POSTERS IN FULL COLOR, edited by Joseph Czestochowski. 46 full-color examples of brilliant school of Polish graphic design, selected from world's first museum (near Warsaw) dedicated to poster art. Posters on circuses, films, plays, concerts all show cosmopolitan influences, free imagination. Introduction. 48pp. 9⅜ x 12¼. 23780-X Pa. $6.00

GRAPHIC WORKS OF EDVARD MUNCH, Edvard Munch. 90 haunting, evocative prints by first major Expressionist artist and one of the greatest graphic artists of his time: *The Scream, Anxiety, Death Chamber, The Kiss, Madonna,* etc. Introduction by Alfred Werner. 90pp. 9 x 12. 23765-6 Pa. $5.00

THE GOLDEN AGE OF THE POSTER, Hayward and Blanche Cirker. 70 extraordinary posters in full colors, from Maitres de l'Affiche, Mucha, Lautrec, Bradley, Cheret, Beardsley, many others. Total of 78pp. 9⅜ x 12¼. 22753-7 Pa. $6.95

THE NOTEBOOKS OF LEONARDO DA VINCI, edited by J. P. Richter. Extracts from manuscripts reveal great genius; on painting, sculpture, anatomy, sciences, geography, etc. Both Italian and English. 186 ms. pages reproduced, plus 500 additional drawings, including studies for *Last Supper,* Sforza monument, etc. 860pp. 7⅞ x 10¾. (Available in U.S. only) 22572-0, 22573-9 Pa., Two-vol. set $19.90

THE CODEX NUTTALL, as first edited by Zelia Nuttall. Only inexpensive edition, in full color, of a pre-Columbian Mexican (Mixtec) book. 88 color plates show kings, gods, heroes, temples, sacrifices. New explanatory, historical introduction by Arthur G. Miller. 96pp. 11⅜ x 8½. (Available in U.S. only) 23168-2 Pa. $7.95

UNE SEMAINE DE BONTÉ, A SURREALISTIC NOVEL IN COLLAGE, Max Ernst. Masterpiece created out of 19th-century periodical illustrations, explores worlds of terror and surprise. Some consider this Ernst's greatest work. 208pp. 8⅛ x 11. 23252-2 Pa. $6.00

DRAWINGS OF WILLIAM BLAKE, William Blake. 92 plates from Book of Job, *Divine Comedy, Paradise Lost,* visionary heads, mythological figures, Laocoon, etc. Selection, introduction, commentary by Sir Geoffrey Keynes. 178pp. 8⅛ x 11. 22303-5 Pa. $5.00

ENGRAVINGS OF HOGARTH, William Hogarth. 101 of Hogarth's greatest works: *Rake's Progress, Harlot's Progress, Illustrations for Hudibras, Before and After, Beer Street and Gin Lane,* many more. Full commentary. 256pp. 11 x 13¾. 22479-1 Pa. $12.95

DAUMIER: 120 GREAT LITHOGRAPHS, Honore Daumier. Wide-ranging collection of lithographs by the greatest caricaturist of the 19th century. Concentrates on eternally popular series on lawyers, on married life, on liberated women, etc. Selection, introduction, and notes on plates by Charles F. Ramus. Total of 158pp. 9⅜ x 12¼. 23512-2 Pa. $6.00

DRAWINGS OF MUCHA, Alphonse Maria Mucha. Work reveals drafts-man of highest caliber: studies for famous posters and paintings, render-ings for book illustrations and ads, etc. 70 works, 9 in color; including 6 items not drawings. Introduction. List of illustrations. 72pp. 9⅜ x 12¼. (Available in U.S. only) 23672-2 Pa. $4.50

GIOVANNI BATTISTA PIRANESI: DRAWINGS IN THE PIERPONT MORGAN LIBRARY, Giovanni Battista Piranesi. For first time ever all of Morgan Library's collection, world's largest. 167 illustrations of rare Piranesi drawings—archeological, architectural, decorative and visionary. Essay, detailed list of drawings, chronology, captions. Edited by Felice Stampfle. 144pp. 9⅜ x 12¼. 23714-1 Pa. $7.50

NEW YORK ETCHINGS (1905-1949), John Sloan. All of important American artist's N.Y. life etchings. 67 works include some of his best art; also lively historical record—Greenwich Village, tenement scenes. Edited by Sloan's widow. Introduction and captions. 79pp. 8⅜ x 11¼. 23651-X Pa. $5.00

CHINESE PAINTING AND CALLIGRAPHY: A PICTORIAL SURVEY, Wan-go Weng. 69 fine examples from John M. Crawford's matchless private collection: landscapes, birds, flowers, human figures, etc., plus calligraphy. Every basic form included: hanging scrolls, handscrolls, album leaves, fans, etc. 109 illustrations. Introduction. Captions. 192pp. 8⅞ x 11¾. 23707-9 Pa. $7.95

DRAWINGS OF REMBRANDT, edited by Seymour Slive. Updated Lipp-mann, Hofstede de Groot edition, with definitive scholarly apparatus. All portraits, biblical sketches, landscapes, nudes, Oriental figures, classical studies, together with selection of work by followers. 550 illustrations. Total of 630pp. 9⅛ x 12¼. 21485-0, 21486-9 Pa., Two-vol. set $17.90

THE DISASTERS OF WAR, Francisco Goya. 83 etchings record horrors of Napoleonic wars in Spain and war in general. Reprint of 1st edition, plus 3 additional plates. Introduction by Philip Hofer. 97pp. 9⅜ x 8¼. 21872-4 Pa. $4.50

THE EARLY WORK OF AUBREY BEARDSLEY, Aubrey Beardsley. 157 plates, 2 in color: *Manon Lescaut, Madame Bovary, Morte Darthur, Salome,* other. Introduction by H. Marillier. 182pp. 8⅛ x 11. 21816-3 Pa. $6.50

THE LATER WORK OF AUBREY BEARDSLEY, Aubrey Beardsley. Exotic masterpieces of full maturity: *Venus and Tannhauser, Lysistrata, Rape of the Lock, Volpone,* Savoy material, etc. 174 plates, 2 in color. 186pp. 8⅛ x 11. 21817-1 Pa. $5.95

THOMAS NAST'S CHRISTMAS DRAWINGS, Thomas Nast. Almost all Christmas drawings by creator of image of Santa Claus as we know it, and one of America's foremost illustrators and political cartoonists. 66 illustrations. 3 illustrations in color on covers. 96pp. 8⅜ x 11¼. 23660-9 Pa. $3.50

THE DORÉ ILLUSTRATIONS FOR DANTE'S DIVINE COMEDY, Gustave Doré. All 135 plates from Inferno, Purgatory, Paradise; fantastic tortures, infernal landscapes, celestial wonders. Each plate with appropriate (translated) verses. 141pp. 9 x 12. 23231-X Pa. $5.00

DORÉ'S ILLUSTRATIONS FOR RABELAIS, Gustave Doré. 252 striking illustrations of *Gargantua and Pantagruel* books by foremost 19th-century illustrator. Including 60 plates, 192 delightful smaller illustrations. 153pp. 9 x 12. 23656-0 Pa. $6.00

LONDON: A PILGRIMAGE, Gustave Doré, Blanchard Jerrold. Squalor, riches, misery, beauty of mid-Victorian metropolis; 55 wonderful plates, 125 other illustrations, full social, cultural text by Jerrold. 191pp. of text. 9⅜ x 12¼. 22306-X Pa. $7.00

THE RIME OF THE ANCIENT MARINER, Gustave Doré, S. T. Coleridge. Dore's finest work, 34 plates capture moods, subtleties of poem. Full text. Introduction by Millicent Rose. 77pp. 9¼ x 12. 22305-1 Pa. $4.50

THE DORE BIBLE ILLUSTRATIONS, Gustave Doré. All wonderful, detailed plates: Adam and Eve, Flood, Babylon, Life of Jesus, etc. Brief King James text with each plate. Introduction by Millicent Rose. 241 plates. 241pp. 9 x 12. 23004-X Pa. $6.95

THE COMPLETE ENGRAVINGS, ETCHINGS AND DRYPOINTS OF ALBRECHT DURER. "Knight, Death and Devil"; "Melencolia," and more—all Dürer's known works in all three media, including 6 works formerly attributed to him. 120 plates. 235pp. 8⅜ x 11¼. 22851-7 Pa. $7.50

MECHANICK EXERCISES ON THE WHOLE ART OF PRINTING, Joseph Moxon. First complete book (1683-4) ever written about typography, a compendium of everything known about printing at the latter part of 17th century. Reprint of 2nd (1962) Oxford Univ. Press edition. 74 illustrations. Total of 550pp. 6⅛ x 9¼. 23617-X Pa. $7.95

CATALOGUE OF DOVER BOOKS

THE COMPLETE WOODCUTS OF ALBRECHT DURER, edited by Dr. W. Kurth. 346 in all: "Old Testament," "St. Jerome," "Passion," "Life of Virgin," "Apocalypse," many others. Introduction by Campbell Dodgson. 285pp. 8½ x 12¼. 21097-9 Pa. $7.50

DRAWINGS OF ALBRECHT DURER, edited by Heinrich Wolfflin. 81 plates show development from youth to full style. Many favorites; many new. Introduction by Alfred Werner. 96pp. 8⅛ x 11. 22352-3 Pa. $6.00

THE HUMAN FIGURE, Albrecht Dürer. Experiments in various techniques—stereometric, progressive proportional, and others. Also life studies that rank among finest ever done. Complete reprinting of *Dresden Sketchbook*. 170 plates. 355pp. 8⅜ x 11¼. 21042-1 Pa. $7.95

OF THE JUST SHAPING OF LETTERS, Albrecht Dürer. Renaissance artist explains design of Roman majuscules by geometry, also Gothic lower and capitals. Grolier Club edition. 43pp. 7⅞ x 10¾ 21306-4 Pa. $3.00

TEN BOOKS ON ARCHITECTURE, Vitruvius. The most important book ever written on architecture. Early Roman aesthetics, technology, classical orders, site selection, all other aspects. Stands behind everything since. Morgan translation. 331pp. 5⅜ x 8½. 20645-9 Pa. $5.00

THE FOUR BOOKS OF ARCHITECTURE, Andrea Palladio. 16th-century classic responsible for Palladian movement and style. Covers classical architectural remains, Renaissance revivals, classical orders, etc. 1738 Ware English edition. Introduction by A. Placzek. 216 plates. 110pp. of text. 9½ x 12¾. 21308-0 Pa. $10.00

HORIZONS, Norman Bel Geddes. Great industrialist stage designer, "father of streamlining," on application of aesthetics to transportation, amusement, architecture, etc. 1932 prophetic account; function, theory, specific projects. 222 illustrations. 312pp. 7⅞ x 10¾. 23514-9 Pa. $6.95

FRANK LLOYD WRIGHT'S FALLINGWATER, Donald Hoffmann. Full, illustrated story of conception and building of Wright's masterwork at Bear Run, Pa. 100 photographs of site, construction, and details of completed structure. 112pp. 9¼ x 10. 23671-4 Pa. $5.95

THE ELEMENTS OF DRAWING, John Ruskin. Timeless classic by great Viltorian; starts with basic ideas, works through more difficult. Many practical exercises. 48 illustrations. Introduction by Lawrence Campbell. 228pp. 5⅜ x 8½. 22730-8 Pa. $3.75

GIST OF ART, John Sloan. Greatest modern American teacher, Art Students League, offers innumerable hints, instructions, guided comments to help you in painting. Not a formal course. 46 illustrations. Introduction by Helen Sloan. 200pp. 5⅜ x 8½. 23435-5 Pa. $4.00

THE ANATOMY OF THE HORSE, George Stubbs. Often considered the great masterpiece of animal anatomy. Full reproduction of 1766 edition, plus prospectus; original text and modernized text. 36 plates. Introduction by Eleanor Garvey. 121pp. 11 x 14¾.   23402-9 Pa. $8.95

BRIDGMAN'S LIFE DRAWING, George B. Bridgman. More than 500 illustrative drawings and text teach you to abstract the body into its major masses, use light and shade, proportion; as well as specific areas of anatomy, of which Bridgman is master. 192pp. 6½ x 9¼. (Available in U.S. only)
22710-3 Pa. $4.50

ART NOUVEAU DESIGNS IN COLOR, Alphonse Mucha, Maurice Verneuil, Georges Auriol. Full-color reproduction of *Combinaisons ornementales* (c. 1900) by Art Nouveau masters. Floral, animal, geometric, interlacings, swashes—borders, frames, spots—all incredibly beautiful. 60 plates, hundreds of designs. 9⅜ x 8-1/16.   22885-1 Pa. $4.50

FULL-COLOR FLORAL DESIGNS IN THE ART NOUVEAU STYLE, E. A. Seguy. 166 motifs, on 40 plates, from *Les fleurs et leurs applications decoratives* (1902): borders, circular designs, repeats, allovers, "spots." All in authentic Art Nouveau colors. 48pp. 9⅜ x 12¼.
23439-8 Pa. $5.00

A DIDEROT PICTORIAL ENCYCLOPEDIA OF TRADES AND IN-DUSTRY, edited by Charles C. Gillispie. 485 most interesting plates from the great French Encyclopedia of the 18th century show hundreds of working figures, artifacts, process, land and cityscapes; glassmaking, paper-making, metal extraction, construction, weaving, making furniture, clothing, wigs, dozens of other activities. Plates fully explained. 920pp. 9 x 12.
22284-5, 22285-3 Clothbd., Two-vol. set $40.00

HANDBOOK OF EARLY ADVERTISING ART, Clarence P. Hornung. Largest collection of copyright-free early and antique advertising art ever compiled. Over 6,000 illustrations, from Franklin's time to the 1890's for special effects, novelty. Valuable source, almost inexhaustible.
Pictorial Volume. Agriculture, the zodiac, animals, autos, birds, Christmas, fire engines, flowers, trees, musical instruments, ships, games and sports, much more. Arranged by subject matter and use. 237 plates. 288pp. 9 x 12.
20122-8 Clothbd. $15.00

Typographical Volume. Roman and Gothic faces ranging from 10 point to 300 point, "Barnum," German and Old English faces, script, logotypes, scrolls and flourishes, 1115 ornamental initials, 67 complete alphabets, more. 310 plates. 320pp. 9 x 12.   20123-6 Clothbd. $15.00

CALLIGRAPHY (CALLIGRAPHIA LATINA), J. G. Schwandner. High point of 18th-century ornamental calligraphy. Very ornate initials, scrolls, borders, cherubs, birds, lettered examples. 172pp. 9 x 13.
20475-8 Pa. $7.95

ART FORMS IN NATURE, Ernst Haeckel. Multitude of strangely beautiful natural forms: Radiolaria, Foraminifera, jellyfishes, fungi, turtles, bats, etc. All 100 plates of the 19th-century evolutionist's *Kunstformen der Natur* (1904). 100pp. 9⅜ x 12¼. 22987-4 Pa. $5.00

CHILDREN: A PICTORIAL ARCHIVE FROM NINETEENTH-CENTURY SOURCES, edited by Carol Belanger Grafton. 242 rare, copyright-free wood engravings for artists and designers. Widest such selection available. All illustrations in line. 119pp. 8⅜ x 11¼. 23694-3 Pa. $4.00

WOMEN: A PICTORIAL ARCHIVE FROM NINETEENTH-CENTURY SOURCES, edited by Jim Harter. 391 copyright-free wood engravings for artists and designers selected from rare periodicals. Most extensive such collection available. All illustrations in line. 128pp. 9 x 12. 23703-6 Pa. $4.95

ARABIC ART IN COLOR, Prisse d'Avennes. From the greatest ornamentalists of all time—50 plates in color, rarely seen outside the Near East, rich in suggestion and stimulus. Includes 4 plates on covers. 46pp. 9⅜ x 12¼. 23658-7 Pa. $6.00

AUTHENTIC ALGERIAN CARPET DESIGNS AND MOTIFS, edited by June Beveridge. Algerian carpets are world famous. Dozens of geometrical motifs are charted on grids, color-coded, for weavers, needleworkers, craftsmen, designers. 53 illustrations plus 4 in color. 48pp. 8¼ x 11. (Available in U.S. only) 23650-1 Pa. $1.75

DICTIONARY OF AMERICAN PORTRAITS, edited by Hayward and Blanche Cirker. 4000 important Americans, earliest times to 1905, mostly in clear line. Politicians, writers, soldiers, scientists, inventors, industrialists, Indians, Blacks, women, outlaws, etc. Identificatory information. 756pp. 9¼ x 12¾. 21823-6 Clothbd. $65.00

HOW THE OTHER HALF LIVES, Jacob A. Riis. Journalistic record of filth, degradation, upward drive in New York immigrant slums, shops, around 1900. New edition includes 100 original Riis photos, monuments of early photography. 233pp. 10 x 7⅞. 22012-5 Pa. $7.00

NEW YORK IN THE THIRTIES, Berenice Abbott. Noted photographer's fascinating study of city shows new buildings that have become famous and old sights that have disappeared forever. Insightful commentary. 97 photographs. 97pp. 11⅜ x 10. 22967-X Pa. $6.00

MEN AT WORK, Lewis W. Hine. Famous photographic studies of construction workers, railroad men, factory workers and coal miners. New supplement of 18 photos on Empire State building construction. New introduction by Jonathan L. Doherty. Total of 69 photos. 63pp. 8 x 10¾. 23475-4 Pa. $4.00

THE DEPRESSION YEARS AS PHOTOGRAPHED BY ARTHUR ROTH-STEIN, Arthur Rothstein. First collection devoted entirely to the work of outstanding 1930s photographer: famous dust storm photo, ragged children, unemployed, etc. 120 photographs. Captions. 119pp. 9¼ x 10¾.
23590-4 Pa. **$5.95**

CAMERA WORK: A PICTORIAL GUIDE, Alfred Stieglitz. All 559 illustrations and plates from the most important periodical in the history of art photography, Camera Work (1903-17). Presented four to a page, reduced in size but still clear, in strict chronological order, with complete captions. Three indexes. Glossary. Bibliography. 176pp. 8⅜ x 11¼.
23591-2 Pa. $6.95

ALVIN LANGDON COBURN, PHOTOGRAPHER, Alvin L. Coburn. Revealing autobiography by one of greatest photographers of 20th century gives insider's version of Photo-Secession, plus comments on his own work. 77 photographs by Coburn. Edited by Helmut and Alison Gernsheim. 160pp. 8⅛ x 11.
23685-4 Pa. $6.00

NEW YORK IN THE FORTIES, Andreas Feininger. 162 brilliant photographs by the well-known photographer, formerly with Life magazine, show commuters, shoppers, Times Square at night, Harlem nightclub, Lower East Side, etc. Introduction and full captions by John von Hartz. 181pp. 9¼ x 10¾.
23585-8 Pa. $6.95

GREAT NEWS PHOTOS AND THE STORIES BEHIND THEM, John Faber. Dramatic volume of 140 great news photos, 1855 through 1976, and revealing stories behind them, with both historical and technical information. Hindenburg disaster, shooting of Oswald, nomination of Jimmy Carter, etc. 160pp. 8¼ x 11.
23667-6 Pa. **$6.00**

THE ART OF THE CINEMATOGRAPHER, Leonard Maltin. Survey of American cinematography history and anecdotal interviews with 5 masters—Arthur Miller, Hal Mohr, Hal Rosson, Lucien Ballard, and Conrad Hall. Very large selection of behind-the-scenes production photos. 105 photographs. Filmographies. Index. Originally Behind the Camera. 144pp. 8¼ x 11.
23686-2 Pa. $5.00

DESIGNS FOR THE THREE-CORNERED HAT (LE TRICORNE), Pablo Picasso. 32 fabulously rare drawings—including 31 color illustrations of costumes and accessories—for 1919 production of famous ballet. Edited by Parmenia Migel, who has written new introduction. 48pp. 9⅜ x 12¼.
(Available in U.S. only) 23709-5 Pa. $5.00

NOTES OF A FILM DIRECTOR, Sergei Eisenstein. Greatest Russian filmmaker explains montage, making of Alexander Nevsky, aesthetics; comments on self, associates, great rivals (Chaplin), similar material. 78 illustrations. 240pp. 5⅜ x 8½.
22392-2 Pa. $7.00

HOLLYWOOD GLAMOUR PORTRAITS, edited by John Kobal. 145 photos capture the stars from 1926-49, the high point in portrait photography. Gable, Harlow, Bogart, Bacall, Hedy Lamarr, Marlene Dietrich, Robert Montgomery, Marlon Brando, Veronica Lake; 94 stars in all. Full background on photographers, technical aspects, much more. Total of 160pp. 8⅜ x 11¼. 23352-9 Pa. $6.95

THE NEW YORK STAGE: FAMOUS PRODUCTIONS IN PHOTO-GRAPHS, edited by Stanley Appelbaum. 148 photographs from Museum of City of New York show 142 plays, 1883-1939. *Peter Pan, The Front Page, Dead End, Our Town,* O'Neill, hundreds of actors and actresses, etc. Full indexes. 154pp. 9½ x 10. 23241-7 Pa. $6.00

DIALOGUES CONCERNING TWO NEW SCIENCES, Galileo Galilei. Encompassing 30 years of experiment and thought, these dialogues deal with geometric demonstrations of fracture of solid bodies, cohesion, leverage, speed of light and sound, pendulums, falling bodies, accelerated motion, etc. 300pp. 5⅜ x 8½. 60099-8 Pa. $5.50

THE GREAT OPERA STARS IN HISTORIC PHOTOGRAPHS, edited by James Camner. 343 portraits from the 1850s to the 1940s: Tamburini, Mario, Caliapin, Jeritza, Melchior, Melba, Patti, Pinza, Schipa, Caruso, Farrar, Steber, Gobbi, and many more—270 performers in all. Index. 199pp. 8⅜ x 11¼. 23575-0 Pa. $7.50

J. S. BACH, Albert Schweitzer. Great full-length study of Bach, life, background to music, music, by foremost modern scholar. Ernest Newman translation. 650 musical examples. Total of 928pp. 5⅜ x 8½. (Available in U.S. only) 21631-4, 21632-2 Pa., Two-vol. set $12.00

COMPLETE PIANO SONATAS, Ludwig van Beethoven. All sonatas in the fine Schenker edition, with fingering, analytical material. One of best modern editions. Total of 615pp. 9 x 12. (Available in U.S. only) 23134-8, 23135-6 Pa., Two-vol. set $17.90

KEYBOARD MUSIC, J. S. Bach. Bach-Gesellschaft edition. For harpsichord, piano, other keyboard instruments. English Suites, French Suites, Six Partitas, Goldberg Variations, Two-Part Inventions, Three-Part Sinfonias. 312pp. 8⅛ x 11. (Available in U.S. only) 22360-4 Pa. $7.95

FOUR SYMPHONIES IN FULL SCORE, Franz Schubert. Schubert's four most popular symphonies: No. 4 in C Minor ("Tragic"); No. 5 in B-flat Major; No. 8 in B Minor ("Unfinished"); No. 9 in C Major ("Great"). Breitkopf & Hartel edition. Study score. 261pp. 9⅜ x 12¼. 23681-1 Pa. $8.95

THE AUTHENTIC GILBERT & SULLIVAN SONGBOOK, W. S. Gilbert, A. S. Sullivan. Largest selection available; 92 songs, uncut, original keys, in piano rendering approved by Sullivan. Favorites and lesser-known fine numbers. Edited with plot synopses by James Spero. 3 illustrations. 399pp. 9 x 12. 23482-7 Pa.$10.95

PRINCIPLES OF ORCHESTRATION, Nikolay Rimsky-Korsakov. Great classical orchestrator provides fundamentals of tonal resonance, progression of parts, voice and orchestra, tutti effects, much else in major document. 330pp. of musical excerpts. 489pp. 6½ x 9¼. 21266-1 Pa. $7.50

TRISTAN UND ISOLDE, Richard Wagner. Full orchestral score with complete instrumentation. Do not confuse with piano reduction. Commentary by Felix Mottl, great Wagnerian conductor and scholar. Study score. 655pp. 8⅛ x 11. 22915-7 Pa. $13.95

REQUIEM IN FULL SCORE, Giuseppe Verdi. Immensely popular with choral groups and music lovers. Republication of edition published by C. F. Peters, Leipzig, n. d. German frontmaker in English translation. Glossary. Text in Latin. Study score. 204pp. 9⅜ x 12¼. 23682-X Pa. $6.50

COMPLETE CHAMBER MUSIC FOR STRINGS, Felix Mendelssohn. All of Mendelssohn's chamber music: Octet, 2 Quintets, 6 Quartets, and Four Pieces for String Quartet. (Nothing with piano is included). Complete works edition (1874-7). Study score. 283 pp. 9⅜ x 12¼. 23679-X Pa. $7.50

POPULAR SONGS OF NINETEENTH-CENTURY AMERICA, edited by Richard Jackson. 64 most important songs: "Old Oaken Bucket," "Arkansas Traveler," "Yellow Rose of Texas," etc. Authentic original sheet music, full introduction and commentaries. 290pp. 9 x 12. 23270-0 Pa. $7.95

COLLECTED PIANO WORKS, Scott Joplin. Edited by Vera Brodsky Lawrence. Practically all of Joplin's piano works––rags, two-steps, marches, waltzes, etc., 51 works in all. Extensive introduction by Rudi Blesh. Total of 345pp. 9 x 12. 23106-2 Pa. $15.95

BASIC PRINCIPLES OF CLASSICAL BALLET, Agrippina Vaganova. Great Russian theoretician, teacher explains methods for teaching classical ballet; incorporates best from French, Italian, Russian schools. 118 illustrations. 175pp. 5⅜ x 8½. 22036-2 Pa. $2.75

CHINESE CHARACTERS, L. Wieger. Rich analysis of 2300 characters according to traditional systems into primitives. Historical-semantic analysis to phonetics (Classical Mandarin) and radicals. 820pp. 6⅛ x 9¼. 21321-8 Pa. $12.50

THE WARES OF THE MING DYNASTY, R. L. Hobson. Foremost scholar examines and iilustrates many varieties of Ming (1368-1644). Famous blue and white, polychrome, lesser-known styles and shapes. 117 illustrations, 9 full color, of outstanding pieces. Total of 263pp. 6⅛ x 9¼. (Available in U.S. only) 23652-8 Pa. $6.00

AN ETYMOLOGICAL DICTIONARY OF MODERN ENGLISH, Ernest Weekley. Richest, fullest work, by foremost British lexicographer. Detailed word histories. Inexhaustible. Do not confuse this with *Concise Etymological Dictionary*, which is abridged. Total of 856pp. 6½ x 9¼. 21873-2, 21874-0 Pa., Two-vol. set $13.00

A MAYA GRAMMAR, Alfred M. Tozzer. Practical, useful English-language grammar by the Harvard anthropologist who was one of the three greatest American scholars in the area of Maya culture. Phonetics, grammatical processes, syntax, more. 301pp. 5⅜ x 8½.　　　23465-7 Pa. $4.00

THE JOURNAL OF HENRY D. THOREAU, edited by Bradford Torrey, F. H. Allen. Complete reprinting of 14 volumes, 1837-61, over two million words; the sourcebooks for *Walden*, etc. Definitive. All original sketches, plus 75 photographs. Introduction by Walter Harding. Total of 1804pp. 8½ x 12¼.　　　20312-3, 20313-1 Clothbd., Two-vol. set $80.00

CLASSIC GHOST STORIES, Charles Dickens and others. 18 wonderful stories you've wanted to reread: "The Monkey's Paw," "The House and the Brain," "The Upper Berth," "The Signalman," "Dracula's Guest," "The Tapestried Chamber," etc. Dickens, Scott, Mary Shelley, Stoker, etc. 330pp. 5⅜ x 8½.　　　20735-8 Pa. $4.50

SEVEN SCIENCE FICTION NOVELS, H. G. Wells. Full novels. *First Men in the Moon, Island of Dr. Moreau, War of the Worlds, Food of the Gods, Invisible Man, Time Machine, In the Days of the Comet.* A basic science-fiction library. 1015pp. 5⅜ x 8½. (Available in U.S. only)
20264-X Clothbd. $15.00

ARMADALE, Wilkie Collins. Third great mystery novel by the author of *The Woman in White* and *The Moonstone.* Ingeniously plotted narrative shows an exceptional command of character, incident and mood. Original magazine version with 40 illustrations. 597pp. 5⅜ x 8½.
23429-0 Pa. $7.95

FLATLAND, E. A. Abbott. Science-fiction classic explores life of 2-D being in 3-D world. Read also as introduction to thought about hyperspace. Introduction by Banesh Hoffmann. 16 illustrations. 103pp. 5⅜ x 8½.
20001-9 Pa. $2.75

AYESHA: THE RETURN OF "SHE," H. Rider Haggard. Virtuoso sequel featuring the great mythic creation, Ayesha, in an adventure that is fully as good as the first book, *She.* Original magazine version, with 47 original illustrations by Maurice Greiffenhagen. 189pp. 6½ x 9¼.
23649-8 Pa. $3.50

ORIENTAL RUGS, ANTIQUE AND MODERN, Walter A. Hawley. Persia, Turkey, Caucasus, Central Asia, China, other traditions. Best general survey of all aspects: styles and periods, manufacture, uses, symbols and their interpretation, and identification. 96 illustrations, 11 in color. 320pp. 6⅛ x 9¼.　　　22366-3 Pa. $6.95

CHINESE POTTERY AND PORCELAIN, R. L. Hobson. Detailed descriptions and analyses by former Keeper of the Department of Oriental Antiquities and Ethnography at the British Museum. Covers hundreds of pieces from primitive times to 1915. Still the standard text for most periods. 136 plates, 40 in full color. Total of 750pp. 5⅜ x 8½.
23253-0 Pa. $10.00

UNCLE SILAS, J. Sheridan LeFanu. Victorian Gothic mystery novel, considered by many best of period, even better than Collins or Dickens. Wonderful psychological terror. Introduction by Frederick Shroyer. 436pp. 5⅜ x 8½. 21715-9 Pa. **$6.95**

JURGEN, James Branch Cabell. The great erotic fantasy of the 1920's that delighted thousands, shocked thousands more. Full final text, Lane edition with 13 plates by Frank Pape. 346pp. 5⅜ x 8½.
23507-6 Pa. **$4.50**

THE CLAVERINGS, Anthony Trollope. Major novel, chronicling aspects of British Victorian society, personalities. Reprint of Cornhill serialization, 16 plates by M. Edwards; first reprint of full text. Introduction by Norman Donaldson. 412pp. 5⅜ x 8½. 23464-9 Pa. **$5.00**

KEPT IN THE DARK, Anthony Trollope. Unusual short novel about Victorian morality and abnormal psychology by the great English author. Probably the first American publication. Frontispiece by Sir John Millais. 92pp. 6½ x 9¼. 23609-9 Pa. **$2.50**

RALPH THE HEIR, Anthony Trollope. Forgotten tale of illegitimacy, inheritance. Master novel of Trollope's later years. Victorian country estates, clubs, Parliament, fox hunting, world of fully realized characters. Reprint of 1871 edition. 12 illustrations by F. A. Faser. 434pp. of text. 5⅜ x 8½. 23642-0 Pa. **$6.50**

YEKL and THE IMPORTED BRIDEGROOM AND OTHER STORIES OF THE NEW YORK GHETTO, Abraham Cahan. Film *Hester Street* based on *Yekl* (1896). Novel, other stories among first about Jewish immigrants of N.Y.'s East Side. Highly praised by W. D. Howells—Cahan "a new star of realism." New introduction by Bernard G. Richards. 240pp. 5⅜ x 8½. 22427-9 Pa. **$3.50**

THE HIGH PLACE, James Branch Cabell. Great fantasy writer's enchanting comedy of disenchantment set in 18th-century France. Considered by some critics to be even better than his famous *Jurgen*. 10 illustrations and numerous vignettes by noted fantasy artist Frank C. Pape. 320pp. 5⅜ x 8½. 23670-6 Pa. **$4.00**

ALICE'S ADVENTURES UNDER GROUND, Lewis Carroll. Facsimile of ms. Carroll gave Alice Liddell in 1864. Different in many ways from final Alice. Handlettered, illustrated by Carroll. Introduction by Martin Gardner. 128pp. 5⅜ x 8½. 21482-6 Pa. **$2.50**

FAVORITE ANDREW LANG FAIRY TALE BOOKS IN MANY COLORS, Andrew Lang. The four Lang favorites in a boxed set—the complete *Red, Green, Yellow* and *Blue* Fairy Books. 164 stories; 439 illustrations by Lancelot Speed, Henry Ford and G. P. Jacomb Hood. Total of about 1500pp. 5⅜ x 8½. 23407-X Boxed set, Pa. **$16.95**

HOUSEHOLD STORIES BY THE BROTHERS GRIMM. All the great Grimm stories: "Rumpelstiltskin," "Snow White," "Hansel and Gretel," etc., with 114 illustrations by Walter Crane. 269pp. 5⅜ x 8½.
21080-4 Pa. $3.50

SLEEPING BEAUTY, illustrated by Arthur Rackham. Perhaps the fullest, most delightful version ever, told by C. S. Evans. Rackham's best work. 49 illustrations. 110pp. 7⅞ x 10¾. 22756-1 Pa. $2.95

AMERICAN FAIRY TALES, L. Frank Baum. Young cowboy lassoes Father Time; dummy in Mr. Floman's department store window comes to life; and 10 other fairy tales. 41 illustrations by N. P. Hall, Harry Kennedy, Ike Morgan, and Ralph Gardner. 209pp. 5⅜ x 8½. 23643-9 Pa. $3.00

THE WONDERFUL WIZARD OF OZ, L. Frank Baum. Facsimile in full color of America's finest children's classic. Introduction by Martin Gardner. 143 illustrations by W. W. Denslow. 267pp. 5⅜ x 8½.
20691-2 Pa. $4.50

THE TALE OF PETER RABBIT, Beatrix Potter. The inimitable Peter's terrifying adventure in Mr. McGregor's garden, with all 27 wonderful, full-color Potter illustrations. 55pp. 4¼ x 5½. (Available in U.S. only)
22827-4 Pa. $1.50

THE STORY OF KING ARTHUR AND HIS KNIGHTS, Howard Pyle. Finest children's version of life of King Arthur. 48 illustrations by Pyle. 131pp. 6⅛ x 9¼. 21445-1 Pa. $5.95

CARUSO'S CARICATURES, Enrico Caruso. Great tenor's remarkable caricatures of self, fellow musicians, composers, others. Toscanini, Puccini, Farrar, etc. Impish, cutting, insightful. 473 illustrations. Preface by M. Sisca. 217pp. 8⅜ x 11¼. 23528-9 Pa. $6.95

PERSONAL NARRATIVE OF A PILGRIMAGE TO ALMADINAH AND MECCAH, Richard Burton. Great travel classic by remarkably colorful personality. Burton, disguised as a Moroccan, visited sacred shrines of Islam, narrowly escaping death. Wonderful observations of Islamic life, customs, personalities. 47 illustrations. Total of 959pp. 5⅜ x 8½.
21217-3, 21218-1 Pa., Two-vol. set $14.00

INCIDENTS OF TRAVEL IN YUCATAN, John L. Stephens. Classic (1843) exploration of jungles of Yucatan, looking for evidences of Maya civilization. Travel adventures, Mexican and Indian culture, etc. Total of 669pp. 5⅜ x 8½. 20926-1, 20927-X Pa., Two-vol. set $7.90

AMERICAN LITERARY AUTOGRAPHS FROM WASHINGTON IRVING TO HENRY JAMES, Herbert Cahoon, et al. Letters, poems, manuscripts of Hawthorne, Thoreau, Twain, Alcott, Whitman, 67 other prominent American authors. Reproductions, full transcripts and commentary. Plus checklist of all American Literary Autographs in The Pierpont Morgan Library. Printed on exceptionally high-quality paper. 136 illustrations. 212pp. 9⅛ x 12¼. 23548-3 Pa. $12.50

AN AUTOBIOGRAPHY, Margaret Sanger. Exciting personal account of hard-fought battle for woman's right to birth control, against prejudice, church, law. Foremost feminist document. 504pp. 5⅜ x 8½.
20470-7 Pa. $7.50

MY BONDAGE AND MY FREEDOM, Frederick Douglass. Born as a slave, Douglass became outspoken force in antislavery movement. The best of Douglass's autobiographies. Graphic description of slave life. Introduction by P. Foner. 464pp. 5⅜ x 8½. 22457-0 Pa. $6.50

LIVING MY LIFE, Emma Goldman. Candid, no holds barred account by foremost American anarchist: her own life, anarchist movement, famous contemporaries, ideas and their impact. Struggles and confrontations in America, plus deportation to U.S.S.R. Shocking inside account of persecution of anarchists under Lenin. 13 plates. Total of 944pp. 5⅜ x 8½.
22543-7, 22544-5 Pa., Two-vol. set $12.00

LETTERS AND NOTES ON THE MANNERS, CUSTOMS AND CONDITIONS OF THE NORTH AMERICAN INDIANS, George Catlin. Classic account of life among Plains Indians: ceremonies, hunt, warfare, etc. Dover edition reproduces for first time all original paintings. 312 plates. 572pp. of text. 6⅛ x 9¼. 22118-0, 22119-9 Pa.. Two-vol. set $12.00

THE MAYA AND THEIR NEIGHBORS, edited by Clarence L. Hay, others. Synoptic view of Maya civilization in broadest sense, together with Northern, Southern neighbors. Integrates much background, valuable detail not elsewhere. Prepared by greatest scholars: Kroeber, Morley, Thompson, Spinden, Vaillant, many others. Sometimes called Tozzer Memorial Volume. 60 illustrations, linguistic map. 634pp. 5⅜ x 8½.
23510-6 Pa. $10.00

HANDBOOK OF THE INDIANS OF CALIFORNIA, A. L. Kroeber. Foremost American anthropologist offers complete ethnographic study of each group. Monumental classic. 459 illustrations, maps. 995pp. 5⅜ x 8½.
23368-5 Pa. $13.00

SHAKTI AND SHAKTA, Arthur Avalon. First book to give clear, cohesive analysis of Shakta doctrine, Shakta ritual and Kundalini Shakti (yoga). Important work by one of world's foremost students of Shaktic and Tantric thought. 732pp. 5⅜ x 8½. (Available in U.S. only)
23645-5 Pa. $7.95

AN INTRODUCTION TO THE STUDY OF THE MAYA HIEROGLYPHS, Syvanus Griswold Morley. Classic study by one of the truly great figures in hieroglyph research. Still the best introduction for the student for reading Maya hieroglyphs. New introduction by J. Eric S. Thompson. 117 illustrations. 284pp. 5⅜ x 8½. 23108-9 Pa. $4.00

A STUDY OF MAYA ART, Herbert J. Spinden. Landmark classic interprets Maya symbolism, estimates styles, covers ceramics, architecture, murals, stone carvings as artforms. Still a basic book in area. New introduction by J. Eric Thompson. Over 750 illustrations. 341pp. 8⅜ x 11¼.
21235-1 Pa. $6.95

GEOMETRY, RELATIVITY AND THE FOURTH DIMENSION, Rudolf Rucker. Exposition of fourth dimension, means of visualization, concepts of relativity as Flatland characters continue adventures. Popular, easily followed yet accurate, profound. 141 illustrations. 133pp. 5⅜ x 8½.
23400-2 Pa. $2.75

THE ORIGIN OF LIFE, A. I. Oparin. Modern classic in biochemistry, the first rigorous examination of possible evolution of life from nitrocarbon compounds. Non-technical, easily followed. Total of 295pp. 5⅜ x 8½.
60213-3 Pa. $5.95

PLANETS, STARS AND GALAXIES, A. E. Fanning. Comprehensive introductory survey: the sun, solar system, stars, galaxies, universe, cosmology; quasars, radio stars, etc. 24pp. of photographs. 189pp. 5⅜ x 8½. (Available in U.S. only)
21680-2 Pa. $3.75

THE THIRTEEN BOOKS OF EUCLID'S ELEMENTS, translated with introduction and commentary by Sir Thomas L. Heath. Definitive edition. Textual and linguistic notes, mathematical analysis, 2500 years of critical commentary. Do not confuse with abridged school editions. Total of 1414pp. 5⅜ x 8½.   60088-2, 60089-0, 60090-4 Pa., Three-vol. set $19.50

Prices subject to change without notice.

Available at your book dealer or write for free catalogue to Dept. GI, Dover Publications, Inc., 180 Varick St., N.Y., N.Y. 10014. Dover publishes more than 175 books each year on science, elementary and advanced mathematics, biology, music, art, literary history, social sciences and other areas.